Industrial Sealing Technology

Industrial Sealing Technology

H. HUGO BUCHTER

Engineering Consultant and Regional Manager
Burgmann Seals America, Inc.
formerly BASF Ludwigshafen, Germany
Monsanto Company, St. Louis, Missouri

A Wiley-Interscience Publication

JOHN WILEY & SONS
New York • Chichester • Brisbane • Toronto

Library of Congress Cataloging in Publication Data

Buchter, H. Hugo.
  Industrial sealing technology.

  "A Wiley-Interscience publication."
  Includes bibliographies and index.
  1. Sealing (Technology) I. Title.

TJ246.B8   621.8′85   78-13548

ISBN 0-471-03184-4

Printed in the United States of America

10 9 8 7 6 5 4 3 2 1

Preface

The material contained in this book is the first compilation of all aspects of seal technology. Included are detailed discussions of theories and design concepts for seals for applications from ultravacuums to pressures above 5000 atmospheres. A wealth of illustrations also is included. The information presented has been accumulated and evaluated critically during forty years' experience in the chemical industries of the United States and Europe.

Because of the ever-increasing severity of industrial, military, and commercial requirements, the search for effective sealing devices has become a major challenge for design engineers. Operating pressures and temperatures are reaching extremes. In addition to meeting operational demands, modern sealing devices must also satisfy environmental, safety, and health requirements.

Although seal technology plays a vital role in modern technology, its importance may not be apparent to the general practicing engineer. Perhaps this attitude has historical roots: before the age of steam effective seals were not required and until the space age the need increased only gradually. Yet who can imagine the wide acceptance of such common items as air conditioners, car engines, vacuum cleaners, and dishwashers without effective seals. In fact, the development of industrial rotating and reciprocating equipment often is limited by the reliability of the sealing devices employed.

In response to the multitude of industrial domestic sealing requirements an entirely new discipline has evolved that up to now has not been properly defined and documented. This book attempts to clarify this situation and to assist the practicing engineer 'in selecting the proper sealing mechanisms and devices from the complex designs and sophisticated configurations available in today's marketplace.

This book is divided into seven chapters. Chapter 1 deals with the fundamentals of gaskets and sealants. Most of the available information comes from company brochures and catalogs and is confusing. This makes it difficult to choose suitable items, particularly when standardization is lacking. This book furnishes explanations of sealing mechanisms and design configurations of gaskets and sealants, and provides data on the properties of materials used. Particular emphasis is placed on solid state metal gaskets for high pressure. For the first time, background information is provided on the fundamental computation of high-pressure flange joints, helping to overcome the standardization lacking in this field.

Chapter 2 covers axial seals, especially mechanical end face seals. The comprehensive range of designs offered provides numerous solutions to rotating seal problems encountered in pumps, compressors, and agitated vessels containing pressures up to 3000 psi and over. Axial seals have been improved to the point that they commonly can provide several years of maintenance-free operation.

Radial seals, which comprise the largest segment of rotating shaft seals, are covered extensively in Chapter 3. Their wide application in combustion engines, refrigerators, driers, and all kinds of motors is possible because of the development of high-temperature-resistant, low-friction plastics, such as Teflon (PTFE), Kel-F, and polyimids and because of the use of a radial lip ring design. Radial lip ring seals have been used recently in agitator shafts operating under mild service conditions and gradually are replacing mechanical end face seals.

Chapter 4 covers circumferential split ring seals, which also allow axial motion.

The important category of packings is treated in Chapter 5, which has in-depth information on packings for rotating and reciprocating shafts operating under both mild and extreme pressures. Answers are offered to the classical question of packing versus mechanical end face seal.

Felt, diaphragm, and piston ring seals, which were developed for both rotating and reciprocating equipment applications, are discussed in detail in Chapter 6. The book concludes with discussions of the numerous types of interstitial seals. It offers specific in-depth information on visco pump, labyrinth, bushing, and ferrofluidic seals.

H. Hugo Buchter

San Francisco
October 1978

Acknowledgments

In preparing this book I was materially assisted by A. O. Fisher, Principal Engineering Specialist at Monsanto's Corporate Engineering Department, St. Louis, Missouri. I gratefully acknowledge his help.

I would also like to thank the following journals and manufacturers for their assistance in obtaining illustrations: *Chemical Engineering*; *Chemie Ingenieur-Technik*; *Hydraulics and Pneumatics*; *Lubrication Engineering*, American Society of Lubrication Engineers; *Machine Design*; *Mechanical Engineering*, American Society of Mechanical Engineers; *Power Magazine*; *Product Engineering*; Condren Corporation, North Brunswick, New Jersey; and Crane Packing Company, Morton Grove, Illinois.

Ferrofluidics Corporation   H.H.B.
Burlington, Massachusetts

Contents

CHAPTER 3
RADIAL SEALS FOR ROTATING SHAFT   235

Industrial Sealing Technology

CHAPTER 1

Gaskets and Devices for Static Sealing

I. Introduction

Seals are mechanical devices used to prevent leakage of liquids, solids or gases. They are also used to prevent penetration of foreign particles into concealed containers or piping systems.

Seals are available in an enormous variety of design configurations utilizing a wide range of diversified sealing principles. The selection of seals from this large variety of modifications, however, is not a matter of picking desirable items from a catalog. On the contrary, for any specific application a careful evaluation must be made before a selection is possible. The selection must consider pressure, temperature, corrosive atmosphere, materials, shaft speed, and so on. Often the only guideline available is the manufacturers' brochures. So making the appropriate choice depends on the designer's experience and thorough understanding of the seal problems.

The first element in arriving at a technically justifiable solution is correct identification of the sealing problem. The second factor is the availability of a suitable seal configuration on the industrial market. The third, but not the least significant, point is definition of the amount of leakage that can be tolerated. This last factor should not be underestimated because there is no such thing as zero leakage.

A. Definition of Zero Leakage

In fabricator literature the concept of *zero leakage* is often referred to. However, this is misleading because an accepted definition of the term is nonexistent. In general practice a zero leakage specification is an indication to use polymeric seats or seals. Metal-to-metal seals generally fail to meet this requirement. An exception is in gaskets for static conditions where metal is plastically deformed to obtain a leakage of less than 10^{-8} atmospheric cm^3/sec helium.

In an extensive study by Advanced Technology Laboratories, General Electric Company, Schenectady, New York, *zero leakage* has been defined to be a leakage of less than 10^{-8} atmospheric cm^3/sec of helium. Another industrial source indicates that although *zero leakage* has no meaning, it may be considered to be in the range of 10^{-4} to 10^{-8} atmospheric cm^3/sec helium. At NASA's Manned Spacecraft Center in Houston, Texas, *zero leakage* is defined as no more than 1.4×10^{-3} standard cm^3/sec GN$_2$ at 300 psig and at ambient temperature. Leakage requirements and specifications for valves for unmanned missions, obtained from interviews with prime manufacturers, varied from 1.15×10^{-5} standard cm^3/sec to 0 for N$_2$O$_4$ and from 8.3×10^{-3} standard cm^3/sec to 1.4×10^{-4} standard cm^3/sec for other gases.

Leakage rate is a relative concept and what rate can be tolerated must be clearly stated. Leakage of extremely expensive, toxic, corrosive, explosive, or flammable fluids must be reduced to a minimum. Tolerable leakage rates are decided from case to case; a general rule does not exist. As a guideline, one should remember that an average drop of fluid amounts to an approximate volume of 0.05 cm^3, which means that it takes about 20 drops of fluid to obtain 1 cm^3 of volume. This figure offers a good guideline for evaluating the leakage rate of a fluid in a system that can or cannot or must be tolerated.

B. Seal Classification

Classification of any subject, technical or nontechnical, serves the purpose of identifying species, making it easier to analyze the problems involved. Accordingly seals may be classified into two major categories, static and dynamic.

Static seals comprise three major groups, known as gaskets, sealants, and direct contact seals.

Dynamic seals can also be subdivided into two basic groups designated as seals for rotating shafts and seals for reciprocating shafts. These two groups make up the bulk of industrial seals by volume and require specific consideration for predominantly custom-made sealing devices. Further in this category trade names must be used to identify the different devices. These trade names must be tolerated, since there is no other system for designating most specific devices.

A general classification chart for all seals discussed in this book is shown in Fig. 1.1.

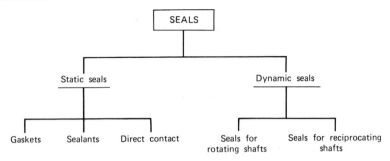

Fig. 1.1   General seal classification.

Rotating shaft seals are given special and extensive consideration since they are the dominating category of major industrial seals. They comprise two principal groups, the interfacial seals and the interstitial seals.

The interfacial seals subdivide into axial seals and radial seals as shown in Fig. 1.2. Interfacial seals represent a very large family of industrial sealing devices that establish a direct contact between the sealing component and the rotating shaft.

Interstitial seals categorize a family of four distinct groups in which the sealing components have no direct mechanical contact with the rotating shaft. The sealing elements allow a certain leakage that retards flow by controlling the clearance gap through which the fluid flow may pass using external forces on the fluid.

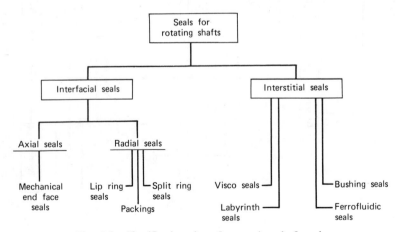

Fig. 1.2   Classification chart for rotating shaft seals.

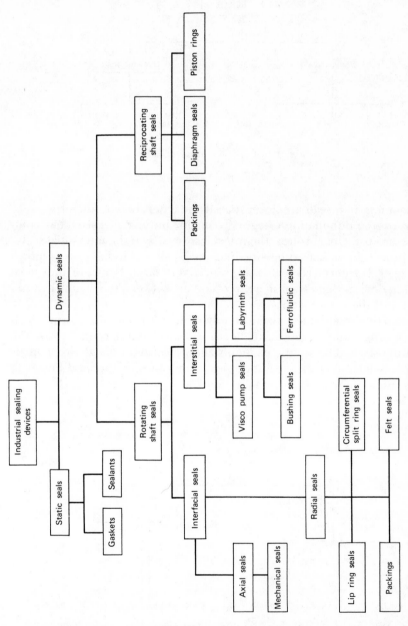

Fig. 1.3 Total classification diagram: industrial sealing devices.

4

The function of interstitial seals is to create a pressure drop of the fluid to be sealed with the least possible flow and simultaneously to permit unrestricted relative motion between moving parts. Interstitial seals are capable of maintaining a pressure differential between the interior of a machine and its surrounding environment by distinctly throttling the escaping fluid. Unlike interfacial seals, no mechanical contact of moving machine parts is intended. This reduces friction to an absolute minimum. To throttle fluid in a controlled manner, however, it is necessary to allow a nominal flow.

Examples of such interstitial seals are visco seals, labyrinth seals, and bushing seals. The family of the ferrofluidic seals, however, is a clear exception. Here the gap for the clearance fluid is filled with a magnetic medium that is held captive without flow with the help of a more or less strong magnetic field. Mechanical parts to establish a seal do not exist. With interstitial seals no rubbing contact exists; hence friction and wear of moving parts in the direct seal environment are practically eliminated.

Figure 1.3 gives an overview of all the seal categories to be described in this book.

II. Gaskets and Their Fundamental Design Concepts

Gaskets are sealing devices designed for systems handling gases, liquids, or solids. They are incorporated between rigid, predominantly metallic sealing areas of machinery, equipment, containers, and piping systems operated under a wide variety of service conditions including pressure, temperature, and corrosive atmospheres.

A. Basic Considerations

Gaskets represent equipment components that are predominantly designed for static working conditions. They are equally suited to seal against dynamic load requirements. They are available in a wide range of materials, such as metals, plastics, and composites. To provide a satisfactorily tight seal gaskets must conform to the mating seal faces of the joint contact surfaces.

With the joint under operating conditions the hydrostatic end force tends to separate the mating sealing faces, thus forcing the bolts to lengthen. As a result the existing seal contact pressure acting on the gasket is reduced and eventually leakage occurs. Thus the proper design of a gasket should be the result of an evaluation of the internal pressure, the gasket contact pressure produced by the bolt forces, and the gasket materials, which are selected to withstand the operating conditions. Fluctuations in both internal pressures and temperatures often cause serious deficiencies, resulting in leakage problems, and must therefore be taken into account.

Gaskets for vacuum service greatly differ from those for atmospheric conditions or high-pressure operations. Likewise gaskets for sealing fluids are different from gaskets used for sealing contained gases.

To prevent leakage in a gasketed joint of any design a bolt force is required to compress the material of the gasket so that the gasket sealing surface fills the asperities of the contact faces of the joint. Leakage is stopped only if the gasket material actually fills all depressions of the seal contact faces. Thus the gasket material plays a very significant role in appropriate sealing devices. Asbestos, plastics, elastomers, metals, and combinations of any of these materials are the essential representative components in sealing devices. A considerable variety of sealed-joint design configurations are available using these materials.

It is obvious that the bolt force must be higher than the effect of the internal pressure, which in turn constantly tends to separate the joint components. Precise data are given in a later section of this chapter. Because most of the seal contact areas are imperfect, the gasket material must "flow" into the surface imperfections of the joint faces to achieve good mating, which then permits a tight seal with practically no leakage.

Gasket materials are therefore designed to flow under compression between the seal contact faces. Consequently the seal effect is improved by using softer gasket materials. The tightening forces can be reduced correspondingly by increasing the softness of the gasket material.

Sealing with gaskets appears at first glance to be a simple procedure. In reality, however, it is extremely complex, and a reliable theoretical design analysis for static gaskets still does not exist. All gasket devices have been developed on a purely empirical basis, and a prediction of their behavior in service, even for new gasket designs, cannot be made. It is customary to define a so-called tightening factor by which the joint design can be judged and appropriately calculated.

The tightening factor expresses what the contact force should be in relation to the internal pressure. For establishing reliable tightness of a system the tightening force must always be greater than the counter effect exerted by the internal pressure. The tightening factor reflects the ratio of tightening force to internal pressure. It represents a safety factor that guarantees leakage-free service in the presence of surface asperities. Specific data on tightening factors is presented later in this chapter. The effectiveness of gasket materials can be judged on the basis of the value of the tightening factor.

B. The Influence of Material on the Effectiveness of Gaskets

An ideal gasket may be defined as one that has two basic sealing properties: one involves the softness, deformability, and memory of the gasket material and the second involves a self-energizing device. Memory behavior requires a high degree of elasticity and flexibility without the material readily hardening or fatiguing. This enables the gasket material to flow and fill surface asperities at low load requiring a tightening factor of maximal 1.50 while still providing an appropriate safety margin without sacrificing safety. A low value of the tightening factor has the advantage of keeping the joint from being underdesigned. A safety margin of 30 to 40% in the tightening factor is theoretically sufficient to prevent leakage.

The provision of a self-energizing effect is a prudent engineering design that utilizes the internal pressure to increase the contact pressure with rising hydrostatic internal pressure without increasing or jeopardizing the tightening force in the closure system. The criterion of the self-energizing sealing effect is discussed in more detail in a later section of this chapter in connection with special design concepts for gaskets.

1. Rubber as a Gasket Material

Rubber, natural and synthetic, is an almost ideal gasket material, allowing very low bolt loads for proper sealing. To help explain the application ranges we present a brief review of rubber chemistry.

Natural rubber is a polymer of the hydrocarbon substance known as isoprene, containing five atoms of carbon and eight atoms of hydrogen.

The first synthetic rubbers, named neoprene, are straight polymers of chloroprene, whereas most current synthetic rubbers are copolymers of two or more constituents combined to provide the desired characteristics. The process of polymerization improves viscosity but reduces flow tendencies.

a. Rubber Processing. When crude rubber with long straight polymer chains is plasticized by milling, the long polymers are broken down and the material becomes soft and is more easily processed. During the processing operation, desirable material characteristics are imparted. For example, elasticity, resistance to rupture and abrasion, durability, rigidity, and particularly mechanical stability are improved as a result of vulcanization. Unvulcanized rubber becomes soft and pliable when heated. Once vulcanized, however, rubber scorches or burns rather than softening. This is an important fact to be considered when selecting rubber gaskets.

b. Durameter. To establish proper sealing of joints with rubber, its hardness must be known. Hardness of rubber or other elastomers is designated by a durameter number. Industrial gaskets are available in durameter numbers ranging from 35 to 95. For comparison, gum rubber has a durameter of 25 and is thus too soft for gasketing. Low durameters require low bolt loads, desirable for minimum flange dimensions. Where higher bolt loads are involved, the durameter number must be increased. Appropriate durameters for normal industrial sealing purposes range from 60 to 90.

c. Buna-S. In Buna-S the letter *S* stands for styrene type, as a copolymer of butadiene and styrene. Buna-S is also vulcanized and can be cured to hard rubber. Carbon black is added to impart optimal physical properties, producing a higher resistance to water than natural rubber has. Aging properties are also improved. In static conditions Buna-S offers greater heat resistance than natural rubber; severe flexing, however, causes heat resistance to deteriorate.

d. Cold Rubber. When Buna-S is copolymerized at low temperature, so-called cold rubber is obtained that is superior to natural rubber in abrasion resistance, but its resistance to gasoline, oils, and concentrated acids is very poor. Resistance to diluted concentrations of acids and alkalis is good.

e. Buna-N. In Buna-N the letter *N* stands for nitrile, comprising copolymers of butadiene acrylonitrile. Typical trade names are Paradril, Hydar, Chemigum, Butaprene.

In general, Buna-N types are compounded to improve abrasion resistance, increase tensile strength, extend heat resistance, and produce a favorable compression set. Buna-N exhibits good nonadhesion properties in contact with metals.

Buna-N is best known for its resistance to petroleum, gasoline, and aliphatic hydrocarbons. It further offers high resistance to high- and low-temperature aging and abrasion. It has poor resistance to light and is soluble in aromatics such as benzene, toluene, naphthalene, and chlorinated solvents such as trichloroethylene. Buna-N cannot be used in the presence of ketones, amines, esters, ethers, and certain organic acids. Consult manufacturer.

f. Neoprene. Neoprene is a chloroprene polymer that is not vulcanized with sulfur; this is an advantage for gaskets that are in contact with flange materials exposed to corrosive atmospheres containing sulfur. With regard to mechanical properties, neoprene is closer to natural rubber than other synthetic components.

Its tensile strength and abrasion properties are practically equal to those of natural rubber. Although its tear resistance is slightly less than that for natural rubber, its dynamic properties are excellent and are retained even at higher temperatures. It also has excellent resistance to sunlight, weathering, and heat.

Neoprene has excellent resistance to oils and nonaromatic gasolines; however, its performance is poor in aromatic atmospheres. Swelling will be observed in mineral oils of low aniline point. Where swelling can be tolerated, this may enhance original physical and mechanical properties. When in contact with a direct flame, it will burn, but it will not propagate when the flame is removed.

g. Butyl. Butyl is a very widely used material for gasketing because of its excellent properties. It is a copolymer of isobutylene and varying small amounts of other diolefines, such as isoprene and/or butadiene. It is an economical product, produced in quantity at ordinary pressures and temperatures above $-6°C$ as a byproduct of petroleum cracking.

Butyl has excellent resistance to gas diffusion. Although its physical properties are not as good as those of natural rubber, it rates high in chemical stability and resistance to deterioration or decomposition.

Its wide technical application is based on its high resistance to chemicals and oxidation, which is considered more important than its tensile strength or resistance to cold flow. Butyl resists attack by vegetable or animal oils and fats; however, it swells like natural rubber when in contact with petroleum or coal-tar solvents. It absorbs negligible amounts of water, provides good heat and low-temperature resistance. Resilience at room temperature is poor, but it is excellent at higher temperatures.

h. Thiokol. Thiokol is a polysulfide rubber available in a variety of grades. All are milled with plasticizers, and not vulcanized with sulfur. They are usually cured with zinc oxide, but cannot be cured to hard rubber consistency.

Grades PR-1 and ST show best overall solvent resistance of the commercial rubbers; however, they are poor against halogenated solvents. They are highly resistant to petroleum, hydrocarbons, esters, ethers, ketones, alcohols, gasoline, aromatic blended fuels, and many others.

PR-1 and ST have good flexibility at low temperatures, and properties remain stable at cold temperatures. They have a tendency to deteriorate at higher temperatures.

Tensile strength, elasticity, tear, and abrasion resistance and compression set are relatively low. Thiokol will be recommended where minimum swell, freedom from shrinkage, and good aging characteristics are required for contact with active solvents and reagents and where physical properties are considered to be of secondary importance.

i. Silicone Rubber. Silicone rubbers are a combination of silicone and hydrocarbons, based on an inorganic skeleton of alternate silicone and oxygen atoms with bonds 50% stronger than the carbon-to-carbon structure in organic rubber.

Silicone rubbers have excellent heat resistance and maintain good flexibility even at low temperatures. They provide high dielectric properties, plus fine oxidation and good weathering resistance. They have good resistance to oils with high aniline points but only moderate resistance to low-aniline-point oils and most gasolines. They deteriorate markedly when exposed to steam under pressure and tend to swell in hydrocarbon solvents.

Silicone rubbers are strongly recommended wherever high or very low temperatures must be encountered. Consultation with manufacturers is recommended, since various types possess different properties.

j. Cellular Rubber. Cellular rubber is a chemically blown rubber having light weight, good flexibility, elasticity, compressibility, and low thermal conductivity. Its insulation property offers excellent resistance to shock, sound, vibration, air, heat, and moisture.

Cellular rubber is preferably used for low-pressure gaskets on openings, ductwork, tanks, and so on, where odd and nonflat surfaces create a sealing problem. It is excellent on air-conditioning units, for gasketing as well as noise abatement. Neoprene sponges are used for sealing against oils; and silicone-blown rubber is used in high-temperature applications.

A brief summary of these elastomeric gasket materials is given in Table 1.1.

The DuPont Company has recently introduced a new kind of seal material known as Kalrez. This material is reported to combine the resistance properties of Teflon with the elastic abilities of Viton and is, therefore, an excellent top candidate for O-ring material.

2. Plastics as Gasket Materials

With the development of fluoroethylene, trifluorochloroethylene, phenolic resins, and polyethylene, an entirely new field of gasket materials was discovered that penetrated into the industrial sealing market, where high temperatures or corrosive atmospheres rapidly destroy natural or synthetic elastomers and rubbers. This has been characterized as the new age of plastics.

a. Teflon.   Teflon, developed by DuPont de Nemours, is a polymer of tetrafluoroethylene that can be compression molded. Teflon gaskets are formed by machining from molded sheets, extruded rods, or tubes.

Teflon has superior resistance to heat ranging from the cryogenic area up to 500°F. By reinforcing it with glass fiber and other embedded materials, its temperature stability can even be raised beyond the 500°F limit.

Teflon will resist chlorides, boron trifluoride, high-boiling-point solvents, ketones, esters, ethers, boiling nitric acid, aqua regia, and sodium hydroxide. The only common environments attacking Teflon are molten alkali metals and fluorine at high temperatures and pressures.

In the range from −300°F to 500°F Teflon provides high mechanical strength, but it becomes very brittle when temperature drops below −300°F. When approaching 500°F (480°F to be precise), Teflon starts to lose most of its tensile and other physical properties. Electrical properties at high frequencies and high temperatures are excellent. Teflon has superior resistance to surface arc-over because it melts and vaporizes instead of carbonizing, thus having no conducting path. Its power factor, low dielectric constant, and high dielectric strength remain unaffected for temperatures up to 400°F and over. Even ultraviolet rays have no affect on Teflon.

Teflon exhibits a smooth antistick surface. Under compressive pressures Teflon undergoes cold flow at any temperature, and at temperatures in excess of 650°F it decomposes, developing toxic and poisonous fumes.

Because of its chemical inertness Teflon is an ideal material for jacketed gaskets of the envelope and molded-shield type. Almost any filler material may be chosen. It is inert against absorption of moisture.

b. Kel-F.   Kel-F, a polymer of trifluorochloroethylene, is a very stable high-temperature thermoplastic developed by the M. W. Kellog Com-

Table 1.1 Basic Physical Properties of Natural and Synthetic Rubbers and

General Classification	Natural Rubber	Buna-S	Buna-N	
			Low Swell Hycar and Butaprene	High Swell Paracril and Chemigum
Specific gravity	0.93	0.94	1.00	1.00
Tensile:				
Pure gum	3000	400	600	600
Reinforced	4500	3000	4000	4000
Tear resistance	Excellent	Poor–fair	Fair	Fair
Abrasion resistance	Excellent	Good	Good	Good
Aging:				
Sunlight	Poor	Poor	Poor	Poor
Oxidation	Good	Fair	Fair	Fair
Heat, max temperature (°F)	300	250	300	300
Static (in storage)	Good	Good	Good	Good
Flex cracking resistance:				
Slow rate	Excellent	Good	Good	Good
Fast rate	Excellent	Poor	Poor	Poor
Compression set resistance	Good	Good	Very good	Very good
Solvent resistance:				
Aliphatic hydrocarbon	Very poor	Very poor	Excellent	Good
Aromatic hydrocarbon	Very poor	Very poor	Good	Fair
Oxygenated solvent	Good	Good	Poor	Poor
Halogenated solvent	Very poor	Very poor	Very poor	Very poor
Oil resistance:				
Low aniline	Very poor	Very poor	Excellent	Fair
High aniline	Very poor	Very poor	Fair	Excellent
Gasoline resistance				
Aromatic	Very poor	Very poor	Poor	Good
Nonaromatic	Very poor	Very poor	Fair	Excellent
Acid resistance:				
Dilute (under 10%)	Good	Good	Good	Good
Concentrated (except nitric and sulfuric)	Fair	Poor	Poor	Poor
Low temperature resistance, max (°F)	−65	−70	−65	−65
Permeability to gases	Fair	Fair	Fair	Fair
Water resistance	Good	Very good	Very good	Very good
Alkali resistance:				
Dilute (under 10%)	Good	Good	Good	Good
Concentrated	Fair	Fair	Fair	Fair
Resilience	Very good	Fair	Fair	Fair
Elongation, max (%)	700	500	500	500

Neoprene		Butyl Nonoil Resistant	Thiokol		Silicone Rubbers
Type GN	Type W		Type PR-1	Type ST	
1.25	1.25	0.91	1.35	1.35	1.2 to 2.6
3500	3500	3000	300	300	200–450
3500	3500	3000	1500	1500	. . .
Good	Good	Good	Fair	Fair	Poor-fair
Excellent	Excellent	Fair	Poor	Poor	Poor
Excellent	Excellent	Excellent	Good	Good	Good
Good	Good	Good	Good	Good	Very good
300	300	300	160	160	450
Very good	Very good	Good	Fair	Fair	Good
Excellent	Excellent	Excellent	Fair	Fair	Fair
Excellent	Excellent	Excellent	Poor	Poor	Poor
Poor	Excellent	Fair	Poor	Poor	Good
Fair	Fair	Poor	Excellent	Excellent	Poor
Poor	Poor	Very poor	Good	Good	Very poor
Fair	Fair	Good	Fair	Fair	Poor
Very poor	Very poor	Poor	Poor	Poor	Very poor
Fair	Fair	Very poor	Excellent	Excellent	Poor
Good	Good	Very poor	Excellent	Excellent	Good
Poor	Poor	Very poor	Excellent	Excellent	Poor
Good	Good	Very poor	Excellent	Excellent	Good
Fair	Fair	Good	Poor	Poor	Fair
Fair	Fair	Fair	Very poor	Very poor	Poor
−50	−65	−65	−40	−65	−120
Very good	Very good	Very good	Good	Good	Fair
Poor	Fair	Very good	Fair	Fair	Fair
Good	Good	Very good	Poor	Poor	Fair
Good	Good	Very good	Poor	Poor	Poor
Very good	Very good	Very poor	Poor	Poor	Good
500	500	700	400	400	300

pany. It shows high electrical resistance and a very low heat conductivity, thus representing an excellent electrical insulator. At very low temperatures it provides high-impact strength with excellent resistance to thermal shock.

Kel-F is nonflammable and provides—as does Teflon—unusually good neutrality and inertness after long exposure to solvents, concentrated mineral acids, alkalis, fuming nitric acid, aqua regia, hydrogen fluoride and peroxide, and other strong oxidizing and reducing agents. It is neutral to water, absorbs no moisture, and offers excellent atmospheric weathering properties. Plasticized Kel-F shows some shrinkage in certain hydrocarbons, solvents, and hot water.

In unplasticized condition Kel-F retains most of its properties in a temperature range from −320°F (liquid nitrogen) to 390°F. The upper limit is reduced to 300°F when Kel-F is plasticized. Kel-F in plasticized form also has the tendency to harden at low temperatures. Cold flow is low and it even exhibits *memory,* a desirable property in gasket materials—under pressure it shows slight deformation but it resumes its original shape when pressure is released.

c. Phenolic Resins.   Phenolic filled resins, known under the trade names Micarta, Synthane, and so on, are formed from layers of fibrous sheet filler that has been impregnated with a thermosetting resin binder bonded under high heat and pressure, forming a strong, solid material through chemical reaction. The layers are formed into a solid mass that will not delaminate again nor resoften at increased heat. The impregnating resin is permanently changed from a fusible and soluble state to an infusible and insoluble state. In this condition the resinous product is said to be *thermosetting* or cross-linked.

Phenolic resins have excellent dielectric strength and are well suited for combination gaskets and insulation of components. They are used in electrical insulating systems where current may be picked up and conducted into wall, piping, underground mains, or stations where electrical current cannot be tolerated.

When applied as material for insulating purposes it is customary to place O-rings of this material between the flange faces to avoid direct metallic contact.

Another method of insulation is to insert layers of variable thickness between the contact faces of the flanges. The thickness of the layer of phenolic resins between the flange faces is governed by the degree of electricity (amperage and voltage) to be insulated against.

To prevent current conductivity across the flange bolts into the flange rings or vice versa it is preferable also to have the bolt shafts insulated to eliminate any trace of possible metallic contact.

In the text to follow a number of seal configurations are discussed that can simply be used as static gaskets or as dynamic seals as well, as is shown for the BAL-Seal, the Omniseal, and the FCS-Seals by Fluorocarbon Company.

3. Metals as Gasket Materials

Metals when used as construction materials for gaskets under static operating conditions serve to combat pressure, temperature, and corrosive atmosphere. In extreme high-pressure service metals are required exclusively for gasketing. Metals are also used for high-temperature conditions where elastomers, plastics, fibrous materials, and their prospective combinations are no longer safe and appropriate.

The category of metallic materials for gaskets ranges from soft metals such as lead, aluminum, copper, brass, Monel, and nickel to various steels, including the highly alloyed, to the noble metals silver and platinum.

All-metal gaskets are available in numerous varieties of forms, both standardized and custom-made: plain, profiled, serrated, cast and corrugated as one piece or multipiece designs, as illustrated later in this chapter. The geometry of the design is governed by the conditions to be met in service.

a. Basic Design Principles for Metal Gaskets. In basic flange joint design the components of the joint determine the geometry of the metal gasket. The majority of gaskets are designed for use with standard flange face configurations. In general, gaskets are used in unconfined, partly confined, or fully confined position. Unconfined gaskets can move horizontally and are free to deform away from and toward fluid pressure. Partly confined gaskets rest in a female-type recess of the shoulder in the face and are also free to deform on either the inside or outside rim. Totally confined gaskets are embedded in a groove of the female flange face and usually have little room in which to extrude.

The unconfined gasket is very practical and is therefore the one most commonly used for piping and numerous other joint configurations. Most metallic gaskets can be considered self-confining. A certain category of gaskets is designed to be semimetallic for unconfined application through use of a retaining ring or similar device either inside or outside, or even on both sides. This is done to limit the gasket compression to a given desirable value, as is shown later for Flexotallic gaskets.

In all unconfined designs consideration must be given to the fact that the gasket can be blown out if the bolt load does not produce satisfactory contact pressure.

The complete lack of a scientifically sound design analysis for selecting gasketing using solid metals as construction materials makes it extremely difficult to evaluate appropriately the wide spectrum of metallic gaskets used in industrial sealing applications. The author therefore undertook an extensive testing program to provide data that could be used for the design of reliable gasket configurations. The results are fundamental and offer a substantial basis for today's high-pressure equipment design. They provide a reliable basis for the development of urgently needed standards for high-pressure flange joint components. The results are discussed in detail. The development was initiated by a paper by W. Seufert (8).

In flange design a series of factors must be considered, such as bolt circle diameter, mean tightening diameter of the contact areas with the gasket, and particularly the bolt force. Here the gasket plays the dominating role. With the gasket in the compact solid form (like lens, octagonal, or delta ring), the bolt force is the most important factor to be analyzed and evaluated properly. This is in high-pressure equipment the basis for the accurate design of a flange joint. In this field relatively little is known about accurately designing joint connections or closures, because basic information on flange-gasket behavior is lacking. Without knowledge of the appropriate bolt force the design of the flange joint components becomes insecure or at least inaccurate.

To determine a reliable value of the bolt force the behavior of the metal gasket must be known when the joint is tightened. In high-pressure service the gasket behavior is characterized by the definition of a tightening factor x, which represents a typical characteristic of the gasket design in conjunction with the gasket material.

W. Seufert (8) reports on fundamental experiments with joints without using gaskets. It was felt that these tests could provide answers about gasket behavior. Seufert's test methods were reproduced and confirmed and then applied to the same flange geometry with the use of gaskets. If there was any difference between the behavior without metal gaskets and a system with metal gaskets then these test results would provide the answer.

The paragraphs to follow analyze precisely joint behavior without gaskets and then compare the results with those obtained from systems with gaskets.

1. DEFINITION OF THE PROBLEM.   The bolt force required to hold the components of a flange joint together to establish a reliable seal against extreme internal pressure is derived from the internal system pressure and is linearly proportional to this pressure. To show this interde-

pendence in a mathematical way we assume a design example representing a cylindrical tube, provided with flanges on both ends and closed by a blind cover on either end.

When the cylinder is filled with pressurized air to establish an internal pressure of p_i, the air volume produces an internal force P that tends to lift up the covers and separate them from the static gaskets, resulting in leakage. This internal force resulting from the internal pressure is computed from the relation

$$P_{p_i} = P_\uparrow = A_T \times p_i \quad [\text{kg}] \tag{1}$$

where A_T is the tightening area with the mean tightening diameter d_T. The bolt force required to hold the cover in space for a tight operating condition is derived from the relation

$$P_B = P_\downarrow = x(A_T \times p_i) \quad [\text{kg}] \tag{2}$$

Equation 1 differs from Equation 2 by the value of the tightening factor x, which must always be greater than one. Since $x > 1$, force $P_\downarrow$ must be greater than $P_\uparrow$.

Comparing the results of numerous publications on this subject it will be noticed that tightening factor x will range from >1 all the way to 10. Such a situation is intolerable in high-pressure design where hazardous gases and fluids are handled. If x is chosen too low the system will leak and the plant environment become a hazard. If x is chosen too high the system will be overdesigned and therefore uneconomical.

The bolt force P_B applied to tighten the system must under no circumstances be smaller than the force developed by the internal pressure p_i; consequently,

$$P_B > P_{p_i}$$

Tightening factor x indicates how much the ratio P_B/P_{p_i} must be to safely prevent leakage, considering all possible operating conditions. This ratio is definitely a function of the design as well as of the construction materials of the solid metal gaskets and the joint components. The mathematical interrelationship can be developed only on an experimental basis.

The confusing range of values for tightening factors found in the literature with a span from one to ten is a clear indication that this problem is not readily understood and this in turn is the result of the intricacy of the behavior of the solid gasket in a high-pressure joint. To clarify this discrepancy the author conducted a comprehensive series of tests to develop experimentally the values of the tightening factor x for system components that did not use solid metallic gaskets. In other

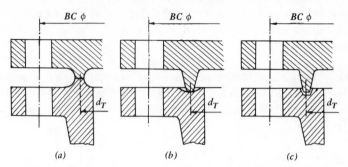

Fig. 1.4 Configurations of seal interfaces subjected to pressure tests.

words, the system components of the joints were designed in such a way that the interface areas were an integral part of the components, thus eliminating the use of solid metal gaskets. If the leakage behavior of those specially designed components then was the same as or comparable to the leakage behavior of joints using solid gaskets, the conclusion could be drawn that the problem was identified and solved.

Derived from a wealth of experience in the high-pressure field, the following flange components were tested, reducing the interfaces to the basic concepts of a flat area as in Fig. 1.4*a*, an interface with a one-line contact as in Fig. 1.4*b*, and a face with a double-line contact as in Fig. 1.4*c*. These somewhat idealized designs yield fundamental results with a minimum of test series. The objective of these tests was to determine the magnitude of the tightening forces and their interrelationship with a variety of interface configurations at various pressure levels for given material characteristics and specific seal area configurations.

2. LEAKAGE CONDITIONS IN THE ELASTIC AND PLASTIC DEFORMATION RANGE. When the bolts of a joint are tightened, a load is transmitted to the flange joint components containing the seal face areas. This bolt load then results in a deformation of the seal contact areas, which can be either in the elastic or in the plastic range. Since relatively little is known about whether the deformation within the elastic range suffices to establish a gastight seal, most engineers are unaware of the kind of deformation they produce when they tighten normal high-pressure flange joints.

To shed scientific light on this significant question preliminary tests were conducted for studying the conditions under which elastic and plastic deformations of the interface areas will occur.

Using flat interface contact areas with a width of approximately 13 mm and a mean tightening diameter of approximately 300 mm ($-\frac{1}{2}$

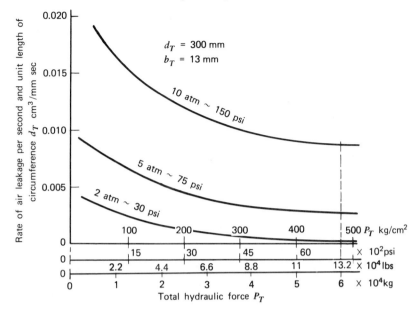

Fig. 1.5 Leakage rate of air at various internal pressures for given test chamber with polished contact areas.

in. width and ~11.80 in. tightening diameter), machined to a surface finish of about 2 μin. it was found that leakage against internal gas pressure cannot be prevented—even with almost perfect contact surfaces—as long as the bolt load produces a deformation that stays within the elastic range.

Figure 1.5 indicates the results of these preliminary tests. At an air pressure of 10 atm (~150 psi) the leakage rate was close to 0.010 cm^3/cm · sec although a tightening force of 60,000 kg was applied by means of a hydraulic press to supply a uniform bolt load across the tightening circumference.

Noticeable leakage was still observed after the gas pressure was reduced to 5 atm (~75 psi). Even at an air pressure of only 2 atm (~30 psi) under 60,000 kg hydraulic contact force leakage did not stop.

As a final conclusion it must be stated that bolt forces that produce elastic deformations in the contact interface areas of the joint components are not satisfactory for preventing leakage in systems that must seal against internal gas pressure. In spite of very good surface conditions there are always microscopically fine surface asperities where leakage channels will develop through which an air jet is possible. Such a

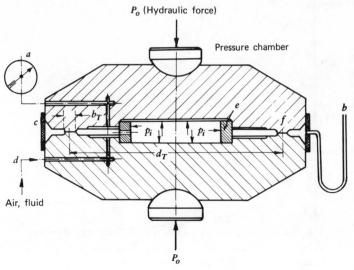

Fig. 1.6 Details of special testing device for static sealing.

leakage rate, although hardly measurable, cannot be tolerated in high-pressure service, particularly in the presence of hazardous gas atmospheres handled in those systems.

The test equipment used to determine leakage behavior is illustrated in Fig. 1.6. Two solid steel components are designed to form a hollow pressure chamber where gases and fluids can be contained to charge the chamber with internal pressure p_i. The interface area for the seal, designated as f, can be modified in width b_T and mean diameter d_T to satisfy a large variety of testing objectives. Centering ring e is used to guarantee perfect alignment of the two halves of the pressure chamber to match the parallelism of the interface areas f.

The test chamber is pressurized through bore d, and the pressure level is measured by manometer gauge a. Elastomer ring c is used to collect leakage across width b_T to be measured by a highly sensitive manometer b. A hydraulic press provides the necessary seal contact force P_0 to establish a uniform distribution of the contact force, required for gas-tight sealing, supplying freedom from human errors when using bolts. The basis for the calculation is interface f with width b_T and mean tightening diameter d_T. The tests were conducted by modifying interface f with width b_T and mean diameter d_T.

Testing was conducted under the following procedure: The joint components were charged by a random hydraulic load; then the chamber was pressurized through bore d. When no leakage was observed the internal gas pressure was raised to a level at which the first bubble of leakage could be seen. Even the slightest amount of leakage was considered critical and pressurization was stopped. At this point the internal pressure was discharged to zero level. In the subsequent pressure cycle the hydraulic contact load was increased, and the test chamber was pressurized again until the first leakage bubble could be observed. Once the pressure of this level was measured, the chamber was discharged down to zero. By repeating these cycles over and over an entire leakage curve was established plotting contact pressure versus internal gas pressure. By using this cycle procedure the three curves of Fig. 1.5 were obtained.

This method is well suited for developing complete leakage characteristics for a given design with b_T and d_T dimensions resulting in families of curves, starting at relatively low loads and moderate pressure conditions that can be raised to any desirable level.

The tightening forces can be computed on the basis of the interface geometry. When the test chamber is pressurized by internal gas or fluid pressure p_i a force P_1 is developed that counteracts the hydraulic tightening force P_0. This force is derived from the relation

$$P_1 = \frac{\pi}{4} \times d_T^2 \times p_i \quad [\text{kg}] \tag{3}$$

where d_T is the mean tightening diameter of the interface area f and p_i is the internal pressure that pressurizes the test chamber. Force P_1 opposes the hydraulic contact force P_0 and tends to separate the two chamber halves, in which case leakage occurs. Consequently for leakage-free test conditions force P_0 must be essentially greater than force P_1; the remaining tightening force in the presence of the internal pressure p_i then is P_T,

$$P_T = P_0 - P_1 = P_0 - \frac{\pi}{4} d_T^2 \times p_i \quad [\text{kg}] \tag{4}$$

Since quite a range of seal interface contact configurations were tested, for flat seal contact areas with a sealing width of b_T it is useful to refer the tightening force P_T to the unit area of the seal contact interface. This makes the results independent of the seal face geometry and permits a comparison of test devices with variable geometry.

The resulting unit tightening pressure p_T for the contact interfaces is

$$p_T = \frac{P_T}{\pi \times d_T \times b_T} \tag{5}$$

where d_T is the mean sealing diameter, b_T the width of interface f, and P_T the implied contact force.

In cases where area contact is being substituted for line contact the real value of the true contact force is very difficult to determine. For this purpose a unit tightening force is defined referring to the line contact along the circumference of the circle with the mean value $(d_T)_m$ of the tightening diameter. Thus the line contact pressure is

$$p_T' = \frac{P_T}{\pi \times d_T} \tag{6}$$

3. SOME METHODS FOR ESTABLISHING A SEAL IN THE PRESENCE OF ELASTIC DEFORMATION OF THE INTERFACE AREAS. Although, as indicated earlier, elastic deformations of interface components due to the hydraulic contact pressure do not suffice to provide a gastight seal, some phenomena can accomplish a seal in the elastic range, but these cases must be considered to be strict exceptions to the standard rule. By using certain kinds of mineral oils with a given viscosity and/or by covering the interface contact areas with a very thin metal foil it is possible to cover the surface asperities and establish a reasonably good seal against internal gas pressure. In emergencies these methods may be very helpful, but they must always be considered tricks that can provide satisfactory leak-free service when quick solutions must be found and other methods are not available. However, they contribute to a better understanding of the phenomenon that elastic deformations of the interface contact areas of the sealing components do not suffice to accomplish a reliably tight seal, particularly against internal gas pressure.

In tests undertaken to investigate the influence of viscous oils on the tightening behavior of metallic interfaces, oils of various degrees of viscosity and molecule size were used. By rubbing the metallic interfaces with oils of viscosities ranging from 1.1° Engler to 97.5°E, leakage characteristics were established as illustrated in Figs. 1.7 to 1.9 for a series of widely modified surface finish conditions.

Figure 1.7 presents the results obtained from a test device in which the contact areas were machined to a surface finish of 0.15 μin. rms. The graph proves conclusively that the leakage behavior improves noticeably with increasing viscosity of the oil film rubbed into the contact surfaces. With the exception of the oil with 1.1°E viscosity all curves are located below the $p_T = p_i$-curve. An oil with 1.1°E is very thin and comes close to having the properties of petroleum.

The testing device of Fig. 1.8 was provided with a slightly rougher contact surface finish, ranging in the order of 0.80 μin. In Fig. 1.9 the curves were obtained in a device with a surface finish of 1.80 μin. rms. In both figures the slopes of the leakage curves increase to a multiple of that of the $p_T = p_i$-curve.

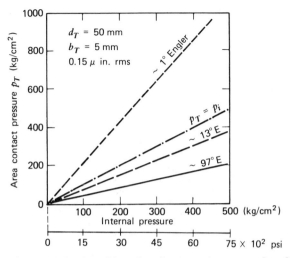

Fig. 1.7 Seal behavior of interface for 0.15 μin. rms surface finish.

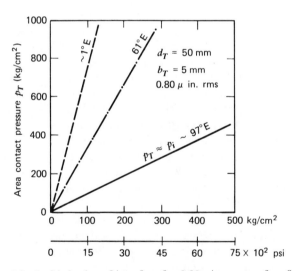

Fig. 1.8 Seal behavior of interface for 0.80 μin. rms surface finish.

23

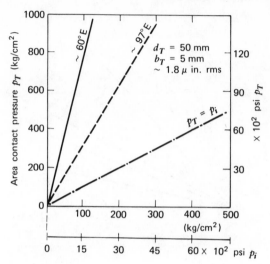

Fig. 1.9 Seal behavior of interface for ~1.8 μin. rms.

As a result the viscous oil film rubbed into the surfaces of the contact interface areas prevents leakage by filling the surface asperities. When the test chamber is charged with internal gas pressure the oil film develops microscopic adhesion and intramolecular cohesion forces that increase markedly as the gap decreases. The forces also increase with rising gas pressure, increasing fluid viscosity, and increasing size of the oil molecules, thus preventing the pressurized gas from leaking out.

However, as soon as leakage has started once at any spot an increase in hydraulic contact pressure will not stop the leak. The test device or joint must be completely dismantled and the interface surfaces must be provided with a new oil film, before the chamber is repressurized. In case of a leakage of any amount the developing gap acts like a specific jet nozzle through which the air (gas) will escape.

The use of oil films of high viscosity for sealing purposes in the presence of metallic interfaces at high internal gas pressures is good only for emergency cases when time is a decisive factor. It should not be considered a solution for general application.

A similar tightening effect can be accomplished by placing a very thin metal foil on one of the interface areas. Suitable metals for foiling of this nature can be silver, aluminum, copper, and so on. On hydraulic contact pressure application the foil easily fills all surface asperities, thus blocking any chance of formation of microscopic leakage paths.

In Fig. 1.10 a test device was used provided with interfaces having a

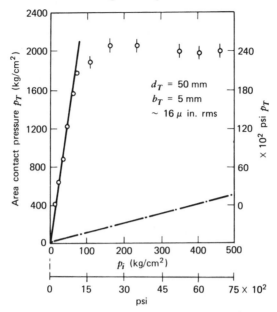

Fig. 1.10 Seal behavior of interface for ~16 μin. rms.

surface finish of approximately 16 μin. rms. With such a surface rough-ness an initially high contact force was required to achieve the initial seal, as shown by the very steep slope of the seal curve. Suddenly, once a certain contact pressure is reached the curve changes slope very dis-tinctly to a practically horizontal position. From now on tightness was achieved and the internal gas pressure could even be raised further without increasing the hydraulic contact force. When the test device was dismantled it was discovered that the interface contact surface areas were permanently deformed. The metals had begun to flow and fill out all surface asperities, thus blocking any possibility of leakage. Now, internal gas pressures were applied that could never be reached when only elastic deformations were present, not even with highly lapped and honed contact surfaces. It was further found that the contact pressure that produced permanent surface deformation was far beyond the yield strength of the construction materials of the interface components.

As a final conclusion it can be stated that an absolutely leakage-free system can be established by using a contact pressure under which the interface areas are capable of being permanently deformed. This in turn means that the bolt forces must be high enough in a practical joint to produce permanent deformation in the contact areas of the joint com-ponents.

4. TESTS TO DETERMINE THE MAGNITUDE OF THE ACTUAL TIGHTENING FORCES. To prove this theory further and to obtain solid numerical data on bolt force requirements for high-pressure service with solid metal-to-metal seal contact, a series of special tests was arranged using various flat contact areas. Devices with line contact with either one-line and/or two-line contacts for high-pressure sealing were also incorporated. The steels of the joint components were chosen from categories with a yield strength of $\sigma_y = 32$ kg/mm² and an ultimate tensile strength of about 70 kg/mm², corresponding to about 45,000 psi yield and $\sigma_u \sim$ 105,000 psi ultimate tensile strength. The tests were conducted in such a way that the yield strengths of the contact components were exceeded with absolute certainty.

The results of these tests are illustrated in the curves of Fig. 1.11, where three test devices were subjected to leak tests. One set of interfaces was provided with a surface finish of approximately 25 μin. rms, another set had a surface finish of about 12 μin. rms, and the third set was prepared to a finish of better than 1 μin. rms. All three devices were pressurized by compressed air.

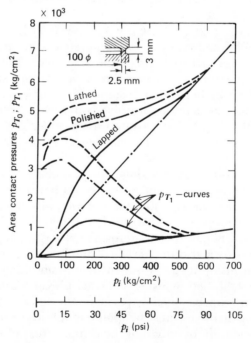

Fig. 1.11 Influence of surface finish (< 1–12–25) on sealing behavior.

The graph shows that the honed surfaces require the lowest relative seal contact forces. For the polished device the initial contact pressures are noticeably higher. Finally, as must be expected, the device with the roughest surfaces requires the highest contact forces to establish the seal.

The graph further shows an interesting phenomenon. Once the permanent deformation of the contact areas has been reached all three curves join to one curve and then follow from this point on a straight line that passes through the origin of the coordinate system. In other words, the seal contact pressures now become linearly proportional to the internal pressures they have to seal.

The area ordinate at which the three seal curves join the straight line corresponds to a hydraulic contact pressure of approximately 6500 kg/cm^2, which in turn is slightly higher than twice the yield strength of the component material forming the interfaces.

In addition to the p_{T_0} curves the graph of Fig. 1.11 also presents the p_{T_1} curves, which indicate the real tightening curves for the condition that the internal pressure is acting. The p_{T_1} curves are computed from the difference between P_0 and P_1 contact forces.

Figure 1.11 proves that sealing metallic contact areas against internal gas pressure is possible only once the interface areas have been permanently deformed. The pressure required to produce this amount of plastic deformation must at least exceed the value of twice the yield strength of the component materials forming the interfaces. Once this plastic deformation has taken place the surface requirements of the contact areas no longer affect the leakage behavior of the interface components of the joint system.

The same type of tests were repeated using the same grades of surface finish in the seal contact areas with the difference, however, that the pressure medium was changed from air to viscous mineral oils.

The results are shown in the graph of Fig. 1.12. The curves are almost a repetition of those of Fig. 1.11. This leads to the conclusion that the surface finish and the pressure medium no longer influence the leakage behavior of the interface areas once these seal areas have been permanently deformed by the hydraulic prestress forces, applied to establish the initial seal.

The defined location on the ordinate of the coordinate system for the value of the required prestress force for tightening is in excess of twice the yield strength of the seal component materials.

5. TESTS USING PRELOADING OF THE FLAT INTERFACE AREAS FOR PERMANENT DEFORMATION.   The next step of interest was to investigate the influence of preloading the test device before pressurizing the chamber.

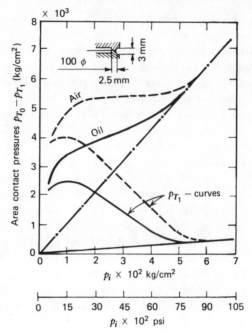

Fig. 1.12 Influence of surface finish (12–25) on sealing behavior.

Three levels of preload were chosen, two levels below the yield strength and one level in close proximity to the yield strength. The actual data were 5000, 5500, and 6500 kg of preload.

The preload procedure was as follows: First a hydraulic load of 5000 kg was applied that was subsequently discharged to zero load. Then a leak curve was established with this preload. The second test device was preloaded with a hydraulic load of 5500 kg, also with subsequent discharge to zero. Then the leak curve was developed. The third testing device was preloaded with a hydraulic pressure of 6500 kg, discharged to zero level and then leak tested.

Once the three steps of preloading were achieved, leak test results were obtained as shown in the graph of Fig. 1.13.

The slope of the test curve with 5000 kg preload is very steep. At 5500 kg preload the slope angle decreases and tends to move closer to the curve with a preload of 6500 kg. This latter curve coincides with the straight line through the origin, as would be expected.

The slope of the p_{T_1} curve through the origin reflects the actual sealing

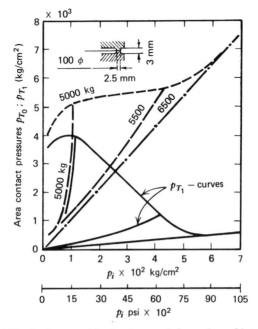

Fig. 1.13 Leak tests with air after predeformation of interfaces.

characteristic of the test devices under consideration. This sealing characteristic can be mathematically expressed as

$$k_T = \frac{p_{T_1}}{p_i} \qquad (7)$$

6. THE INFLUENCE OF THE WIDTH OF THE SEAL INTERFACE ON SEAL BEHAVIOR. To study the influence of the width b_T on the seal behavior of flat interface areas three devices were tested designed with a width of 1 mm, 2.5 mm, and 6 mm. Each device was preloaded with a load of 6500 kg. The test results are illustrated in Fig. 1.14. All p_{T_0} curves are straight lines through the origin, as was expected. All p_{T_1} curves fall together into one single straight line, as was also expected. This summary p_{T_1} curve reflects the tightening characteristics of all test devices with flat contact interface areas and can be expressed as

$$k_T = p_T = \frac{p_{T_1}}{p_i} = 1.50 \qquad (8)$$

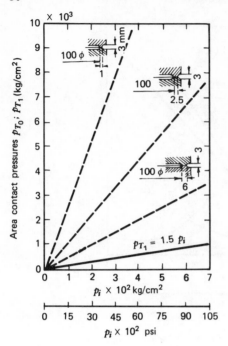

Fig. 1.14 Influence of width on interface and leakage behavior, tested with air (b_T = 1–2.5–6 mm).

Thus flat seal areas can be sealed against internal gas pressure if the preload is high enough to produce permanent deformation of the contact areas. To be exact, this load must be at least twice the yield strength of the contact materials. Since this load is a direct function of the width of the seal areas, this width should be designed for a minimum. The BASF Company of Germany uses the following relationship, which agrees with the author's experience,

$$\frac{d_o}{d_i} \geqslant \frac{1}{\sqrt{1 - P_0/\sigma_u}} \tag{9}$$

with d_i, d_o to be the internal and external diameters of the flat faces, σ_u the ultimate tensile strength of the contact materials, and P_0 the bolt force. The curves of Fig. 1.15 prove that a certain optimal width is required for satisfactory sealing under reasonable bolt forces.

7. SEALING BEHAVIOR FOR DEVICES WITH LINE AND DOUBLE-LINE CONTACTS. Wide and flat sealing areas require considerable bolt force to establish satisfactory sealing conditions when high-pressure fluids and/or gases are handled. This forces the designer to reduce the sealing width.

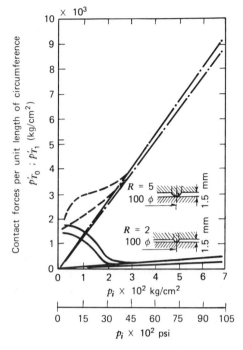

Fig. 1.15 Influence of the profile radius of the tongue on the leakage behavior (test medium air, $R_1 = 5$ mm, $R_2 = 2$ mm).

The question comes up of whether a sharp line of contact would be ideal to permit a concentrated sealing force thus reducing the total seal contact pressure to a minimum. For the solution of this problem a series of tests were conducted using curved shoulders with a spherical contour to be in contact with a flat area for sealing. Two devices having a spherical shoulder were investigated, one with a radius of 2 mm and the other with a radius of 5 mm. The profiles are shown in Fig. 1.15 together with the corresponding test results.

The tests prove that for sealing of gas pressures in the presence of metal-to-metal contact a certain minimum contact area is essential. Sharp edges will always result in leakage. Figure 1.15 indicates clearly that the 5 mm radius surface requires a smaller contact force than the device with the 2 mm radius in the curvature of the contact face.

With line contact the seal curve also converges with the straight line through the origin as soon as a permanent deformation of the contact surfaces is achieved.

b. Summary Conclusions. Before summarizing the results of these tests it may be stated that a complete series of tests was also conducted under the same basic conditions, but using solid metal gaskets (lens rings). The results obtained are practically the same as those previously described. To avoid repetition the curves are not shown here, but the results of Fig. 1.15 are fully applicable to joints using solid metal gaskets.

In a comprehensive series of tests to clarify the conditions for sealing metallic interface areas against internal gas pressure for the purpose of establishing fundamental concepts for high-pressure flange joint design, significant results were achieved that may be summarized as follows.

1. Contact forces or contact pressures that produce elastic deformations of the seal contact areas of a metallic interface do not suffice to establish a gastight seal at high pressure. Even in the presence of perfect contact interfaces with supersurface finish, leakage will still occur.

2. Sealing under elastic deformation conditions is possible only if the surface finish conditions are extremely high and the interface contact areas are rubbed with a viscous oil that provides an oil film on the seal areas. Similar results are obtained with very thin metal foils placed between the contact areas and then pressurized by the hydraulic load or a corresponding bolt force. The metal foil or the oil film easily fill the surface asperities of the seal contact areas and thus prevent leakage.

3. For metal-to-metal seal interfaces an unconditional seal can be achieved only if the contact areas are permanently deformed by the seal contact pressure. The metals fill the surface asperities in the moment of plastic flow and form a tight seal. The contact pressure required to produce the permanent deformation that prevents gas leakage must be at least twice the value of the yield strength of the materials of the components forming the interfaces.

4. When gaskets are used they can preferably be preloaded, again using a load of twice the yield strength. Beyond this critical point the seal contact pressure becomes directly linearly proportional to the internal pressure to be sealed.

5. When solid metal gaskets are used for sealing of high internal gas pressures, an optimal width of interface area is mandatory that must be wider than a sharp line. The minimum width required can be mathematically computed. As a rule of experience the minimum contact area should be no less than $\frac{1}{16}$ to $\frac{3}{32}$ in. As an example the German BASF-style lens ring joint may be mentioned, to be discussed in a later section. Here the line contact for sealing high-pressure gases may range in width from $\frac{1}{16}$ to $\frac{3}{32}$ in.

6. Line contact with a sharp distinguished line in one of the components is not desirable. This is a theoretical concept only. A perfect seal with such a design cannot be achieved.

7. Although a high degree of surface finish of the contact components in high-pressure joints is not required, a minimum of 16 μin. rms should be used to keep the contact forces within reasonable limits.

4. Tightening Factor

With regard to the tightening factor as a basis for the computation of the bolt forces for the joint, the following guidelines, derived from over 40 years of experience in the high-pressure field are recommended:

1. A tightening factor of $x = 1.50$ as determined by comprehensive experimentation is satisfactory and will provide reliable service for all static pressure conditions up to 1000 atm, ambient temperatures, and moderate sizes, not exceeding 2 in. in ID.

2. For pressures in excess of 1000 atm and up to 2000 atm and in the presence of fluctuations in pressure and temperature, and for designs with cylinder diameters of more than 2 in. Ø, the tightening factor should be raised to $x = 2.0$.

3. For pressures exceeding the level of 2000 atm and larger cylinder diameters for service with pulsating pressures and temperatures, the tightening factor x should be chosen in the range between 2.0 and 3.0, depending on specific plant site service conditions.

4. For pressures in excess of 3000 atm and pulsations of pressure and temperature, regardless of size of the cylinder diameter, the tightening factor should range between $x = 3.0$ and 3.5.

5. In using solid metal gaskets such as lens ring, octagonal ring, and the like, a difference in hardness between the contacting components is mandatory. This difference should amount to at least 50 Brinell units, with the metal gasket being the softer component. Gaskets are easier to remachine once they have been permanently deformed when pressurized by the bolt force.

The increase of the tightening factor in excess of the value of 1.5, as determined by reliable experiments, is based on a variety of factors learned about through the experience of many years in high-pressure service in the field and in all kinds of industrial production. Bolts, nuts, washers, and flange rings show noticeable losses in their assembly conditions after a certain exposure time. Other factors to be considered are material inhomogeneities, errors in assembly accuracy and machining, and the usual human errors.

Once the tightening factor has been established an economical and accurate design of the joint components can be made. The tightening factor is the cornerstone for the evaluation of the joint components. Once the tightening factor is known the computation of the bolt force is no longer a problem. The determination of the flange dimensions in connection with the bolt circle diameter and the mean tightening diameter involve nothing but ordinary design routine.

C.   Design Configurations of Gaskets of Metal and Their Combinations

The design of metallic gaskets is governed exclusively by the requirements of the sealing mechanism. The tests discussed in Section II.B.3a reveal that in metal-to-metal contact sealing there are two factors to be considered: the area contact and the line contact. In both cases sealing is established by permanent deformation of the interfaces. In configurations with line contact the final sealing effectiveness is achieved when the line is first widened to a small area and then "pressure welded" as a result of the permanent deformation. The so-called line wedge principle widens to an area contact principle with small cross-sectional area, large enough to establish a seal.

Now a third concept must be added that is applied in resilient metal gaskets, which utilize resilient components that act as springs. To increase the seal effectiveness further, the gaskets are placed in restraint cavities and the spring-type arms are inserted into specific grooves where they undergo a prestress load when the joint is tightened, thus sealing effectively and still offering some space for flexible movements of the springy arms. The configurations illustrated in the following tables are self-explanatory.

The category of resilient metal seals is largely used in applications where sealing efficiency is combined with the extended temperature capability of metallic gaskets. The basic structural element is usually a high-strength metal, sometimes combined with a soft coating of metal or plastic to provide the actual sealing. Like O-rings, these seals are self-energizing, have small cross sections, and require relatively low loading forces. They are generally reusable and thus offer almost indefinite life. Unlike O-rings, however, they are expensive and their availability is limited as far as diameter of joint is concerned. They should not be chosen in place of O-rings. Resilient metal seals and even O-rings become very costly once they exceed certain optimal size limitations—in other words, once they must be custom-made and are no longer shelf items.

Taking into account the complexity of the seal analysis and further considering the lack of essential information other than experimental data, the wide range of possible design configurations for static seals may be grouped roughly into major categories as follows:

1. Metallic gasket seals, nonenergized by internal pressure.
2. Metallic gasket seals, energized by the internal pressure.
3. Combination of materials, energized by internal pressure.
4. Combination of materials, nonenergized by internal pressure.
5. Nonmetallic gasket seals, energized by internal pressure.
6. Nonmetallic gasket seals, nonenergized by internal pressure.

1. Metallic Gasket Seals, Nonenergized by Internal Pressure

Metallic seals that are not designed to take advantage of the existing internal pressure to improve the sealing effect are basically designated as nonenergized gasket seals. A typical example is a flat gasket placed in between two raised-face flanges. Nonenergized gaskets require an external force at all times to establish and maintain contact in the interface areas to tighten the system. Since the theoretical and actual contact pressure is directly proportional to the internal system pressure, a high force is required to provide leak-free service. Since the interface areas must be permanently deformed, it becomes mandatory to design for minimal contact area width. The best approach, therefore, is the near-line contact, which provides optimal seal conditions with minimal contact forces.

Nonenergized gaskets are strongly susceptible to pressure and temperature pulsations in connection with the relaxation effects of the bolt-nut combination, resulting in bolt load reductions and subsequent leakage. Designs considering these gaskets must provide high preloads with an increased tightening factor and must incorporate compensation devices in the form of Belleville springs or similar spring devices to establish a safe joint counteracting the pulsation action.

Surface topography in the interface areas is considered a function of the internal pressure and the nature of the construction materials that provide the seal. Since permanent deformation of the contact areas is mandatory, a surface finish of better than 16 μin. rms is not required. Wherever machine marks are still visible they should be concentric with the bore of the gasket. Radial marks are critical.

Using ground and superpolished contact surfaces without a gasket in between is not only costly and absolutely uneconomical, regardless of the pressure level to be sealed, but it is also ineffective, as was shown in Fig. 1.5.

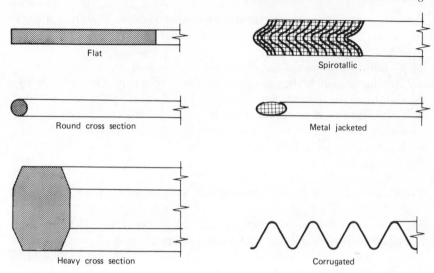

Fig. 1.16 The six basic static gasket configurations (From *Condren Gasket Catalog*, courtesy of Condren Company).

Gaskets based on the nonenergized design principle are not suited for service with high temperatures in the creep range.

Looking at the design components, metallic seals can be derived from six basic design configurations: flat metal gaskets, round cross section, heavy cross section, corrugated gaskets, spirotallic gaskets, and metal-jacketed gaskets. These basic components are presented in Fig. 1.16.

a. Flat Metal Gaskets and Their Design Configurations. In general industrial practice the flat metal gasket is perhaps the most misunderstood gasket in view of scientific computation. It is generally applied in excessive widths although the required bolt load is directly proportional to the width of the gasket. Increasing the width does not improve the sealing efficiency of the joint. Further, when the width of the gasket seal face area is increased, a higher degree of machining is required that may become intricate and costly when heavy equipment is involved.

A suitable flat gasket design therefore calls for a design configuration of minimal width, as can be derived from Table 1.2, which illustrates a variety of modifications of the flat gasket.

The numbers given with the figures indicate the gasket designations for identification in the Condren catalog.

Gasket #940 represents a plain solid and flat ring gasket, easy to

Table 1.2 Flat Metal Gaskets and Their Modifications

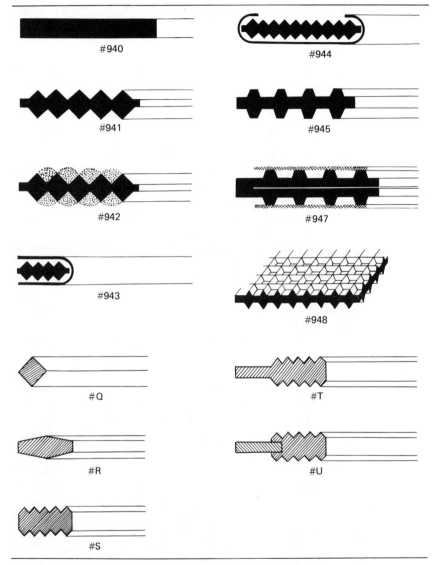

#940

#944

#941

#945

#942

#947

#943

#948

#Q

#T

#R

#U

#S

Source: Condren Gasket Catalog, courtesy of Condren Company.

fabricate in any desirable width. The material of which the gasket is constructed must be softer than the material of the contacting flange faces. High bolt load is required to establish plastic deformation of the gasket interface areas. This gasket is not suited for large-scale equipment.

Gasket #941 illustrates a profile gasket that provides solid metal and a multitude of line contact areas with differential pressure spaces in between by offering V-shaped ribs in the seal surface contact areas. Specifically suited for screwed attrition closures where low friction is mandatory. Available in standard dimensions for raised face flanges, large tongue and groove flanges, and male-female flanges.

Gasket #942 is a profiled gasket with V-shaped grooves filled with twisted and treated asbestos cord cemented between the grooves on both sides, providing good sealing effectiveness in high-temperature and high-pressure service. Easy to apply even on rough and uneven contact surfaces.

Gaskets #943 and #944 are profiled gaskets with a metal jacket of either the French type (narrow gaskets) or the single-jacketed type for wider gaskets. The V-shaped ribs subdivide both contact areas into a series of subspaces to improve sealing efficiency and reduce contact force. Smooth surface prevents scoring of flange contact areas. Consultation with the manufacturer is recommended.

Gasket #945 is a serrated gasket consisting of solid metal with spaced concentric ribs. Well suited for high-pressure service and for high temperatures and corrosive atmospheres. Smooth contact areas in the flange faces are required.

Gasket #947 is a bellowseal gasket. Similar to the #945 gasket but designed as bellowseal, utilizing the self-energizing sealing principle for increased tightening without increasing the bolt force. A ring of thin, compressed asbestos sheet packing is placed on top of the serrations on both sides. This gasket is made of two metal discs, each serrated on one face and welded together on the inner rim. This gives the bellow construction sufficient flexibility to allow it to "breathe" in the presence of abnormal thermal movements of the joint components. However, the gasket is not designed to be used as a tube expansion joint.

Gasket #948 illustrates a typical multiseal device, made of solid metal with raised cross-sectional ribs providing a wafflelike surface pattern. The seal area offers a multiple interface design. Gasket metals should be soft, such as aluminum, copper, and soft steels. Ribspacing is in $\frac{1}{16}$ in. squares.

Gasket Q has a single diamond as a cross section that matches a square groove. With slightly differential angles in the contact peaks excellent sealing efficiency can be achieved.

Gasket R is a flattened diamond. Service conditions similar to those described for gasket Q.

Gasket S has a regular comb profile. Similar advantages to those described for gasket #941.

Gasket T is similar to gasket S; however, one piece has centering ring.

Gasket U is the same as gasket T, but the centering ring is separate.

b. Round Cross-Sectional Gaskets of Metal.   The round cross-sectional gasket is an approach to line contact design for lowering bolt load without sacrificing sealing efficiency. Made from round wire of the desired solid diameter, the gasket is designed for insertion either in special grooves or in male-female flange face combinations. The use of flat flange faces without special recess face configurations is not recommended for this style of metal gasket.

The three basic designs of round cross-sectional gaskets are presented in Table 1.3.

Gasket #903 half illustrates a half-sectional view of the round gasket partially wrapped with a jacket of the same or softer metal for application between nongrooved flanges. The external edge of the jacket is used for centering purposes. The metal core is usually hard, to allow radial reinforcement, whereas the jacket is the softer part for improved sealability.

Gasket #903 full represents the full-sectional round gasket using two concentric metal core rings enclosed and spaced by a soft jacket. The internal wire normally has a cross section 0.020 in. wider than the wire of the external wire ring. The inner wire ring assures full seating load at the gasket ID and the external ring provides a limitation of the maximum possible flange deflection. The bolt holes are located in the web of the jacket.

Table 1.3   Round-Cross-Section Gaskets

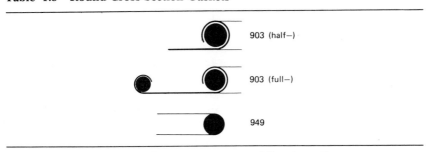

	903 (half–)
	903 (full–)
	949

Source: Condren Gasket Catalog, p. 5, courtesy of Condren Company.

Gasket #949 illustrates the plain round nonsectional wire ring gasket. The flanges should be adequately grooved to assure accurate installation of the gasket. Not recommended for flat face sealing contact.

Sealing efficiency can be markedly improved by plating the contact areas with metals or coating with Teflon, Kel-F, and similar plastic materials where temperature is not a problem. Plating of gaskets with a thin layer of soft metals such as silver, copper, or nickel permits good sealing of the joint, reduction of the bolt contact force, and size of joint geometry.

When gaskets are plated the plating thickness must be chosen on an individual basis, varying from case to case. The layer thickness may range from two thousandths to six thousandths of an inch, depending primarily on the particular conditions. The plating layer usually deforms by itself without affecting the base material underneath. Consequently the bolt force for establishing seal contact can be reduced.

2. *Metallic Gasket Seals, Energized by Internal Pressure*

A gasket design is termed self-energizing when the internal pressure provides an increase in sealing effect without requiring additional bolt force for tightening. As a result an increase in internal pressure assures improved sealing efficiency without calling for higher bolt force. The bolt load must, however, suffice to maintain a tight seal at all times and prevent separation of the flange interface areas. A self-energizing gasket design principle does not assure complete tightness of the system. The prestress of the bolts must satisfy all possible and conceivable service conditions, including eventual pulsations in pressure and temperature. This in fact is just one of the reasons why in an earlier section a tightening factor of 3.50 is given for superpressures in spite of the use of self-energizing heavy-sectional metal gaskets.

The category of self-energizing metal gaskets may be subdivided into two basic groups distinguished by function. One group represents heavy-sectional gaskets and the other comprises the resilient metal gaskets.

a. The Rigid Heavy-section Self-energized Metal Gaskets. The category of rigid metal gaskets with heavy cross sections represents the major family of gaskets predominantly used in all kinds of equipment serviced in general and specialized high-pressure. Because of the nature of their application in chemical and related process industries they are designed as self-energizing gaskets. The major and basic components are illustrated in Table 1.4 as they are preferably used in the field all over the world.

Table 1.4 Heavy-Cross-Section Gaskets

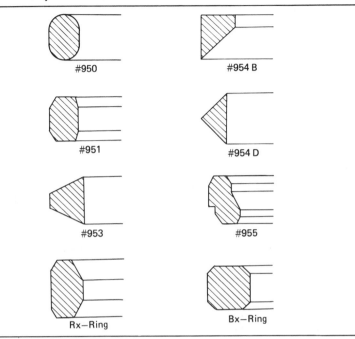

#950	#954 B
#951	#954 D
#953	#955
Rx—Ring	Bx—Ring

Source: Condren Gasket Catalog, courtesy of Condren Company.

Gaskets #950 and #951 illustrate solid metal gaskets designated as oval and octagonal ring gaskets. They have to match both standard and special grooves for applications almost entirely in the oil and petrochemical industries and liquid hydrogen and oxygen services. They are applied preferably in equipment operated at pressures up to 10,000 psi. Octagonal rings, often referred to as V-Tite gaskets, are interchangeable in flanges of the newest flat and bottom groove design; however, only the oval cross section can be used in the old type round-bottom groove.

Gasket #953 reflects the so-called English style lens ring gasket, which seals theoretically by a line contact. Details are shown in a later section of this chapter.

Gasket #954, the Bridgman ring gasket, is of a special design that also utilizes the wedge sealing principle. Well suited for joints where pressure-temperature pulsations and shock must be encountered.

Gasket #954, the delta ring gasket, is an elegant design providing a seal with dual line contact. Details and application are discussed later in this chapter.

Gasket #955 illustrates the V-Tite Transition ring. Used for ring-type joints in which the internal diameters of the joining tube faces are different. Often made with oval or octagonal contact facings.

Gaskets #951 BX and #951 RX show special modifications of the oval and octagonal seal rings with similar sealing characteristics.

Tables 1.2, 1.3, and 1.4 are from a catalog published by the Condren Corporation of North Brunswick, New Jersey. These gaskets were formerly fabricated by the Johns-Manville Company of Denver, Colorado. Their metal gasket facilities have recently been sold to the Condren Corporation.

The lens ring in general is one of the most congenial gasket designs in the history of the development of high-pressure equipment sealing. It is used for piping, tubing, machinery, and all kinds of equipment. The application of the lens ring, designated as #953 in Table 1.4, is characterized by a straight line as sealing slope face. This type of lens ring gasket configuration is shown in a special application to seal a high-pressure tube joint in Fig. 1.17. For comparison we show the conventional German style lens ring gasket, which has a seal face of spherical shape with a defined radius in relation to the tube ID, illustrated in Fig. 1.18. The difference in the design of the two lens rings is obvious. In the straight slope configuration the contacting tube face has a sharp edge to provide some sort of a line contact for seal, establishing the seal at the ID of the tube. The sharp edge of the tube face "bites" into the seal face of the lens ring forcing the lens ring to have a smaller ID. This corner can be a detrimental flow restriction depending on the condition of the

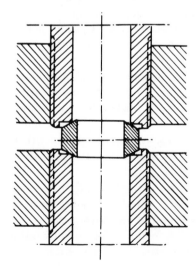

Fig. 1.17 Joint using British-style metal lens ring.

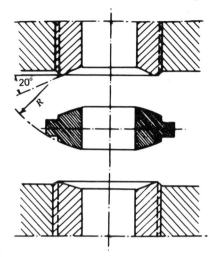

Fig. 1.18 Joint using German-style metal lens ring.

system fluid. High viscosity of the fluid can have a significant influence on the development of the flow pattern at this restrictive corner.

The design of Fig. 1.18 represents the more common lens ring as developed by the BASF of Ludwigshafen/Rhein in the early phases of high-pressure syntheses with large-scale heavy equipment in the first decade of this century. The ID of this lens ring is the same as the ID of the adjacent tube faces, providing a flow passage area without restriction of any kind. The seal interface is formed by two planes. The tube face is a straight cone with a 20-degree slope contacting a spherical plane of the lens ring, thus establishing a theoretical line contact that is in reality a small plane to satisfy minimal sealing requirements. This design allows repeated applications before refinishing of the lens ring surface becomes necessary.

For pulsating pressures in the presence of elevated temperatures the bellows-type lens ring should be used, shown in Fig. 1.19. This lens ring is provided with an interior gap and a bleed hole for the pressure. After machining is completed the ring is compressed by a certain predetermined plastic strain to a shape shown in the illustration. This kind of ring is specifically suited for utilization of the self-energizing sealing principle. Pressure of any level will not blow out the ring, but will increase the sealing efficiency instead.

The delta ring is a simple sealing device with excellent self-sealing effectiveness, as can be concluded from Fig. 1.20. Part (*a*) reflects the condition of assembly at zero bolt load. When bolt load is applied (*b*) the Delta ring is compressed so that the external flange sections contact each

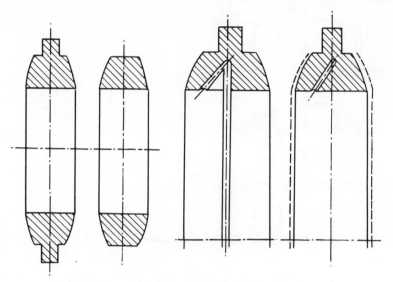

Fig. 1.19 Standard German-style lens ring designs.

other. At this stage the delta ring functions as a spring, following the flange movements should pulsations of the pressure occur. If the pressure still increases markedly, the bolts stretch slightly; the flanges move apart by a minute amount. Leakage, however, does not occur since the delta ring follows the flange movement, releasing some of its compression but still maintaining a tight seal, as illustrated in Fig. 1.20c.

An application of the delta-ring gasket is shown in Fig. 1.20d, where it seals an autoclave closure. The transition device in the autoclave closure serves as a flow reducer.

b. Resilient Metal Gaskets.   Resilient metal gaskets comprise an enormous range of metallic design configurations with highly diversified sealing principles. The simplest type of flexible metal gaskets are the corrugated devices, followed by the family of flexotallic gaskets. Another significant category may be classified as special design configurations developed to satisfy specific sealing purposes.

1. CORRUGATED METAL GASKETS. Corrugated metal gaskets actually represent a family of serrated gaskets applicable for moderate pressure ranges not exceeding the 600 psi level. Three basic gaskets representative of this category of design are illustrated in Fig. 1.21, from the *Condren Gasket Catalog.* Gaskets similar to these are manufactured by all major metal gasket fabricators in practically all industrial countries.

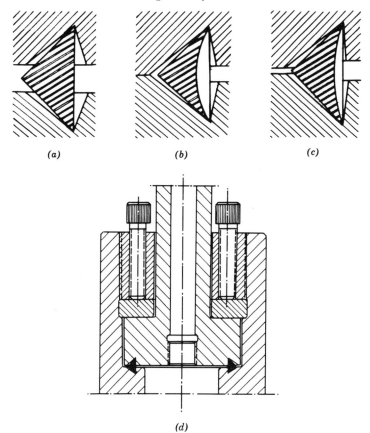

(a)   *(b)*   *(c)*

(d)

Figure 1.20   (*a,b,c*) Designs of delta ring gasket, (*d*) Delta ring gasket in autoclave closure.

Gasket #900 represents an all-metal, plain, corrugated gasket used for low-pressure service conditions, requiring a thin line contact type of seal because of space limitation and weight. The metal thicknesses may range from 0.010 to 0.031 in., depending on the metal and the nature of the corrugation pitch. The average design suggests a minimum of three corrugations; however, there are applications with one or two. A slight flat inside the inner and outside the outer corrugation contributes to the stiffness of the gasket.

Gasket #904, the Flexseal gasket, has corrugations covered with a jacket of woven asbestos. The cloth combines the conformability of the soft jacket and the line contact labyrinth reinforcing characteristics of the corrugated metal filler. The asbestos cloth may be either plain or rubberized.

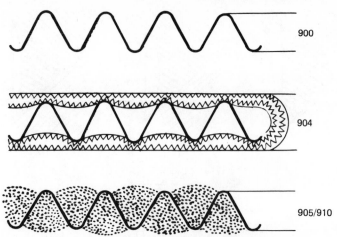

Fig. 1.21   Corrugated metal gaskets (From *Condren Gasket Catalog*, courtesy of Condren Company).

Gaskets #905 to #910 consist of a #900 design with asbestos cord cemented into the V-shaped corrugations. The metal is generally 0.019 to 0.020 in. thick with a corrugation pitch of $\frac{5}{32}$ to $\frac{3}{16}$ or $\frac{1}{4}$ in., depending on the gasket width or kind of application.

Temperatures of up to 1150°F and pressures not exceeding 600 psi are tolerated.

2. FLEXOTALLIC METAL GASKETS.   Flexotallic (also termed *spirotallic*) metal gaskets are multipurpose sealing devices, spiral-wound and consisting of preformed V-shaped metal strips wound into a spiral. The various metal layers are separated by a filler of any kind, ranging from asbestos strips to fibers and Teflon. The basic customary designs are illustrated in Fig. 1.22.

Spirotallic gaskets are available in five major designs with and without centering devices and inner and outer rings for the control of maximum compression during tightening. Materials of construction range from carbon steels to all types of stainlesses, Inconel, Hastelloys, Carpenter 20, titanium, Monel, nickel, copper, aluminum, and even silver.

The sealing mechanism is the result of a combination of the yield and flow of the metal and soft filler plies, when the gasket is compressed during the tightening operation to a predetermined degree. Metal-to-metal plies making up the inner and outer edges must be under compression.

Spirotallic gaskets provide greater resiliency than any other metal-

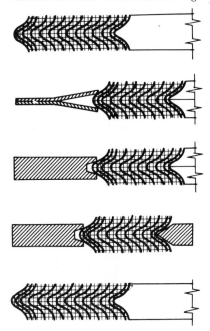

Fig. 1.22 Spirotallic metal gaskets (From *Condren Gasket Catalog*, courtesy of Condren Company).

asbestos gasket type. They offer high mechanical strength. Further details are comprised in the *Condren Gasket Catalog*.

They can be used for joint sealing with flat face, raised face, male-female combinations, or even tongue and groove designs. Their elastic and springy behavior makes them excellent candidates for high-temperature service conditions.

3. SPECIAL DESIGNS OF SELF-ENERGIZING RESILIENT METAL GASKETS. Resilient metal seals are offspring products, resulting from aerospace developments. They are now strongly penetrating general commercial applications at all levels wherever extreme service conditions must be met. Their sealing mechanism is characterized by resilient members of a metal gasket, providing springy action of a high degree.

Resilient metal seals combine the practicality of elastomeric component seals with the extended high-temperature capability of flexible metals, utilizing the internal system pressure for self-energizing sealing purposes. This in turn results in a reduction of the closing forces for the joint. Resilient metal gaskets can be repeatedly used without remachining.

In Fig. 1.23 the basic shapes of resilient metallic gaskets are illustrated.

Fig. 1.23 Basic shapes of resilient metal gaskets.

From these initial designs practically all other design configurations can be derived.

The sealing mechanism of typical resilient metal gaskets may best be described by the discussion of the Bobbin Seal, which has been developed for rocket propulsion systems. This seal exhibits all the basic sealing principles expected from a resilient metal seal.

Bobbin Seal. The Bobbin Seal actually reflects the principle of two integrated Belleville springs attached in reversed position to a short tube section, combined to one solid piece of metal. The cavity of the connector components is smaller in height in assembled condition when the faces are in mutual contact than the height of the Bobbin Seal before compression is applied. When the joint is tightened to its final service condition the Belleville sections are compressed, forming an interference fit with the cavity of the connector components, thus establishing a perfect seal that maintains its spring capability at all times, regardless of pressure or temperature pulsations. The final position of the connector faces is achieved after tightening is completed and when flange faces are in solid contact with the central cylindrical section of the Bobbin Seal, shown in Fig. 1.24.

In final sealing position the contacting interface areas are plastically deformed, achieving a perfect seal, and the contact force is independent of cyclic loading pulses.

Initial clearance is preferably held between 0.002 to 0.008 in. First contact during assembly with the flange faces and the Bobbin springs is established at point A. By increasing the bolt load, face B of the Bobbin spring finally establishes contact with the connector wall and reaches interference condition along the sealing areas. As deflection further increases contact, stress on face B surfaces exceeds the compressive yield limit of the construction material of the seal. When the seal is fully

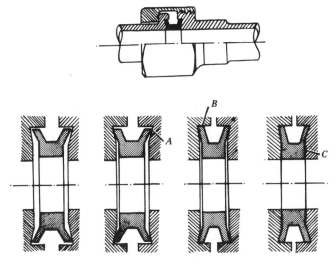

Fig. 1.24 Bobbin seal, phases of assembly.

closed, reaching the final assembly stage, contact at faces C is made and a solid joint is reached between connectors and the cylindrical body of the Bobbin Seal.

Resilient metal gaskets are frequently used and can be applied with or without constraining groove in the contact faces. Wherever this groove is geometrically possible it should be applied, since it facilitates plastic flow of the gasket material at low bolt load, practically reflecting the same seal relationship as described for the Bobbin Seal.

Clam Seal. The Clam Seal has been developed in the United Kingdom for application in aerospace and nuclear equipment. The sealing mechanism is simple and unique and establishes its own sealing surfaces. The gasket has sharp edges that bite into the softer countercontact areas, resulting in subsurface grooves. This is a complete reversal of the generally accepted sealing principle of metallic gaskets, where the interface environment surfaces are harder than the gasket itself. The Clam ring is assembled so that it can act as a spring, utilizing the self-energizing sealing principle. Details of the seal are illustrated in Fig. 1.25.

During assembly and tightening of the bolts the seal is compressed and bows slightly inward. When internal pressure is applied, the gasket is forced back toward its initial shape, using the self-energizing seal principle when the gasket edges are forced to penetrate deeper into the seal groove.

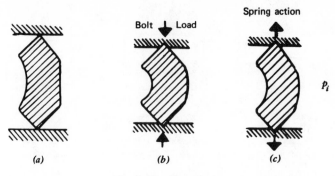

Fig. 1.25 Clam Seal.

Major applications of the Clam Seal are for fittings using widely modified seating configurations. These designs are particularly suited for sealing threaded connections of all kinds.

A modification of the Clam Seal is presented in the sketch for the Boss Seal of Fig. 1.26, representing a NASA development used to seal aerospace fluid systems. NASA claims maximum leakage rates in the order of 10^{-9} cm³/sec of helium for temperatures ranging from -450 to $1500°F$. Reports show that these devices have been used for pressures up to 30,000 psi. The gasket can be used repeatedly. The Boss Seal was developed by NAVAN Inc., a subsidiary of North American Aviation, Inc., El Segundo, California.

Expandable Seal—"Mott" Principle.   The Mott Seal is a gasket designed to utilize the toggle expansion principle, which creates a seal that takes

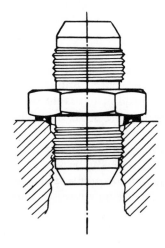

Fig. 1.26 Boss Seal as modified Clam Seal.

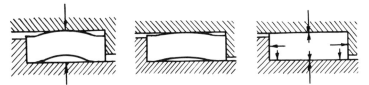

Fig. 1.27 Resilient metal seal, "Mott" Principle.

advantage of its own pressure. Gasket design and sealing mechanism of the Mott Seal are illustrated in Fig. 1.27.

In the initial condition before compression the gasket is higher than the cavity into which it will be forced when the joint is tightened. By tightening the bolts of the joint, the gasket flattens and tends to expand, thus increasing the contact forces against the corresponding seal faces, providing a tight interference fit with a strong locking effect between the joint components. As a result of the toggle design principle, the ratio of the radial force to the clamping force increases significantly as the toggle approaches its final flat position. When fully tightened, the high radial sealing force provides a perfect seal.

When the gasket is made of aluminum, which exhibits a coefficient of thermal expansion twice that of steel, the seal improves its effectiveness as temperature rises. To achieve appropriate seal efficiency it is mandatory to provide adequate dimensional tolerances for both mating face diameters. Tolerance requirements are usually offered by the seal manufacturer.

This basic seal principle can be modified in a number of ways, as by inserting flat rings into sloped grooves of special flanges. It is significant that the ring always acts as a spring and the volume of the gasket is slightly bigger than the volume of the seal groove cavity. This requires a given prestress to fit the seal into the cavity for sealing. Gaskets of this nature are generally not reused.

Grayloc Joint. The Grayloc joint is a unique design in several ways. Figure 1.28a illustrates a cross-sectional view of the joint, showing all components. As part of a piping system the Grayloc joint serves the same purpose as the standard flange joint. The hub element, which is the equivalent of a flange, is preferably welded to the pipe.

The clamps, which consist of two halves, hold the joint together. They are designed with a C-shape profile providing massive material concentration to satisfy maximum strength requirements. Four bolts establish the necessary rigidity. The claws of the clamp must be strong to take the axial forces developed by the internal system pressure. The bolts must establish a tight contact between gasket and clamp.

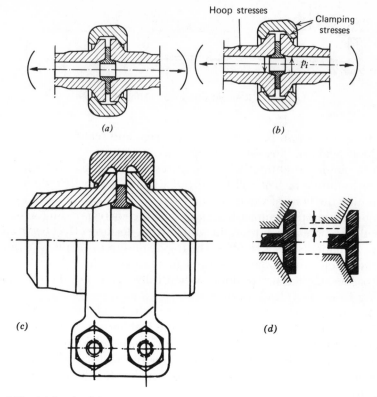

Fig. 1.28 (a) Grayloc joint, cross section. (b) Stress distribution. (c) Grayloc joint total. (d) Alignment faults.

The gasket actually represents a double-cone design with a shoulder ring permitting a wedge-type sealing contact with the sloped face of the tube hub, requiring minimum clamping force for theoretical line contact.

In Fig. 1.28b the forces acting on the joint are illustrated. This arrangement allows a minimum of bolts with smaller dimensions than is customary for standard flange joints without sacrificing stability.

Maintenance is not required during normal operation. Assembly, shown in Fig. 1.28c, must be done with care to prevent misalignment with the gasket as indicated in Fig. 1.28d. Cleaning the hub seat and the contacting gasket seal areas prior to assembly provides leak-free operation. Polishing of all contact surfaces with steel wool is mandatory.

A typical example of a Grayloc flange joint is shown in Fig. 1.29, representing the closure of a high-pressure vessel with an internal

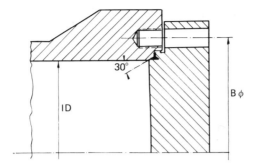

Fig. 1.29   Closure using Grayloc gasket.

diameter of 32 in. to be operated at a pressure of 4500 psi and a temperature of 430°F. This design is similar to the double-cone seal configuration developed by the BASF of Ludwigshafen some 70 years ago in conjunction with the famous ammonia synthesis, giving rise to the start of modern large-scale high-pressure technology.

3. Combination of Materials, Nonenergized by Internal Pressure

By combining metals with a soft material, combinations can be obtained that will do sealing jobs not possible with either one of the materials alone. Combinations of this kind are generally referred to as *jacketed gaskets*. They are made of a soft, easily conformable filler that is partially or completely enclosed within a metal jacket. When corrugated metals are used, the internal structure of the metal provides the stability support and the soft material surrounds the metal. The combination then establishes the final seal.

The primary seal to prevent leakage is the inner metal lap, where the gasket is thickest when compressed. This section flows, effecting the seal. The entire inner lap must always remain under compression. If there is an outer lap this then provides a secondary seal when compressed between the flange faces. Intermediate corrugation metals are often added for strength and resiliency, acting as labyrinth seals.

Jacketed gaskets allow more compression than the plain corrugated gasket group. They impart considerable compensation for flange irregularities and joint misalignments.

A series of major jacketed gaskets customarily used in all branches of industry is shown in Table 1.5. To summarize the details of Table 1.5 the gaskets consist of the following components:

a. Metals Used in Combination Gaskets.   A wide range of metals is used for the fabrication of jacketed gaskets, including lead, grade 1100 alumi-

Table 1.5 Metal Jacketed Gaskets

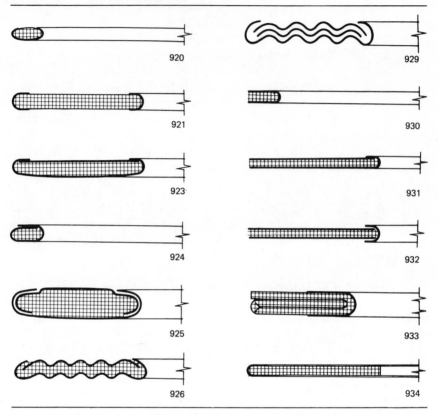

920	929
921	930
923	931
924	932
925	933
926	934

Source: Condren Gasket Catalog, p. 4, courtesy of Condren Company.

num, copper, the brasses, soft steels, nickel, Monel, Inconel, Hastelloy, and stainlesses of both the #300 and #400 series.

b. Fillers for Jacketed Gaskets.   Fillers are generally recruited from asbestos, either millboard or asbestos paper or both. Their application is normally limited in temperature to 850 to 900°F. In noncritical service the temperature limit may be raised to 1200°F.

Other fillers are preferably chosen from the Teflon family as virgin Teflon or filled with glass fiber as reinforcement to combat corrosion and temperatures up to 500°F, depending on the type of Teflon used.

c. Metals as Filler Material for Jacketed Gaskets.   When temperature is a problem, metals should be used to replace nonmetallic fillers in jacketed gaskets, especially #924 and #925. With metals as filler materials the upper temperature of operation may far exceed the 900°F limit. Consultation with the fabricator is recommended.

4. Nonmetallic Gasket Seals Energized by Internal Pressure

The nonmetallic gasket seals energized by internal pressure are characterized basically by gaskets that use a resilient nonmetallic member with a high degree of elastomeric memory. The memory is used to establish a seal under pulsating conditions of pressure and/or temperature, making the joint effectiveness independent of flange deflection and thermal expansion within a certain tolerable range. High resiliency is the prerequisite for perfect sealing, since the elastomer fills the surface asperities completely at relatively low contact forces. A compression of the gaskets within 15 to 20% of their initial cross section suffices to achieve a perfect seal within the elastic deformation range without imparting excessive load forces. A higher degree of compression is detrimental to the elastomer and will not improve the sealing efficiency.

The resilient elastomeric gaskets described in this section do not include the conventional O-ring designs, which are discussed in a later section of this chapter, with vacuum as the operating condition.

Using elastomeric sealing components automatically imposes a temperature limitation dictated by the chemical component composition. Consultation with the manufacturer is needed, since a wide range of elastomeric materials is involved. Their natural elastic behavior and their high degree of resiliency give them the ability to conform with the mating surfaces of the interface components, which in turn permits less stringent surface quality requirements. Surface finishes of 64 μin. rms are satisfactory.

a. Nonmetallic Solid Elastomeric Gaskets.   Nonmetallic elastomeric gaskets are available in two major groups, either solid or hollow.

The family of solid elastomeric gaskets comes in a large variety of cross sections. Some of the more frequently used gaskets, listed in Fig. 1.30, are made of Viton, rubbers of all kinds, and/or silicone rubber. Their sealing function is self-explanatory.

b. Nonmetallic Hollow Elastomeric Gaskets.   Hollow elastomeric gaskets usually have.a hose-type design to meet specific service requirements. They are generally internally pressurized by a gas or a fluid to give a high degree of sealing effectiveness. Where temperature is involved, air or nitrogen as pressure media can be replaced by a fluid, either water or

Fig. 1.30 Elastomeric solid gaskets, basic profiles.

a heat transfer fluid for cooling. The addition of suitable pressure control devices for the fluid or the gases is no problem.

This category of gaskets provides ideal sealing solutions to a variety of severe service conditions where other conventional devices are not available.

Figure 1.31 shows typical elastomeric hollow gaskets using additional secondary fluids to improve the seal requirements by introducing internal fluid pressure into the interior cavity.

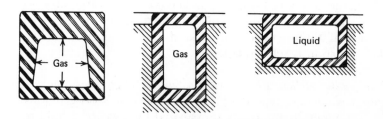

Fig. 1.31 Elastomeric gaskets with hollow cross section.

5. Teflon-Base Seals with Springs, Self-Energized

The Teflon-base seal is a spring-actuated seal of virgin or filled Teflon. A helical flat or round wire spring is placed within the circumferential cavity machined or molded inside the seal ring. Omniseal, BAL-Seal, and Fluorocarbon-FCS-Seals are typical representatives of this category of lip-type sealing configurations. Their sealing mechanisms are practically alike and function with high efficiency.

a. Omniseal Design. The Omniseal gasket was formerly a product of the Aircraft Division of Aeroquip Corporation. The seal fabricating facilities have recently been sold to the Fluorocarbon Company of Anaheim, California.

The design reflects a flat, helical spring incorporated in a C-shaped ring, made of Teflon. The springs provide after installation a constant contact force against the seal areas, because of compression of the springs during assembly. The installation is accomplished with an interference fit that results in springy counteraction of the spring to maintain automatic contact force for the seal, Fig. 1.32. The Omniseal must be installed in the seal cavity in such a way that the C opens against the internal pressure of the system in order to take advantage of the self-energizing sealing principle. Omniseal design configurations for dynamic load service conditions are described in Chapter 5.

b. BAL-Seal Design. The BAL-Seal is a U-shaped ring with the opening always directed toward the system pressure. A special metal spring, embedded inside the U-opening, tends to spread the U-legs constantly to establish automatic contact with the corresponding seal contact area, thus imparting a resiliency with memory into the system. The dynamic applications are discussed in Chapter 5 (see rotating shaft seals).

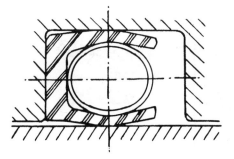

Fig. 1.32 Omniseal.

Fig. 1.33 BAL-Seal.

The edges where the seal contact takes place are strengthened by an increased cross section. The seal body is available in virgin Teflon, glass-filled Teflon, graphite-filled Teflon, or combinations with graphite or molybdenum disulfide. The springs can be chosen from any metal compatible with the process. A cross section of the conventional static seal is presented in Fig. 1.33. More details.are given in a discussion of dynamic sealing devices in Chapter 5. The BAL-Seal is a product of the BAL-Seal Engineering Company, Tustin, California, which offers a wide range of seal configurations of this type.

c. FCS–Fluorocarbon Seal.   The FCS-Seal fabricated by the Fluorocarbon Company, Anaheim, California, is also a Teflon-base ring seal with a metal spring actuation used for imparting resiliency with memory effect. Contrary to the BAL-Seal and Omniseal designs the spring of the FCS-Seal is not helical but made of a flat metal ribbon with special winding. The spring either is fully embedded in the Teflon material or is openly inserted into the spring cavity. Details of the FCS designs are presented in Fig. 1.34. The helical spring has limited resiliency and is formed of one piece of flat stainless strip, wound to form a U-shaped cross section. It exhibits an automatic load that provides continuous contact in both interface areas because of precompression at assembly.

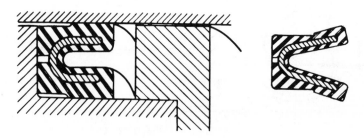

Fig. 1.34   Fluorocarbon-FCS Seal.

The FCS-Seal gasket is available in two basic types of sealing alloy materials:

1. Fluoroloy S is predominantly used, essentially consisting of Teflon. Resistance to wear for dynamic seals is 10 to 100 times that of virgin TFE without the abrasive character, normally found in wear-resistant compounds.
2. Fluoroloy SL is a TFE mixed with carbon and graphite, providing up to 1000 times the seal resistance of unalloyed TFE, according to the manufacturer.

More specific data on the dynamic behavior of Fluorocarbon-Seal SL is presented in the section on dynamic seals for rotating shaft equipment, Chapter 5.

D. Gaskets for Specific Closure Designs under Internal Pressure

The best, simplest, and most secure closure design is achieved when the head is welded to the vessel. However, this is accomplished at a considerable sacrifice, for the vessel can never be opened again without destruction of the weld.

The majority of vessels must be able to be opened, for a number of reasons. Many are designed with internal agitators, catalyst beds, heating and cooling systems, gas and liquid introduction pipes and instrument lines of all kinds, and internal heating and cooling, thus making the removal of the head mandatory once the life expectancy of the catalyst is terminated. Where cylinder heads are impractical, manholes must be used to make the interior of the cylinder accessible.

The quality of a closure design is judged by a variety of factors: simplicity of geometry, minimum labor in shop, transportation and in-field assembly, maintenance, and particularly the force required to tighten the closure properly, to permit long-term periods of safe continuous operation. The pressure level governs the size of the closure hole.

1. Closures with Gaskets Not Utilizing the Self-Energizing Principle

Closures that do not utilize the self-energizing sealing principle are not recommended for vessels where high pressures are applied. In services with small- to medium-sized autoclaves flat gaskets are often erroneously used, as shown in Fig. 1.35. Here the flat gasket is inserted between flat faces with and without an external support ring, in a conventional

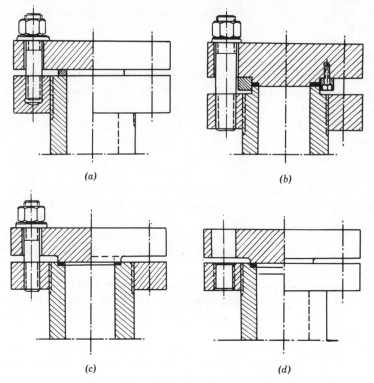

Fig. 1.35 Closures using flat metal gaskets.

male-female joint and in a male-female closure where the gasket is completely constrained. By using combination-type gaskets, closures can be used with considerably reduced contact loads for sealing. The designs of Fig. 1.35 show that these kinds of sealing devices can be costly without being specifically efficient.

2. Closures with Tongue-Groove Modifications

The conventional tongue and groove design basically uses a small and flat gasket for sealing. The tightening efficiency can be markedly improved by modifying tongue and groove geometry with serrations either in the face components or in the gasket or both. Use of the wedge principle in the tongue matching sloped grooves provides line contact with noticeably reduced contact force and good seal effectiveness. Figure 1.36 shows a series of groove and tongue modifications frequently used in equipment for pilot plant operations.

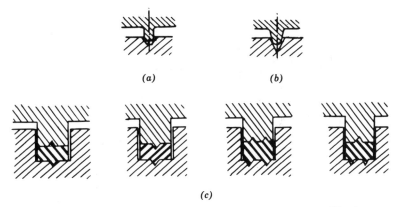

Fig. 1.36 (*a,b*) Profiled contact areas. (*c*) Tongue-groove modifications.

The softer these gasket materials are, the better the sealing efficiency. Plating the gaskets with soft metals enhances seal effectiveness and reduces contact force.

3. Closures Utilizing the Self-Energizing Wedge Principle

An elegant design that takes advantage of both the wedge-type sealing contact and the self-energizing sealing principle is based on the wedge ring as a metallic gasket. The closure may be designed for one wedge ring, two wedge rings, or a combination of two wedge rings with an O-ring where temperature permits on O-ring.

The sealing principle using the wedge ring gasket is illustrated in Fig. 1.37. The design of Fig. 1.37*a* uses a solid cover inserted on the top of the autoclave. A wedge ring gasket is placed between the cover and the autoclave wall. When the bolts of the cover are tightened, the wedge ring closes the gap and provides a perfect seal.

Another approach is shown in Fig. 1.37*b* utilizing two wedge rings inserted in the gap with an O-ring between the wedge rings. By screwing the cover into the top of the autoclave, the final position of the seal rings is reached. Hand-tight arrangement suffices to establish initial sealing. If internal pressure is applied it can be increased to extreme levels without additional tightening of the head component. O-ring and double-wedge ring arrangement guarantee any pressure desired without jeopardizing the seal effectiveness.

The author has operated such autoclaves in high-pressure service for pressures up to 45,000 psi without experiencing any leakage. The tightening of the bomb head is not a problem.

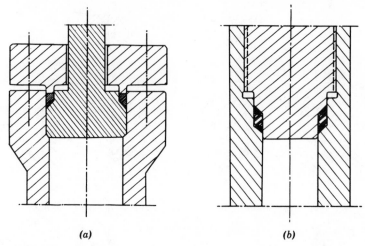

(a)   *(b)*

Fig. 1.37   *(a)* Closures with wedge and *(b)* multiple-wedge gaskets.

The autoclave in Fig. 1.38 uses a so-called double-wedge gasket in one seal ring. Initial tightening is accomplished by screwing the head component into the top of the autoclave. The double-wedge ring then achieves the necessary seal. An additional possibility of retightening is provided by the top closure bolts in the head ring. When they are tightened the system tends to increase seal efficiency. This latter device is not required, since initial tightening with the head in combination with the self-energizing effect due to the internal pressure offers a complete seal.

A combination of two wedge rings on top of each other in one seal groove is another possibility to provide a perfect seal in the presence of extreme pressures. Wedge and self-energizing sealing principle combine to establish a perfect seal effectiveness. Temperature and even pressure pulsations create no problems.

The autoclave of Fig. 1.39 utilizes the double-wedge ring arrangement. Once the closure nut is screwed into the autoclave head the bolts can be tightened to establish any degree of desirable contact force. Double-wedge rings represent ideal sealing devices, although the seal rings may be costly in fabrication.

4.   *Wave Ring Gasket*

The wave ring gasket is a development of ICI (Imperial Chemical Industries, United Kingdom). The gasket, shown in Fig. 1.40, seals when

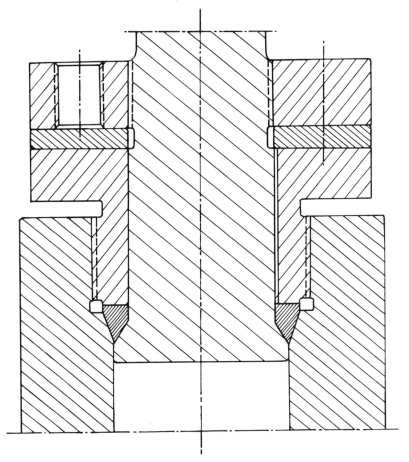

Fig. 1.38   Closure using special wedge seal.

the system pressure deforms the gasket by pressurizing the entire internal surface and deforming the wall so that all external surfaces of the gasket are pressed against the walls of the cavity of the joint components.

   The sealing effect can be markedly improved by providing an initial oversize of the gasket OD in the order of 0.005 in. Sometimes metal plating of the external gasket surface may be used (silver, nickel, copper, aluminum), depending on compatibility with the process medium. When operations are under high temperatures, it is useful to choose stainless materials with a high coefficient of thermal expansion. Once the system comes up to temperature, the sealing efficiency is even greater than when influenced by pressure alone.

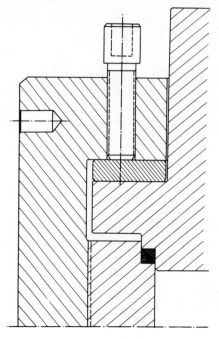

Fig. 1.39 Closure with double-wedge seal arrangement.

The wave ring sealing principle has been utilized by J. Pickup & Sons, Ltd., in Marple, Cheshire, U.K., to develop the sliding expansion joint, illustrated in Fig. 1.41. The device consists of a short tube with the ends belled, strongly similar to the wave ring gasket and incorporating the same sealing mechanism. The spherical sections of the joint match the cavities of the tube ends with an interference fit of up to 0.005 in. when

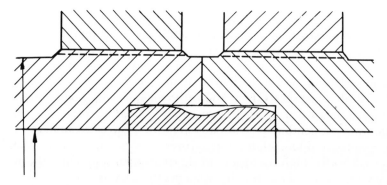

Fig. 1.40 Wave ring gasket.

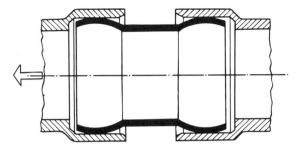

Fig. 1.41 Wave-type seal as expansion joint.

under temperature. The spheres of the joint permit certain amounts of misalignment. Since the joint was designed as an expansion joint for hot exhaust gases, the joint material needs to be selected from the family of stainless steels with a coefficient of expansion considerably greater than that of the pipe system. The stainless predominantly used is 18/8, titanium stabilized, fully softened, and scale-free, capable of providing resistance against corrosive exhaust gases, frequent expansion and contraction cycles, and the wear of the rubbing contact surfaces.

5. *Pressure Coupling*

Harvey N. Pouliot of Sandia's Engineering Division in Livermore, California, developed a coupling device that can be tightened by hand to seal even against high gas pressures. The device is reported to have been tested at pressures up to 25,000 psi and at temperatures ranging from −320 to 500°F.

Unlike most other joint devices in which the internal system pressure tends to separate the contact faces of the joint components, this all-metal seal configuration becomes tighter with rising internal pressure. The thread of the connector prevents separation of the joint through the internal pressure.

The device shown in Fig. 1.42 consists of three parts and is made of beryllium copper.

One part of the joint tubing ends with a sphere, designated as the *inner ball*, the sphere shaping out with a thin, tapered lip. The second tube ends in a female spherical cavity, designated as the *outer ball*, at the mouth of the enlarged threaded end. These two parts are inserted into a nut and held in position by a closure nut. The inner ball has a slightly smaller contact radius than the outer ball.

When it is coupled together, the inner ball fits tightly in the outer ball,

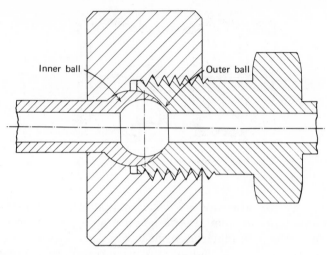

Fig. 1.42   Pressure coupling without gasket.

with the lip of the inner ball making the initial contact. With tightening, the nut forces the lip against the inner wall of the outer ball. Once the system pressure is applied and increased, the flexible inner ball is pressed against the outer ball, thus creating a tighter seal by self-energizing action. A few degrees of misalignment is not critical.

Tests with helium gas at high pressures and temperatures provided leak-free service, in spite of repeated reuse of all parts. The contact areas of the spheres must be lapped to a fine finish to seal helium properly. Construction materials are not a problem as long as process compatibility is guaranteed.

6.   Bridgman Modifications

Static sealing devices cannot be discussed without paying tribute to the extraordinary contributions P. W. Bridgman has made to modern sealing technology. For these unusual achievements Bridgman was awarded the Nobel Prize in 1933. Three applications of Bridgman seal designs are illustrated in Fig. 1.43. As will be noticed the cone sealing principle is utilized in two configurations (*a* and *b*), whereas (*c*) uses a soft, flat gasket. The efficiency of the cone seals is obvious and self-explanatory. The tube faces are conical to match conical cavities of the gasket. There are no alignment problems (Fig. 1.43*a*). In the cone design of Fig. 1.43*b* the wedge seal principle is utilized. Figure 1.43*c* applies a flat gasket of soft metal that deforms easily when contact pressure is applied.

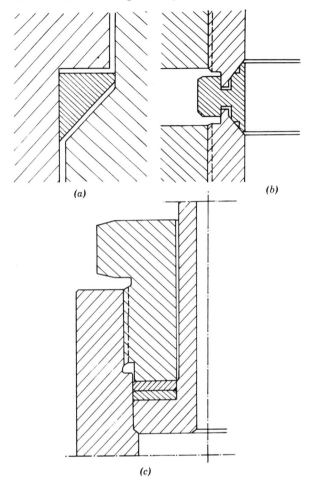

Fig. 1.43 Bridgman gaskets.

7. *Closures Using Solid Wire Rings as Gaskets*

Gaskets designed as solid wire rings provide excellent seal efficiency working on the principle of two-line contact in each groove, as is the case for octagonal oval ring gaskets described earlier for heavy-cross-section sealing configurations.

The seal efficiency can further be improved by plating the ring gasket surface with metal, such as silver, copper, and the like. Another improvement can be reached by suitably lining the contact surfaces in the seal grooves. These designs are well suited for medium pressure ranges. In high-pressure service the oval ring should be preferred.

8. General Application of the Cone Seal Principle

In the initial phase of the development of high-pressure heavy-wall equipment design the cone was the dominating component for establishing a reliable seal. The cone was utilized in an enormous range of design modifications, used in all kinds of laboratory, pilot plant, and large-scale production equipment. Today the cone is still the predominant seal component. It verifies the line contact principle in an elegant manner, providing a seal with minimal contact force. Despite the lack of analytical design methods, the cone has become a standard seal component of the first degree with absolutely reliable and predictable seal behavior as a result of many years of empirical experience in field service.

Before heavy-sectional high-pressure gaskets were developed and welding of thick-wall equipment was not yet available as a reliable fabrication method because of the lack of high-power X-ray machines, the cones were designed as integral parts of the vessel and the seals were accomplished without the use of any gasket. This made the application of heavy flange sections and enormous cover plates mandatory to supply means for arranging the bolts to produce the required seal contact force. With the perfection of modern welding methods in conjunction with reliable welding seam testing machines by radioactive procedures and with the perfection of multilayer design, high-pressure vessel design has completely changed in the past 20 years of steady progress in vessel construction.

a. Single-Cone Design for Closures of Large-Scale Vessels. The single cone was used as a sealing element for high-pressure vessels 5 to 6 ft in diameter with wall thicknesses of up to 10 in. The cone was an integral part of the cover, with the male component matching the female cavity in the face of the cylinder wall.

Two examples of cone seal designs for heavy vessels without gaskets are illustrated in Fig. 1.44. To seal the closure of vessel (*a*) the two mating surfaces were provided with differential slope angles with 2 to 3 degrees of slope variation. This guarantees a close approach to line respectively small-area contact for sealing at minimal contact force. The interface areas must be free from machine marks, with a surface finish of no less than 16 μin. It is further important that the hardness in the contact surface of the cover cone be at least 50 Brinell hardness units

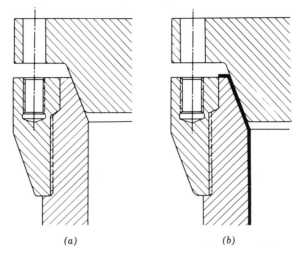

(a)   (b)

Fig. 1.44 Single-cone vessel closures.

softer than the hardness of the matching female surface in the vessel face.

Where the achievement of this difference in surface hardness is a problem, it is preferable to apply a liner to either the cone surface or the vessel. Where liners are used they should be soft. A surface finish of 20 μin. suffices to establish a seal. Vessels of this type are still in operation in Germany today.

b. Double-Cone Seals for Large-Scale Vessels.   The double-cone seal device in high-pressure service was a considerable step forward in the development of static sealing technology. The design was derived from small laboratory seals developed for high-pressure syntheses some 60 years ago in the plants of the BASF in Ludwigshafen.

The double-cone gasket is a solid steel ring provided with two conical slopes for contact with the corresponding female seal areas in the cover and in the vessel end face. Details of a typical double-cone gasket are illustrated in Fig. 1.45. Seals of this nature are in use in numerous vessels all over Europe with diameters up to 6 ft or more, operated at pressures in the order of 700 atm (or 11,000 psi) and temperatures reaching the 600°C (~1100°F) limit.

The vessel cover is provided with a groove for the insertion of the double-cone gasket. A support ring underneath the cone is bolted to the bottom of the cover to hold the cone ring in place. The male-female

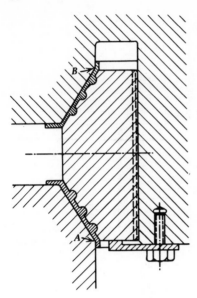

Fig. 1.45   Double-cone gasket.

slope areas are provided with a differential angle to facilitate line contact when the bolts are tightened.

When the bolts are tightened the contact force bends the cover toward the vessel flange. Contact is first established at points *A* and *B* at the corners of the gasket slopes with the interface components. With increased degree of tightening the contact lines widen to small contact areas with slight permanent surface deformation for the seal.

By lining the slope contact areas of the female seal surfaces the seal efficiency can be markedly improved without increase in bolt tightening force. The author has been repeatedly successful in lining the slope areas of the gasket with very thin aluminum or other metal foil, which can be placed on the cone surfaces without difficulty. The foil, however, should be compatible with the process atmosphere. Any other metal foil with suitable thickness (~0.0005 to 0.001 in.) can be applied to accomplish a perfect seal.

For application of the self-energizing sealing principle the double-cone gasket is provided with gas channels on the ID in an axial direction, allowing the internal pressure to enter the female cavity, thus pressurizing the gasket further into the closure and effecting a higher contact force.

An application of a double-cone seal device for a large-scale reactor vessel inserted in the cover is shown in Fig. 1.46.

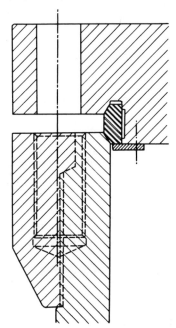

Fig. 1.46 Closure with double-cone gasket.

The illustration shows that the flange has its upper section shrunk to the vessel while the lower section is screwed. By arranging the bolts halfway in the cylinder wall and halfway in the flange, the bolt circle diameter can be kept at a minimum, which facilitates vessel design by reducing the weight of the flange considerably.

Experience over the years has shown that a double-cone metal gasket requires less contact force for initial sealing than is necessary to seal a ¼ in. flat gasket in standard autoclave configurations. It has further been found that the double-cone design requires 25% less tightening force than must be used for a design with a single-cone arrangement, for vessels of equal size and operating conditions.

E. Joints and Closures for Vacuum Connections with Elastomers

In vacuum technology the static seal is an entirely different concept. Contrary to the widely held opinion that a vacuum system is easier to seal than a pressure vessel, vacuum equipment is many times more difficult to seal than a system operated under extreme pressures. This is true despite a maximum pressure differential of only 14.7 psi in vacuum, whereas a pressure system may encounter a differential of tens of thousands of pounds per square inch.

In high-pressure design the internal pressure can be utilized to facilitate sealing, whereas in vacuum sealing this principle is only conditionally applicable in the presence of elastomers. In vacuum systems the pressure differential is not great enough to be of noticeable help. The tendency of trapped gases in the sealing materials to leak into the vacuum, designated variously as *degassing* or *outgassing*, creates the problem. It can be minimized by the selection of suitable gasket materials and proper groove design and by the application of bakeout procedures.

1. Definition of Vacuum Ranges

By definition vacuum pressure is measured in torrs, with 1 torr representing the pressure of 1 mm of mercury. The word *torr* is derived from the name Torricelli.

Three basic vacuum ranges have been defined as follows:

Low vacuum	10^{-3} to 10^{-6} torr
High vacuum	10^{-6} to 10^{-9} torr
Ultrahigh vacuum	10^{-9} to 10^{-12} torr

Outer space—called *hard vacuum*—is defined to be in excess of 10^{-12} torr.

Pressures (forces per unit area) of this magnitude are meaningless, because they are so small that they cannot be measured accurately. As a vacuum approaches absolute conditions, the physical quantity to be measured is expressed by the number of molecules hitting a system surface per unit of time. As an example, at a vacuum of 10^{-6} torr the surface is still completely covered with a new layer of gas molecules in one second. Therefore, at a pressure of 10^{-7} torr it takes 10 seconds to cover the surface with a new layer of molecules. At ultrahigh vacuum with a pressure of 10^{-12} torr it takes 10^6 sec (27.78 hr).

In outer space a clean surface remains clean. Oxidation cannot take place since there are no oxygen molecules present. When two clean surfaces are brought together, interdiffusion occurs and the surfaces actually "weld" together. In most cases they cannot be separated without crude force, often by destruction only.

2. Materials for Vacuum Gasketing with Elastomers

There are actually hundreds of variations of types and compounds used for gaskets in vacuum technology. We do not attempt to describe the

Table 1.6 Permeation Properties of Seal Materials

Properties	Neoprene	Buna-N	Butyl	Viton-A	Silicone	Teflon
Air permeability*	1.0	1.3	0.2	0.8	115.	0.7
Heat resistance	2	2	3	1	1	1
Cold resistance	3–2	2	2	3	1	1
Abrasion resistance	2	2	3–2	2	4	1
Cold flow	2	2–1	2	3	2	4
Oil resistance	3–2	1	4–3	1	3	1
Ozone resistance	2–1	4	2–1	1	1	1

Source: From NASA publications.

* Air permeability = $\dfrac{cc \times cm}{sec \times cm^2} \times 10^{-8}$ for ΔP of 760 torrs at 24°C.

modifications of base materials compounds and specialized treatments because they are so numerous and each supplier is continually modifying his own elastomers to establish very specific characteristics. Because of the lack of uniform specifications in the elastomer industry, it is practically impossible to compare similar gasket materials with any assurance that their chemical and physical properties are uniform or even closely comparable.

Gasketing materials generally used are neoprene, Buna-N, Butyl, Viton-A, silicone rubber, Teflon, and Kel-F elastomer. These materials have been described earlier in Section II.B. A few details on vacuum technology may be added. Table 1.6 compares significant properties together with air permeation rates.

3. Outgassing Rate for Elastomers

Neoprene has the highest rate of outgassing of most commonly used vacuum gasket materials. With the appearance of better materials on the industrial market, neoprene is losing increasingly more of its initial significance.

Butyl has a lower rate of outgassing, but has outstanding resistance to permeability. Its poor oil and heat resistance in connection with a high rate of outgassing limits its use for general vacuum service.

Buna-N is probably the most commonly used elastomer in vacuum technology today, in spite of its high outgassing rate. By vacuum baking Buna-N at 100°C for 4 hr its outgassing tendency is reduced by a factor of at least 10. Vacuum baking does not deteriorate its properties for

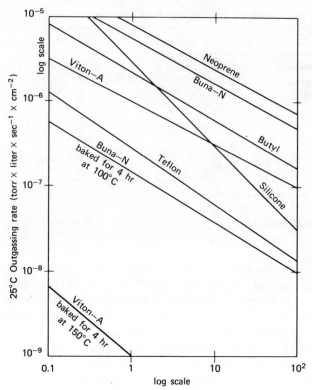

Fig. 1.47 Outgassing rate of typical vacuum gasket materials (NASA).

vacuum gasketing. Subsequent exposure to atmosphere does not affect the lower outgassing rate obtained from vacuum baking, since the gases released during vacuum baking are primarily plasticizers and not water.

Figure 1.47 indicates outgassing rates of materials frequently used for desirable gaskets in vacuum technology.

Viton-A has outstanding resistance to heat and a lower outgassing rate. However, it does have a high rate of cold flow. Its outgassing rate can be further improved by a factor of 1000 by vacuum baking at 150°C for 4 hr. Great care must be used when applying a vacuum-baked Viton-A gasket. Any exposure to atmosphere after baking will cause it to reabsorb gases again, thus eliminating the outgassing effect achieved by the baking process.

Silicone has excellent heat resistance but an extremely high rate of permeability. It is, therefore, often replaced by Viton-A. Its rate of outgassing is relatively good.

Teflon has the best rate of outgassing. Its lack of memory and high degree of cold flow create problems in maintaining vacuum over long time intervals. Teflon also produces other problems for vacuum technology because of its high permeability to helium. After long periods of service in vacuum it cannot be reused because of tendencies for permanent set.

4. Design Factors for Elastomeric Vacuum Gaskets

The simplest design of vacuum joints is possible with elastomeric gasket materials. The most popular gasket configuration is the conventional O-ring. Deviations are the flat gasket for horizontal flat flange faces and the rectangular gasket inserted in a groove of one of the flange faces. The basic configurations are shown in Fig. 1.48.

The flat gasket requires the highest bolt force because of the many surface irregularities in the flange faces.

The rectangular gasket seal offers a variety of advantages. The width is an optimum, exhibiting enough area to fill all the surface asperities of the contacting flange face with relatively low contact force. The optimal protrusion of the gasket above the face of the flange for sealing should be about 20%.

Elastomeric O-rings satisfy a wide range of vacuum sealing requirements. They are the most commonly used sealing gaskets in vacuum

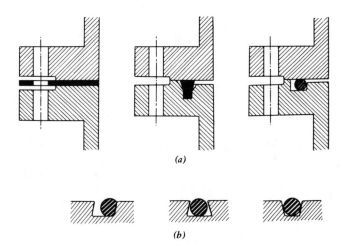

(a)

(b)

Fig. 1.48 Design configurations of elastomeric vacuum seals. (a) Basic seal designs. (b) Groove modifications.

technology, largely because their sizes are standardized and they are available in many configurations and dimensions, materials, and durometers. The compression should be 15% minimum, but not more than 20% of the initial diameter and no less than 10% of the uncompressed cross section.

The groove is usually rectangular but can also be machined as a parallelogram, trapezoid, or double-dovetail to hold the O-ring inside the groove when the flange face is vertical. The groove angles are then preferably 75 degrees. A slight gap between the flange faces should exist to facilitate gas evacuation. This gap is usually 0.003 to 0.005 in. maximum. Toward the atmosphere the flange faces are in contact (see Fig. 1.48) to guarantee the internal gap and prevent excessive compression of the O-ring.

A common and very practical vacuum seal using an elastomeric O-ring is the so-called CVC-con O-ring seal. Two aluminum rings are used to fit the sides of a commercial elastomeric O-ring. When flanges are tightened to establish metal-to-metal contact with the aluminum rings, the protruding portion of the gasket is compressed in the free space established between the aluminum rings. The tendency for degassing practically does not exist (see Fig. 1.49).

This same principle can also be used by having the O-ring inserted in the conventional groove and then tightening the joint to establish full metal-to-metal contact. When the flange faces have a good surface finish

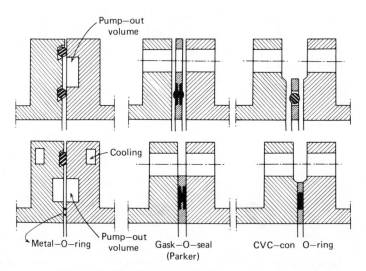

Fig. 1.49 Modifications with O-ring arrangements.

and are parallel, a degassing volume is not required. Using this design the author has established vacuum conditions in the order of 10^{-4} torr without any difficulty in numerous diversified installations.

The Parker Gask-O-Seal, also illustrated in Fig. 1.49, is a molded combination with a metal container ring. The void is designed to be filled when the joint is fully tightened. The rest of the system does not see the rubber. The seal is available in many design configurations, operating effectively.

For large-scale vacuum volumes a double O-ring arrangement is often used. This is also presented in Fig. 1.49, with a pump-out volume between the two elastomeric O-rings. A leak in this system is almost impossible to detect.

By replacing the inner elastomeric O-ring with a hollow metallic O-ring, a joint is formed, developed by NASA, that permits a high degree of vacuum on the order of 10^{-10} and better (see Fig. 1.49).

The design of systems for ultrahigh vacuum operating with pressures of 10^{-9} to 10^{-12} or better requires very special seals. The outstanding characteristic of such systems is that no bolts are required, since the pressure of the outer atmosphere is utilized to seal the existing connections and closures. Specific details on this subject are available from NASA.

Two O-rings are used in the seal design. One seal is the elastomeric ring used as an external device; the other ring is a metallic O-ring applied internally with a pump-out volume between the two rings. The external ring is preferably made of Buna-N, which may be baked out before application. The internally used metal ring is loosely placed between the flat mating faces of the metal flanges without a groove but externally wrapped with thin Teflon tape, as is customary for sealing threaded pipe joints. The outer elastomeric ring is compressed 5 to 10% and no more, leaving the inner metal ring with the Teflon tape exposed to the major force; the flange faces should never touch each other. By establishing this condition with the proper balance of forces between the elastomeric and metal rings, the gaskets can be reused.

Surface finish of the interface areas in metal should be of the order of 8 to 15 μin. rms. It is best if the pump-out volume is evacuated to 10^{-5} torr.

5. *Basic Rules for Design and Behavior of Elastomeric Static Seals*

Appropriate design of elastomeric seal components takes advantage of their desirable properties and minimizes their weak ones. Some rules apply to all designs whether pressure or vacuum is involved.

1. The O-ring must be inserted in a protective groove having a volume larger (~40%) than the volume of the O-ring. Elastomers are basically incompressible and therefore need suitable volume space when deformed. They should not have to take the full tightening load.

2. The O-ring must be preloaded (15 to 20% of its initial volume) while the flanges are tightened to establish face contact. This compression is the prerequisite for the sealing mechanism of O-rings as static gaskets.

3. The gap created when internal system pressure is applied, which tends to separate the flange faces, must be a minimum to prevent extrusion of elastomer into the gap.

4. When face separation becomes excessive, backup rings must be used. These backup rings can be separate components or they can be designed for direct metal support, as shown in Fig. 1.50. The design without the metallic backup reflects clearly the danger of extrusion.

By reversing the metal backup ring having the U shape of the metal support open to the external atmosphere, the backup device can also be used for high-vacuum service. The elastomer provides the sealing action and the steel spring ring provides rigidity and prevents extrusion.

The seals division of the W. S. Shamban Company, West Los Angeles, California, developed a Teflon seal with metal backup designated as the Dryerseal. This design seals rough, damaged, warped, or wavy surfaces, minimizing the requirement for surface finish. Teflon acts as the seal device and the metal prevents extrusion. Under pressure in excess of its compressive yield point, the Teflon acts as a contained fluid and fills the surface asperities of the contact faces, thus establishing a leak-free joint. This seal is also suited for cryogenic and vacuum service. The Dryerseal is illustrated in Fig. 1.51.

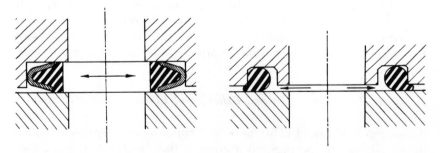

Fig. 1.50 Elastomeric seals with and without backup ring under high pressure.

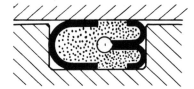 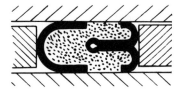

Fig. 1.51 Dryerseal design. Teflon-metal combination.

F. Joints and Closures for Vacuum with Metal Gaskets

For vacuum service all metal-to-metal gaskets can be installed, providing plastic or permanent deformation in the seal contact areas, as in high-pressure service. This includes all gaskets of the heavy-section design, discussed in Section II.C.2a, and also the resilient metal gaskets with springy members, described in Section II.C.2b.

1. Resilient Metal Gaskets

The rigid heavy-section gaskets (lens ring, octagonal ring, oval ring, delta ring, etc.) permit use of extreme pressures and very high vacuum in excess of 10^{-6} torr.

The category of resilient metal gaskets is also well suited for vacuum service. The spring-type resilient arms are either coated with Teflon or plated with a soft metal. The same effect or even better results can be obtained by installing the arms with preload by restraint of their flexibility. The spring capacity of the resilient member is not needed to compensate for pressure or temperature pulsations and this property is very helpful in providing preload with low contact force.

2. Clam-Style Metal Gasket Configurations

All resilient metal seals discussed earlier, including the designs illustrated in Fig. 1.23, the Bobbin Seal, the Clam Seal, and particularly the Boss and the Mott Seal, are excellent devices for sealing vacuum systems. It is common practice to use seals that utilize the principle characterizing the Clam Seal, or the English-type heavy-sectional lens ring. The fundamental idea is that a sharp edge bites into the gasket as in the English-type lens ring, or the gasket bites into the contact area of the flange face, as is typical of the Clam Seal design. In NASA and aerospace service designs are frequently used as illustrated in Fig. 1.52. The configurations are self-explanatory.

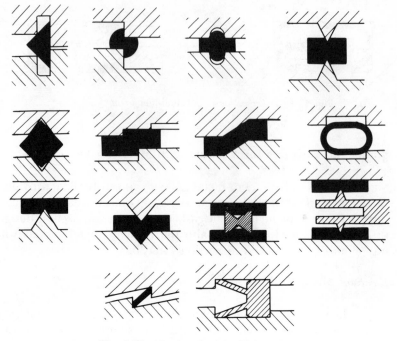

Fig. 1.52 Metal gaskets for high vacuum.

The resilient and highly springy metal gaskets are used in a variety of ways. They may be installed either as plain metal gaskets without restraint or as plain metal gaskets assembled with restraining preload, or as plastic- or Teflon-coated metal gaskets. When they are used without restraint during assembly, the contact width is of considerable significance. Figure 1.53 shows the major members of the springy metal gaskets compared with O-ring modifications. The contact width is shown as quantity W.

3. Joints Using Welds for Sealing

Joints are frequently welded to achieve sealing. This method was initiated in the early phases of high-pressure development. The initial idea was to provide an absolutely tight seal without the necessity of controlled tightening or the deformation of any of the contact interfaces. The areas to be welded together are matched to be parallel and a subsequent internal or external weld establishes the seal. The second advantage is that faces are joined without intricate machining preparations to achieve a tight seal.

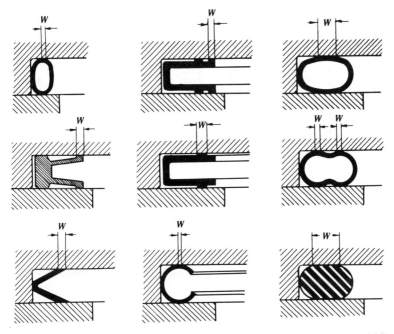

Fig. 1.53 Special metal gaskets for vacuum and pressure (*W* is contact width).

The use of welds as a seal against internal pressure has not been found practical. The method requires costly machining and the weld is not necessarily reliable. The welding process has therefore been restricted predominantly to equipment where absolute tightness is mandatory and the pressure is atmospheric or vacuum.

From a large number of possible design modifications, we show some weld joint methods in Fig. 1.54 for sealing vessels operated under vacuum. In Fig. 1.54*a* the vessel wall is prepared to be used for the face-to-face joint. The prospective contact areas are machined to join as parallel faces, provided with a thin lip for the weld at the vessel OD. This method is acceptable as long as the wall's thickness remains relatively moderate and the pressure is below atmospheric.

Figure 1.54*b*, a better approach, uses two membrane rings, which are very easy to machine to match the faces of the vessel walls. The membrane rings are welded together on the OD to prevent penetration of air between them. The weld does not reach either of the wall faces. At the ID the membrane rings are slightly larger than the ID of the vessel. This leaves a small annular cavity, which is filled by the weld, connecting both the rings and the vessel walls. This weld then is responsible for the

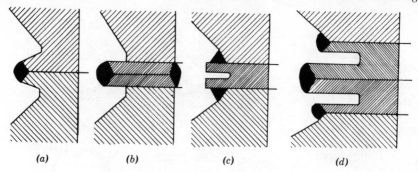

(a) (b) (c) (d)

Fig. 1.54 Joints using welds for sealing.

tightness of the joint. This configuration permits moderate pressures and gives excellent results for operation under a vacuum.

Figure 1.54c represents a modification of the normal membrane design. The double ring is replaced by a single ring with a thickness greater than two rings of Fig. 1.54b. For feasible welding and for imparting greater flexibility of the weld lips the membrane ring is provided with a groove on the OD, thus providing two thin, flexible lips welded to the adjacent contact faces of the walls of the joint components. On the ID of the membrane no welding seam is provided. This design permits only moderate pressures but is excellent for vacuum service conditions.

Figure 1.54d uses two membranes with a groove in each ring to provide thin lips for the welding seams. The upper and lower lips are welded to the faces of the walls. The lips on the membrane contact faces are welded on their OD. This design functions well in vacuum service even for large-scale equipment, although it is costly and time-consuming when the joint must be reopened.

In conclusion, welded joint seals are not practical for pressure vessels of large-scale equipment and higher pressure levels. The designs become too intricate and welding is too complicated to be reliable. The weld design for joints is preferably used for vacuum services of all design configurations, especially vacuum chambers for the space, chemical, petroleum, textile, and related industries, to name just the more important ones. Welded pressure vessels are preferably of multi-layer design.

G. Gaskets and Joints for Cryogenic Service Conditions

With the development of aerospace vehicles the seal industry was faced with new concepts and encountered problems involving service media such as liquid hydrogen, liquid nitrogen, and helium under most unusual conditions, such as ultrahigh vacuum and cryogenic temperatures. The solution to these new problems was found with the development of bimetallic seal configurations.

1. Aerospace Seals

The development of aerospace seals started with the metallic O-ring. Soon the K-seal, V-seal, C-seal, E-seal, and Naflex-seal were added. Their design generally utilizes the self-energizing principle and the various configurations using coatings, platings, and hoods of softer metals to achieve better performance with a simultaneous reduction of the clamping force. The chief characteristic of these seals is flexibility.

a. Ring Spring Design as a Basis for Seals. In the laboratories of General Dynamics/Astronautics of Advanced Products Company, North Haven, Connecticut, a seal was developed consisting of a series of inner and outer rings with double-cone interface contact areas, arranged as a column, as shown in Fig. 1.55, reported by W. A. Prince (7).

A series of inner and outer rings provided with double-conical surface areas is alternately assembled to a column-type spring arrangement. By applying an axial load the inner rings tend to expand the outer rings while being compressed at uniform stress, thus establishing a tight seal, since the outer rings strongly oppose the expansion of the inner rings.

This spring principle utilizing the cone effect is applied for a seal, as indicated in Fig. 1.56. During assembly an axial force is introduced, compressing the spring by radial expansion of the external ring. Simultaneously the internal ring is compressed by this external force acting uniformly along the entire circumference. By using different materials, advantage is taken of the variations in thermal contraction and expansion during the existing temperature pulsations. This design, reported by J. W. Hull (5), further permits greater flange deflection. Both rings are made of high-strength steel with relatively small cross-sectional areas.

b. Temperature-Actuated Seal. For operations having very low temperature and handling cryogenic fluids, S. E. Logan (6) describes another development, illustrated in Fig. 1.57. An elastomeric rubber-type material is formed to a rectangular cross section and is inserted into the female cavities of two flange faces. The rubber ring is supported by an Invar ring. For initial sealing, the rubber ring, compatible to cryogenic fluids, establishes the contact in the groove surfaces by compression

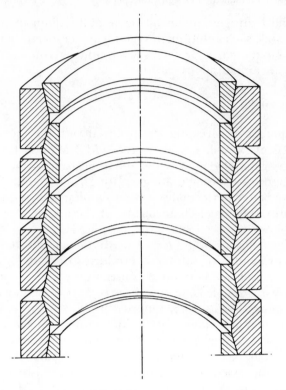

Fig. 1.55 Ring springs with double-cone contacts as sealing mechanism.

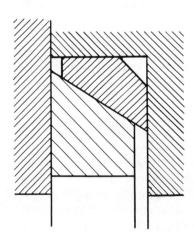

Fig. 1.56 Conical ring springs as seal devices.

Fig. 1.57 Temperature-actuated seal using rubber-metal combination.

when the joint is tightened. The flanges are preferably made of aluminum or stainless steel. On deep cooling, the flanges freeze and subsequently contract by a differential amount producing an additional squeeze on the rubber ring. For temperatures down to −40°C the rubber ring remains elastic and reasonably resilient, like a conventional O-ring would. When cooled below −40°C, the rubber becomes glassy but still maintains a seal.

The ring spring combined with a temperature-actuated seal is another step forward in seal development. It is obvious that surface considerations are secondary to understanding the principle of the sealing mechanism.

2. Bimetallic Seals

The bimetallic seal concept, developed by General Dynamics/Astronautics, provides a wide range of new and possible design configurations for metallic components for sealing cryogenic fluid systems. The basic idea is to match two ring components with conical slopes of 45 degrees in a horizontal arrangement and place the two dissimilar metal spring rings into a cavity of two joint flange faces. The outer ring is preferably of aluminum, the inner ring of stainless steel. When cooling occurs by passage of cryogenic fluid through the system, the outer aluminum ring contracts to a higher degree than the stainless ring. Moving along the slope, the metal spring rings tend to move away from each other, increasing the contact pressure against the cavity interfaces and thereby establishing a better seal as temperature decreases.

A reduction of friction and sealing contact pressure results when both metal rings are coated with a layer of 2 to 4 mils of PFE fluorocarbon.

W. A. Prince (7) describes two applications of bimetallic spring cone seals, shown in Figs. 1.58 and 1.59.

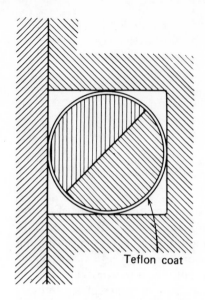

Teflon coat

Fig. 1.58 Bimetallic O-ring with Teflon coating.

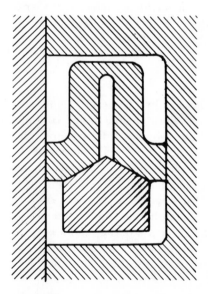

Fig. 1.59 Stainless cone ring as seal actuator.

In Fig. 1.58 an O-ring-type gasket is made of two metal rings joined in a 45-degree slope contact area and the total ring is coated externally with Teflon. The outer ring section consists of an aluminum alloy; the inner ring section is made of stainless steel. On cooling, the outer aluminum ring section contracts to a higher degree than the stainless inner ring section. As the contraction causes the outer ring to move inward, the sealing efficiency is improved. The differential contraction results in an increase of the contact pressure along the groove sealing walls, thus reducing leakage tendencies.

Figure 1.59 also illustrates a bimetallic seal design using a double cone to increase the seal contact pressure. This design could further be modified in a number of ways to provide conical surfaces within the groove itself.

These seal configurations can also be used in high-temperature service by inverting dissimilar materials, or for both temperature extremes by compounding the geometry. These designs have frequently been utilized for sealing aerospace fluid systems.

Additional information on aerospace cryogenic static seals is given in a paper by C. M. Daniels (3).

III.   The O-Ring as a Static Sealing Device

The conventional O-ring has gained such a prominent position that today's industry is inconceivable without it. For static sealing purposes, involving chemical resistance, the O-ring offers the most unusual range of ideal solutions of any kind of static sealing devices for an unusually extended range of service conditions.

A.   The Elastomeric O-Ring

The effectiveness of the elastomeric O-ring is based on its outstanding property of memory. Once compressed, the ring always tends to restore its initial cross section, thus producing the automatic tightening force effect. Every deformation of the housing surrounding the O-ring in service position is compensated by the automatic motion of the O-ring material until the initial compression effect is eliminated; the ring ceases to seal when it loses its memory characteristics or the initial compression.

The elastic material "flows" easily under compression and fills all surface asperities of the contact areas, thus establishing a tight seal with low contact force. The elastomeric O-ring permits a tightening factor of less than 1.50 (1.10 theoretically, established by statistical tests).

Elastomeric O-ring materials are incompressible, so that volume remains constant during compression and only the shape is changed. The most favorable service life is obtained at a rate of compression not exceeding 15 to 20% of the initial cross section. The groove must therefore be rectangular, with the depth corresponding to the height of the O-ring when under compression and the width of the groove being larger than the width of the O-ring under compression, thus leaving enough space for the ring to expand. With a compression of 15 to 20% for the elastomeric ring, the contact mating surfaces for the interfaces with the O-ring do not require extreme surface finish conditions. A surface quality of 16 to 32 μin. rms is completely satisfactory to meet any sealing requirements.

Pressure pulsations have little influence on the seal effectiveness and reliability of the joint, as long as the compression of the elastomeric ring is not eliminated or in any other way affected.

Once assembly is completed and the suitable prestress is established, the flange mating faces should be in mutual contact with each other without the O-ring "seeing" the pressure fluid to be sealed. Because of machining marks—always present—and surface irregularities as a result of eventual surface waviness, a bubble-tight seal is difficult to achieve. This is the requirement the O-ring has to satisfy. These conditions can always be satisfied as long as the O-ring is providing standard specifications.

The elastomeric standard O-ring is equally suited for pressure as well as vacuum service. The author has used elastomeric O-rings in static and dynamic sealing devices for pressures as high as 30,000 psi without difficulty. If the pressure exposure is too long, the ring tends to show a permanent set and, therefore, cannot be reused. Great care must be given to the calculation of the bolts to determine the proper bolt extension for tightening. Antiextrusion devices in the grooves should be used to prevent permanent ring damage. The manufacturer should be consulted for extrusion prevention. Figure 1.60 gives a series of sketches for O-rings used for static and dynamic pressure sealing. These basic designs can be modified in many ways. The sketch configurations are self-explanatory.

Where temperature allows, the O-ring can be used for all designs described earlier in which the wedge or double-wedge gaskets can be replaced by the elastomeric O-ring, offering a seal that requires a tightening factor noticeably below the value of 1.50. The author has sealed gas storage tanks operated at pressures in excess of 1000 atm ($\sim$15,000 psi), inserting the O-ring in a groove machined into the spherical-type metallic lens ring. Regardless of pressure the design can be based on a

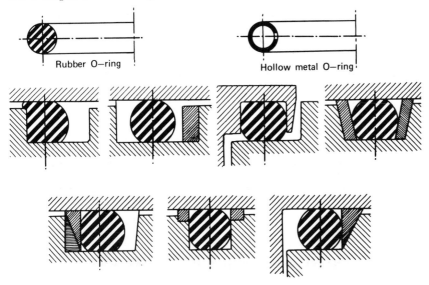

Fig. 1.60 Elastomeric O-rings for static sealing.

maximal tightening factor of 1.50 offering a highly efficient design on a favorable economical basis. The lens ring provides a minimum tightening diameter, resulting in a noticeably reduced tightening force at extreme internal pressures. The O-ring–lens ring combination is shown in Fig. 1.61, compared with a design of a straight-sloped lens ring used by NASA for operating hydrogen, nitrogen, oxygen, and helium.

Combinations of elastomeric O-rings with suitable metallic static gaskets are not limited to the lens ring designs alone. These gaskets have the advantage that the machining operation for providing the groove is considerably less costly when the groove is an integral part of the gasket instead of the end face of the cylinder wall or the cover of heavy equipment. Accordingly it is up to the skill and the imagination of the designer to provide suitable seal combinations using the O-ring, keeping in mind the basic requirements for static sealing and the compatibility of the O-ring with the environment.

For elastomeric static O-ring seals, the following design rules should be observed:

1. The groove for the O-ring must have a volume larger than the volume of the O-ring. Elastomers are incompressible and will not change their volume when subjected to hydrostatic compression, resulting in

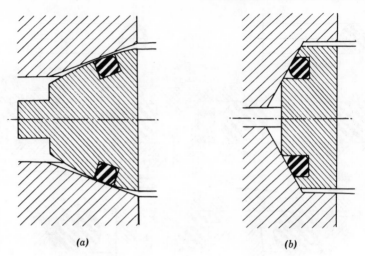

Fig. 1.61 Metallic lens rings with elastomeric O-rings.

cross-sectional variations. The elastomer is not capable of carrying any bearing load for extended periods of exposure time.

2. The prestress applied must suffice to establish and maintain contact between the metal faces at all times. Maximum compression of the elastomer should not exceed a value of 20%. The elastomer is expected to seal the microscopic leakage gaps between the metal faces, and the O-ring must be prevented from being extruded by the internal pressure.

3. When separation of the metal faces occurs while under high pressure, the prestress should be sufficient to keep the gap between the faces low so as not to permit any extrusion of the elastomer. This separation, resulting from the elastic bolt stretch, can be computed and measurements can be made that will keep bolt extension within elastic and controllable limits, thus avoiding damage of the elastomer and preventing leakage.

4. As the internal pressure increases, the elastomer should continue satisfying self-energizing sealing principles. With rising pressures the flange faces should not separate and the seal must be maintained mainly by interfaces between the bottom of the groove and the opposite metal face and the O-ring.

More information on O-rings is presented in Chapter 5 for dynamic seals.

B. The Plastic O-Ring

O-rings made of plastic material, usually Teflon with or without fillers, Kel-F, Viton, and silicone, are also used extensively. Their range of application, however, is quite limited because of their geometric instability and tendency to cold flow. Experience with many pressure applications shows that Teflon O-rings should not be reused.

Where Teflon or Kel-F are needed for compatibility with the process fluid, elastomeric O-rings externally coated with Teflon or Kel-F elastomer may be used. The elastomeric O-ring provides the flexibility and the Teflon or Kel-F impart the corrosion protection. The manufacturer can provide dimensional details.

C. The Metallic O-Ring

The metallic O-ring is usually a hollow, tubular ring with vent holes at the ID. It is used for applications where the elastomeric O-ring is not suitable. The metallic O-ring is available in a variety of basic designs, with round, elliptical, diamond, or double-diamond cross sections, to mention just the basic, most conventional shapes. Some of these major O-ring types are illustrated in Fig. 1.62.

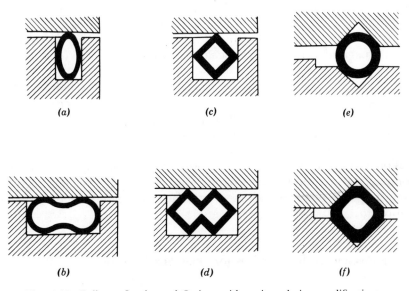

Fig. 1.62 Fully confined metal O-rings with various design modifications.

To offset the inherent loss of strength at elevated temperature, the metal O-ring can be filled with gas under pressure, often about 600 psig. When temperature is rising the gas expands and pressurizes the O-ring internally, thus increasing the resiliency of the O-ring. For O-rings consisting of stainless steels and/or Inconel, pressurizing by gas is recommended for service temperatures above 1000°F. For temperatures above 1200°F with Inconel X and Haynes No. 25 alloy, pressurizing should also be specified.

The vented hollow circular O-ring tends to achieve the same purpose as the pressurized O-ring provided with internal gas under pressure. For rings containing gas pressures exceeding the 600 psi limit the wall thickness may range from 0.010 to 0.080 in., depending on the ring diameter chosen.

The oval metal O-ring (Fig. 1.62a) is a modification of the circular cross-sectional O-ring that is basically installed in a vertical groove arrangement.

Figure 1.62b gives an exploded view of a standard circular O-ring after being prestressed for initial sealing.

Diamond-type O-rings (Fig. 1.62c) generally uses grooves deeper than those for circular O-rings. They establish seal contact at four points simultaneously. Rings with a diamond cross section usually require a prestress with a value 10% below that for circular conventional O-rings of metal. The more complicated ring geometry requires more machining and the rings are therefore more costly.

Double-diamond metal O-rings (Fig. 1.62d) have wider grooves than single-diamond rings require. Sealing is accomplished at four contact points.

It is common practice to use triangular grooves for hollow metal O-rings (Fig. 1.62e). When the ring is fully tightened, the cross section assumes the shape illustrated in Fig. 1.62f.

As indicated earlier, the seal efficiency of metal O-rings is markedly improved by surface plating with such metals as silver and copper. Teflon coating is also suited for application, depending on chemical compatibility requirements. Coating with Teflon has become increasingly popular in recent years, particularly as coating methods have improved.

IV. Joints and Sealing Devices for Small-Sized Tubing

In chemical or petrochemical plants tubing represents a very significant group of components, involving a high percentage of capital investment in operating equipment. The cost of tubing within a plant may range from 25 to 40% of the overall investment in equipment, depending on the product.

Conventional tubing and the corresponding joints with standardized connections are not discussed here, because information on them is readily available in standards books. This discussion is confined to nonstandard joints, which are custom-made. These joints are specifically designed for easy handling and quick opening and closing with high sealing efficiency.

A. High-Pressure Tubing Joints for Laboratory, Pilot Plant, and Instrumentation

The design, assembly, and handling of laboratory equipment involves tubing and fitting problems completely different from those experienced in the general process field. Here tube and equipment connections must be made quickly with an optimal degree of versatility and flexibility because these connections must often be dismantled repeatedly without excessive downtime or loss of expensive material and production. Laboratory equipment is always in a state of change and it is important that all parts be reusable. Since laboratory tubing is generally limited to small sizes, a large number of fittings must be available. Flanges can be cumbersome, heavy, expensive, and often impossible to use because of space limitations. Fittings are therefore the vital components enabling the designer to achieve optimal flexibility of the system within limited space.

In high-pressure technology it has become general laboratory practice to use stainless steel tubing in increments of $\frac{1}{16}$ in. All fittings are designed to match the tube sizes available in both external and internal diameters. Welding is seldom used for joining laboratory tubing. Most laboratory fittings utilize the grip principle, which means that certain components establish a tight grip on the tube surface at both ends. A union-type nut is then used to combine the two tube ends.

1. Sealing Principle

Joints made of conventional thick-wall, high-pressure tubing use the tube faces as seal contact areas, and seal on the application of specially designed metallic gaskets. Metal gaskets are customarily not used for laboratory tubing. The final design of the seal is a function of the wall thickness of the tube. In regular and medium wall tubing the tube ends are introduced into the bore of the fittings without being specially prepared for the final seal, which is accomplished along the external tube surface by the use of additional ferrules. In heavy-wall tubing the tube faces are prepared as cones, which in turn are used within the corresponding bore of the fitting, thus establishing direct sealing contact. The resulting sealing effect is self-explanatory, as is shown in the following discussion.

2. Historical Development

High-pressure laboratory and pilot plant installations were functioning with a high degree of efficiency long before the small-sized stainless steel tubing of today became readily available. The first basic components used in Europe were developed at BASF-Ludwigshafen, Germany, the birthplace of modern plant-scale high-pressure technology in connection with the industrial production of ammonia, synthetic gasoline from coal, methanol, and many other significant products. Two of the first fundamental joint components are shown in Fig. 1.63. Each joint utilizes either the single-cone or the double-cone principle to establish a reliable seal. These components are reusable and are still applied today, especially for extreme pressures, surviving a period of more than 75 years of successful application.

Component (a) was used by the author for many years of service for systems operating at pressures in the order of 100,000 psi. Component

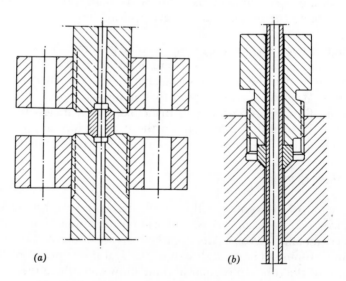

(a)   (b)

Fig. 1.63 High-pressure tube joints using cone gaskets.

(*b*), although also suited for extreme pressures of the same magnitude, was basically used in systems operating at pressures in the 20 to 30,000 psi range, particularly for instrument lines, limited to small sizes of tubing.

3. Major Fittings Primarily Used Today

Fittings for high-pressure tubing are available in a confusing number of design variations. They are not standardized in the normal sense of standardization, and are generally known by their trade names. In the following discussions these trade names are used for identification. There is no wish to favor any manufacturing company by mentioning their products, nor is there any intended discrimination against companies whose products are not discussed. The author has selected those products that demonstrate basic principles and also those that he has used personally with success in plant or laboratory services either in the United States or in Germany.

a. Swagelok Fitting. Swagelok fittings for high-pressure laboratory tubing are perhaps the fittings most frequently used in this country as well as in Europe. They employ the external grip sealing principle based on the permanent deformation of both tubing and corresponding components.

Figure 1.64*a* shows the basic components used to establish a high-pressure tubing joint. Two ferrules, designated as front and back ferrules, are slipped over the end of the tube to be joined. A closure nut is later applied to fasten and then tighten the joint. When the closure nut is tightened, the front ferrule is forced to penetrate into the external surface of the tube, establishing a tight seal by permanent deformation. When disassembled the ferrule can no longer be separated from the tube.

Figure 1.64*b* illustrates how ferrules are combined with a connector. The welded high-pressure tube, however, is not shown. The tube can be attached to endless kinds of modifications and the number of possible connections is literally unlimited. In Fig. 1.64*c* the tube is inserted into a male connector that is prepared for welding either to a thick-wall tube or to any type of adaptor, a vessel, or the like.

Swagelok fittings are simple in design and easy to handle. They permit fastening and tightening without twisting the tube. If handled correctly by tightening the nut $1\frac{1}{4}$ turns the joint is absolutely leakproof. The author has used numerous Swagelok combinations for pressures of 30,000 psi and over without any leak failure. For higher pressures it is

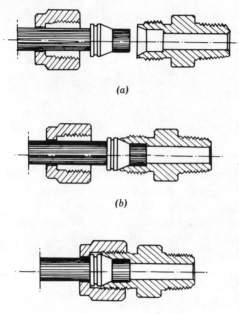

(a)

(b)

(c)   **Fig. 1.64** Swagelok fitting.

mandatory to have correct alignment, particularly for tubes of ⅛ in. and smaller. The versatility of this joint design is obvious. This design exists in many modifications with other trade names, and they are often difficult to distinguish from each other. Specific details may be found in company bulletins or catalogs.

b. Parker-Hannifin Fitting.   The Parker-Hannifin fitting uses only one ferrule, which is designed to seal on two points, thus providing a double sealing action. A model of such a fitting is illustrated in Fig. 1.65. The sealing mechanism is self-explanatory. The fitting consists of three pieces, including the connector, the ferrule, and the nut. The connector is provided with a male pipe thread, suited for both heavy-wall tubing and regular standard tubing. The nut moves the ferrule into final position in the cavity provided in the connector. The cutting or gripping edge of the ferrule is deformed and directed toward the external surface of the tube wall, establishing a perfect and reliable seal. The cavity of the connector provides self-centering action for the ferrule to achieve a concentric grip with adequate seal. In the final tightening motion of the nut the opposite edge of the ferrule is also bent toward the tube surface, establishing a second tightening grip contact with the tube joining surface.

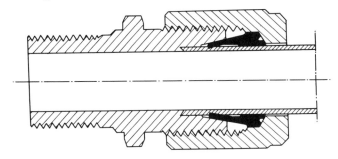

Fig. 1.65   Parker-Hannifin fitting.

These fittings are also available in many design configurations, sizes, and construction materials. The connectors are available in many different designs to permit a high degree of flexibility for joint connections.

c. Hi-Seal Fitting.   The Hi-Seal is a tube fitting using a ferrule that has great similarity to the Parker-Hannifin fitting. A major difference is that the end face of the ferrule does not provide the biting effect. The ferrule is designed with a special toothlike edge that penetrates into the external tube surface for sealing. The second distinct difference is that the fitting usually undergoes a special presetting operation before being used in a tubing system, and this requires a special presetting tool.

Nut and ferrule are slipped over the tube end, which is then placed into the prepared recess of the tool, representing a stepwise assembly with gradual plastic deformation of the tube end before the final grip is achieved. Figure 1.66 shows the finished assembled joint with the ferrule in seal position.

d. Braze Seal Fitting.   The braze seal fitting does not utilize the external grip and bite-in method for sealing but uses instead a ferrule that is brazed on the tube face. By tightening the nut the ferrule is squeezed

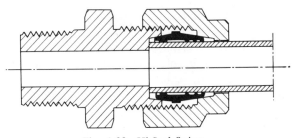

Fig. 1.66   Hi-Seal fitting.

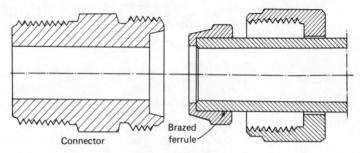

Fig. 1.67   Braze seal fitting.

along its cone into the matching cavity countercontour of the connector. The seal is established by utilizing the wedge and cone principle, which provides ideal sealing action once the sealing components are centered.

An example of a braze seal fitting is given in Fig. 1.67. Certain precautionary measures must be taken to produce a satisfactory brazed joint. The reliability of this joint depends solely on the quality of the brazing operation, which is not particularly difficult. Tolerances of ferrule and tube end must agree fairly well, since excessive deviations will result in immediate tube seal failure.

As a general rule, best results are obtained when the joint is heated rapidly and kept at the brazing temperature for the minimum time required for proper flow of the brazing alloy. When using resistance heating devices the electric current must be kept low enough to prevent severe burning at the important contact points.

This fitting is presently available in tube sizes of $\frac{1}{4}$ in. OD up to $1\frac{1}{2}$ in. The sealing principle is sound. The ferrule is a simple device and can easily be attached by the application of heat. The tube need not be plastically deformed to establish a reliable seal. The same tube end can be used over and over again, and the ferrule can be removed by heat without damaging the tube face. The pressure the fitting is capable of withstanding actually depends on the reliability of the brazing effect and the tube size.

e. Gyrolok Fitting.   The Gyrolok fitting greatly resembles the Swagelok grip fitting, using a front ferrule and a back ferrule. The sealing principle is also the same. The ferrules deform plastically and then bite into the external tube surface to establish a concentric grip. The similarity to the Swagelok design is obvious; thus an illustration of this design configuration is not needed.

As is the Swagelok design, this fitting is available in numerous configurations providing a wide range of modification possibilities for tubing systems. Size range variations come in increments of $\frac{1}{16}$ in.

f. Aminco Fitting. An elegant fitting design that incorporates the single-cone seal principle is known as the Aminco fitting, developed by the American Instrument Company, Silver Springs, Maryland. Sealing is accomplished along the tube face itself, which is machined as a cone. The fitting is capable of handling extreme pressures, depending solely on wall thickness of the tubes involved. Preferable cone angle is 60 degrees. For effective sealing the contact faces are provided with differential angles of 2 to 3 degrees. This assures line contact at the first approach with perfect sealing after stronger tightening. This joint is leakproof at any pressure level through establishment of the reliable cone-wedge sealing principle.

Details and assembly of the Aminco fitting are presented in Fig. 1.68. The sketches are self-explanatory.

For tightening, a sleeve is threaded over the cylindrical section of the tube end. For safety reasons the thread is left-hand. A nut with an external thread forces the sleeve with the tube into the female cavity of

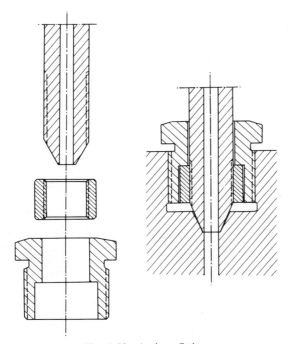

Fig. 1.68 Aminco fitting.

the counterpart of the fitting, with the differential angle in the cone slope. Misalignment is practically nonexistent. The fitting can be reused many times before the seal faces must be remachined. The Aminco fitting is available in a wide range of sizes and construction materials.

g. Sno-Trik Fitting. In recent years a new design has been added to the series of effective fittings that combine the ferrule grip with the principle of the cone contact of the tube face. The idea is to eliminate the external thread on the tube for the sleeve and replace the sleeve by ferrules. The fitting actually represents a combination of the Swagelok design and the Aminco cone principle. A cross-sectional view of the Sno-Trik fitting is given in Fig. 1.69. The illustration is self-explanatory. The ferrules prevent the tube from being pushed back by eventually leaking gases. If it is properly handled the tube face cone will supply the sealing effect. In Fig. 1.69 the Sno-Trik design is compared with the Swagelok configuration. The difference is obvious. The Sno-

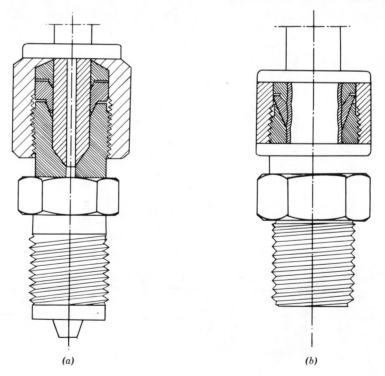

(a)   (b)

Fig. 1.69   Sno-Tric fitting compared with Swagelok joint.

Trik fitting is preferred for tubes with thicker walls than customary for Swagelok fittings.

h. Conomaster Pipe Joint. A rather interesting pipe joint, exhibiting the same sealing principle as the "Mott" gasket (discussed in detail in Section II.C.2b), is designated the *Conomaster pipe joint.* Details are illustrated in Fig. 1.70. Although this joint is not necessarily designed for laboratory tubing, it is presented because of its uniqueness for all tube sizes. The normal tube ends are provided with flanges for butt welding. The tube faces are machined and grooved for double V-band contours. The gasket is highly elastic and flexible and acts like a spring, being inserted with prestress. When the flanges are mated, the edges of the gasket interfere with the sealing surfaces of the flange grooves and are subject to prestress in a radial direction, imparting permanent deformation to the gasket edges and thus creating the sealing effect with radial locking force. This plastic deformation insures intimate intermeshing of the surfaces of the gasket and flanges at the seal areas, providing a 100% circumferential contact.

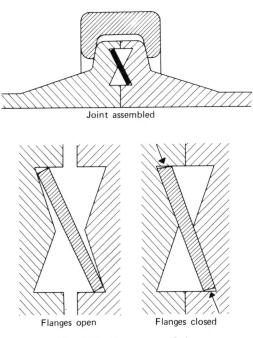

Joint assembled

Flanges open   Flanges closed

Fig. 1.70 Conomaster fitting.

The preload of the gasket further provides some sort of a spring memory, causing the gasket to follow should the flanges undergo thermal expansion. It is mandatory that the gasket be designed with proper interference fit and that the gasket material be some 50 Brinell units softer in hardness than the flange material. Spring effect produced by interference fit and hardness differential of the contacting materials are the fundamental prerequisites for this joint to function with zero leakage.

The flange design with coupling used to fasten the overall joint utilizes the Grayloc clamping principle (described in Section II.C.2b). This specific coupling design makes the joint practically independent of the prestress of the flange bolts.

The Conomaster pipe joint is available in sizes from 1 to 6 in. with pressure ratings up to 2500 lb.

The manufacturer claims that the joint is suited for high-pressure steam systems, nuclear installations, and cryogenic plant facilities.

i. Flare-Type Fittings. Flare-type fittings have been widely used in the hydraulic field for many years. In this design the connector has a conical face that matches with the flared ends of the tubes. Flared fittings are available in various designs in the form of three components, two components, or as inverted-type fittings. These three major categories are presented in Fig. 1.71.

In the three-piece joint the fastening nut has a conventional shape and the tube flare is matched by an additional component. The two-piece design provides conical contact areas in connector and nut and matches the flare of the tube. The inverted-flare type is a simple modification of the two-piece fitting. No gaskets are used and the seal is produced by the direct participation of tube components. In the self-flaring fittings a special flaring operation is not required. In general, the seal effect is accomplished by torquing the nut, which presses a wedge-shaped sleeve against the tubing end and a mating female part of the fitting. This creates a flare-type enlargement on the participating tube end.

These devices work best on thin-wall tubing where large torques are not required to produce the flaring action. The joint is mechanically strong and vibration resistant and can be used repeatedly without repair or maintenance work. Soft materials should be used.

j. Threaded-Sleeve Fittings. Tubes can also be joined for leak-free service by a tightening action between nut and tube without the use of ferrules. The seal is accomplished on the external surface of the tube by

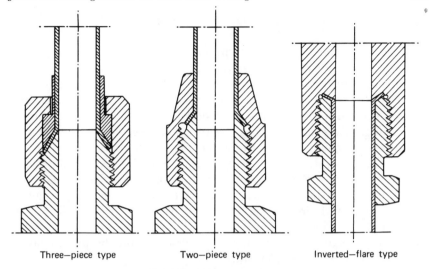

Three—piece type   Two—piece type   Inverted—flare type

Fig. 1.71   Flare-type fittings.

providing the nut with a corresponding shape that forces the face of the nut, similar to a biting edge, into the tube surface, producing permanent deformation by grip action.

An example of this joint, which is designated as a *threaded-sleeve fitting*, is illustrated in Fig. 1.72. The connector contains a recess for the tube face. By tightening the nut, the cone-shaped face of the nut is made to force the tube end into a corresponding recess and produce a permanent deformation, establishing the seal by grip-type action.

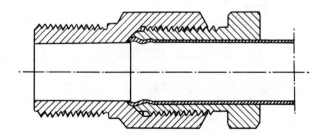

Fig. 1.72   Threaded-sleeve-type fitting.

B. Tube Joints with Elastomeric Seal Components

Laboratory tubing systems can employ many fitting categories when temperature is not a consideration. For all such occasions elastomeric seal devices can be used, providing a wide selection for tube joints. As illustrated in Fig. 1.73, O-rings and other thermoplastic backup rings form a seal through compressive action without deforming the tube. The fittings seal effectively and are reusable, provided only moderate pressures are encountered in the system. The elastomers, Teflon, nylon, Kel-F, silicone, and the like, can replace all ferrule-type metal designs, offering numerous joint combinations that provide perfect seals.

At this point it should be mentioned that most fittings using single or double ferrules of metals are also available with ferrules of Teflon, Kel-F, and other plastic materials. Cold flow of Teflon or Kel-F is not a problem, since the ring cone is totally enclosed and the plastic has no place to go. A perfect seal results.

An interesting joint developed for NASA is shown in Fig. 1.74. The manufacturer claims it can withstand up to 10,000 psi. A Teflon or Kel-F cone ring seals against a flared tube on one side and a beveled metal face of the connector on the other.

The Yarway grip seal joint, presented in Fig. 1.75, uses a Teflon ring of special design between the tube faces. The Teflon is fully confined and cold flow cannot occur. The beveled contact slopes of the Teflon ring with the collets on both sides provide multicontact areas for safe sealing operation. Although a series of extra components is required to establish the joint, the tube ends need no special machining preparation. The collets are split-spring-type cone rings that establish the joint grip

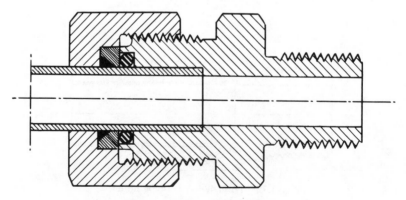

Fig. 1.73 Thermoplastic compression fitting.

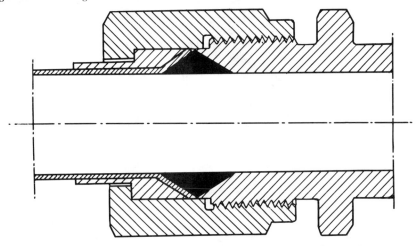

Fig. 1.74 Flared tube with additional Kel-F cone ring.

against the external tube surface when the nut of the union is tightened. The components of the joint can be reused without refinishing. After a series of repeated applications, only the Teflon ring need be replaced. The seal ring may be chosen from any type of elastomeric or plastic material compatible with the flow medium. Accordingly, a wide range of applications is possible.

A combination of metallic ferrules and Teflon or elastomeric seal rings is illustrated in Fig. 1.76. The ferrule is used against a flared tube end. Teflon rings inserted in the conical sections of the ferrules ensure perfect extra seals at relatively low sealing force. The Teflon, Kel-F, rubber O-rings, or other plastic materials cannot be extruded, since they are fully restrained in the cavity. The same is true for the prevention of cold flow of Teflon rings. This fitting is customarily applied to tubing ranging in size from $\frac{1}{8}$ to 1 in., with seal effectiveness claimed for pressures as high as 10,000 psi.

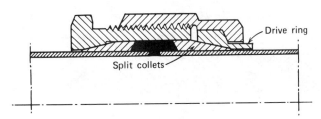

Fig. 1.75 Yarway grip seal joint.

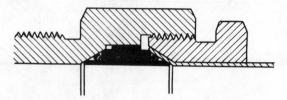

Fig. 1.76 Joint using metal ferrule with O-rings.

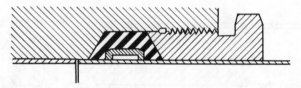

Fig. 1.77 Sealastic fitting with metal ferrule and elastomeric cone ring.

A similar type of fitting also utilizing the double-compression seal principle is shown in Fig. 1.77 and is known as *Sealastic fitting*. A combination between a metal ferrule and a rubber ferrule is suited for any tube-flange combination, providing a perfect seal for a very wide range of joint applications. A plastic deformation of the tube does not take place and shop preparation of the joint components for final sealing is not required. This fitting is also highly suited for vacuum service.

Figure 1.78 shows replacement of the ferrule with a double wedge to establish a leakproof joint for vacuum and pressure service. This design is designated as the *Uni-Frik joint*. The device was developed by Henry Wallenburg and Company, Stockholm, Sweden. In this concept a friction-locked wedge produces a pressure-vacuum fitting without flaring the tube end. When the nut is tightened into the connector, a spring washer is compressed and establishes a locking effect, resulting in a leakproof joint. The washer then squeezes the wedge ring against the external tube surface, deforming it permanently and thus locking it in place with high friction.

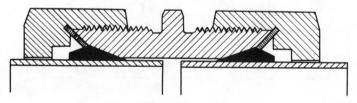

Fig. 1.78 Uni-Frik joint (Swedish design).

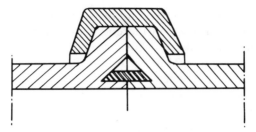

Fig. 1.79 Double-truncated cone seal. *(NASA)*

In an interesting NASA device, illustrated in Fig. 1.79, a cone shaped Kel-F washer is clamped between two matching flange faces. The sketch shows a joint using a 3-in.-diameter double-truncated cone ring of Kel-F material, clamped into a cavity, formed by two closely matching parallel flange faces. The clamping function is accomplished by a Marmon clamp modified with wing nuts. NASA claims service for cryogenic conditions at −320°F. Operation under vacuum is no problem.

This configuration can be hand tightened for use on cryogenic equipment, however, when it is limited to seal diameters of ¾ in. and less. When it is used in ambient temperature ranges, there is no diameter limitation.

This seal can be used in laboratory instrumentation, flow bench hook-ups, and all kinds of hydraulic service connections. The insert design provides sufficient compression of the Kel-F ring cone to allow a metal-to-metal fit between the cavity slopes of the joint faces. Because of a constraint arrangement of the Kel-F seal, cold flow does not occur if Teflon is used.

The Tylok fitting is also a device resembling those fittings that use a ferrule and establish the seal by utilizing the external grip principle. The Tylok fitting produces permanent deformation on the tube surface and the parts cannot be separated again. Tylok fittings are applicable for vacuum and pressure as well, preferably applied to systems in operation for long periods of service. The sealing principle has great similarity to the design configurations of the Swagelok and Gyrolok fittings.

C. Special Joints of Small Tubing for Vacuum Service

Laboratory service requires both vacuum and pressure applications. Many, if not most, of the fittings described in this chapter are suited for both uses. As a practical rule, all fittings relying on plastic deformation of the metal components for establishing a pressure seal are also suited for vacuum service.

Tubing joints using elastomers and plastic material rings on the external tube surface can also be considered good vacuum joint components. However, in most cases plastic deformation of ferrules and tubing is not required in vacuum joints. Most vacuum seal joints can be established by hand without using a wrench, whereas plastic deformation of metal ferrules and tubing can only be accomplished with wrenches using excessive force. Elastomeric vacuum seal joints made simply by manual tightening are often called *quick-connects*.

Typical vacuum fitting joints of this nature are shown in Fig. 1.80. They are frequently used in laboratory service for all levels of vacuum requirement. Here the O-ring represents the major seal component tightened manually. In Fig. 1.80a the O-ring is inserted in a groove that is part of the face of the connector. In Fig. 1.80b the O-ring is inserted in the conical face of the male connector to establish a perfect seal. In Figure 1.80c the face of the connector is flat. The O-ring is placed around the tube and located in a flat metal ring with groove surrounding the O-ring. The metal ring also limits the tightening compression of the O-ring. All fittings for vacuum service using elastomeric O-rings as

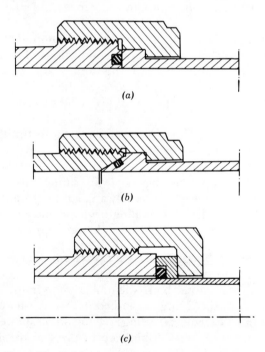

(a)

(b)

(c)

Fig. 1.80 Quick connects using O-rings of elastomers.

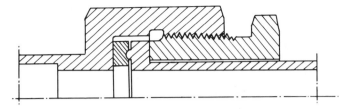

Fig. 1.81   Cajon metal joint for vacuum.

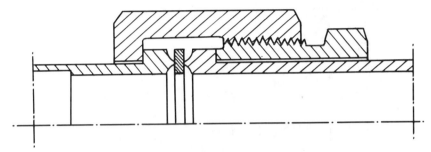

Fig. 1.82   Modified Cajon fitting.

sealing components can simply be tightened by hand to provide a satisfactory vacuum seal.

Cajon has marketed and developed a series of joints with sensitive metal gaskets, both for pressure and vacuum service. One of the Cajon joints is presented in Fig. 1.81. It uses a metal gasket that is deformed from one side by a shoulder in the face of one of the connectors. For effective sealing the gasket must be relatively soft to facilitate plastic deformation before the threads of the closure nut are permanently deformed.

A modification of the joint of Fig. 1.81 is given in Fig. 1.82 in which both connector faces show a raised-face shoulder. The gasket should be aluminum or the like when these materials can be tolerated. The connectors are designed with sockets on the external ends for brazing the tubing.

D.   Typical Cone Joints for General Application

Tube joints are required in countless numbers of applications, including tube-to-tube, vessel-to-tube, vessel-to-instrument, and many others. Whatever the design there are joints sealed externally with ferrules, sealed with extra gaskets, or having the tube end formed as a cone. With

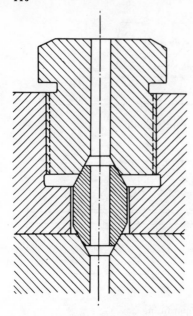

Fig. 1.83 Joint with special double-cone arrangement.

metal-to-metal seals with and/or without gasket the best seal is achieved if the contact area for the final seal is a cone with a differential angle providing line contact. Line contacts require a minimum seal force, and the stability of such a joint is ensured if the seal force is a minimum with low or zero distortion from torque.

Whenever ferrules or tube flare designs cannot provide a satisfactory solution, a separate cone component can always be used, as shown in Fig. 1.83. As indicated in the sketch, this cone actually uses a piece of thick-wall tubing with a conical slope on both ends. The only requirement is that this cone segment be introduced in a bore where it has little chance to buckle. The close clearance then assists in centering the cone and establishing perfect alignment for proper sealing.

These separate double-cone components also should be provided with a differential angle against the female cone cavity. Preferred angles are 58 to 60 versus 60 to 62 degrees. The penetrating depth should be kept at a minimum. Differential angles of the cone permit initial line contacts for sealing at reduced tightening force.

Figure 1.84 presents a series of applications for cone seals of all types. These seals can easily be made in the shop and offer numerous opportunities for solving most connection problems between all types of equipment and for any level of pressure service.

When manufactured in stainless steels, preferably of the #300 series,

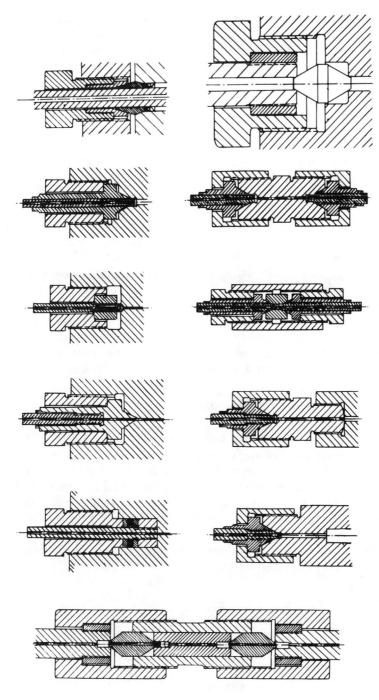

Fig. 1.84 Modifications of cone seals for high-pressure laboratory tube fittings.

these seals can be used without heat treatment. By using high-strength low-alloy steel of the #4340 category plus a heat treatment to $R_c \sim 35$, a high-durability seal will be obtained, since the steel is undergoing strain hardening when it is plastically deformed and it becomes better with repeated applications.

E. Weld Adapters

The chapter on gaskets and tube joints cannot be concluded without a brief discussion of the weld adapters. In research and process development versatility is much more desirable than in a production plant, where conventional standards suffice. In both laboratory and pilot plant operations special parts are used that will never be required in the large-scale plant. Although welding is the simplest and generally the least expensive method of joining tubing, this method is seldom used on very small-sized tubing. The aim is to be able to assemble and disconnect joints as quickly as possible and to reuse parts without machining.

Where welding is required, usually for tube sizes greater than $\frac{3}{8}$ in., it is customary not to weld the tubes by butt welding. It is safer to use adapters, as indicated in the sketch of Fig. 1.85. These adapters are available from the shelf in all types and grades with sizes up to 2 in. They facilitate design and field construction and improve versatility. They are readily attached to any kind or part of process equipment.

V. Summary Considerations for Static Sealing Devices

Gaskets are the most influential factor that governs the design of a joint or a vessel closure and piping system. This becomes even more obvious as pressure increases in the system. Leak-free service of a closure is achieved by applying enough bolt force to provide the required interface contact pressure in the seal component. The bolt force is a function of the mean tightening diameter, the tightening factor, and the internal pressure. Both tightening factor and mean tightening diameter are interrelated functions of the design configuration and the construction material of the gasket.

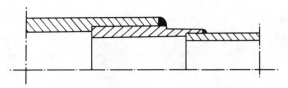

Fig. 1.85 Weld adapters.

A variety of modes must be identified, such as pressure, either static or dynamic; pressure and temperature, both static or pulsating; pressure, temperature, and corrosion; and finally cryogenic conditions in the presence of or without vacuum as encountered in outer space.

A. Design for Pressure Only

With pressure as the sole variable in a system, a distinction must be made between, first, closures for large-scale equipment and, second, small joints in tubing and laboratory equipment and connections to instrument lines.

1. Sealing Devices for Large-Scale Equipment under Pressure

In large-scale closures for high-pressure equipment, such as reactors, columns, containers, separators, surge tanks, pressurized gas bottles, and the like, it is customary to use rigid metallic gaskets for sealing. For pressures exceeding the 5000 psi level, it is preferable to use heavy-sectional gaskets as described in Table 1.4. Very large vessels are sealed with lens rings, double cones, and systems developed by BASF or any other comparable device. The design must aim at the line contact sealing principle when metal contact is involved. As long as temperature is not a factor, the metal gaskets can be combined with O-rings, as indicated in Fig. 1.61. This combination provides a simple design with high efficiency and low sealing contact pressure.

For medium pressure ranges the category of the flexotallic gasket may be considered, as illustrated in Fig. 1.22. The Graylok device is also a candidate for this pressure level.

In the presence of medium to low pressures, the combination-type gaskets are suitable and offer an enormous range of design configurations, shown in Table 1.5. Numerous solutions are available from the category of elastomeric and plastic sealing components, which doubtlessly represent the largest and most significant family of gaskets. The elastomers are available in almost any desirable cross section and material combination.

2. Sealing Devices for Small-Sized Tubing, Instrument Lines, and Equipment

Laboratory equipment requires the greatest versatility of components for quick handling, assembly, dismantling, and operation. An extensive supply is available, provided with highly reliable joints. With the use of stainless steel tubing, preferably of the 300-series assembly, handling and operation are easy and can be achieved from the bench, often without the use of machine tools.

Depending on the selection of the wall thickness, pressures up to 60,000 psi are commonly handled with ease using conventional pressure fittings. The category of the Aminco fittings has practically no upper pressure limitation; they can often be used for pressures higher than the pressure rating of the tubing for which they are chosen.

B. Design for Pressure below Atmospheric (Vacuum)

Metals are best suited for the design of gaskets in vacuum service. Self-energizing principles are of little value, since the pressure differential is too low. Heavy-cross-section gaskets must be permanently deformed to establish satisfactory vacuum, with the same conditions being valid as for extreme internal pressure.

Resilient metal gaskets provide excellent vacuum service if handled properly. Suitable preload with spring action is a basic prerequisite. Plating with soft metals and coating of the metal gaskets facilitates sealing considerably for both pressure and vacuum conditions.

Perhaps the best gaskets for vacuum service are the elastomers, available in countless configurations of cross-sectional areas and construction materials. Design of the gasket cavities for elastomers has to consider the property of outgassing. For ultrahigh vacuum two O-rings should be used with a pumpout volume between the O-rings.

C. Design for Pressure and Temperature

Metallic gaskets with heavy cross section are perfectly suited for pressure of high levels in the presence of elevated temperatures. If pressures and temperatures pulsate, the category of resilient metal gaskets should be chosen. The resilient gaskets are designed to provide spring action and therefore can compensate for bolt expansion without sacrificing contact force needed for sealing. Self-energizing principles should be applied wherever feasible.

For cryogenic conditions bimetallic gaskets are used that provide substantial differences of contraction when cooled or of expansion when heated. The differentials in thermal expansion of the gasket materials are responsible for producing additional contact forces for the gaskets in excess of the preload by the bolt force. In the presence of vacuum and temperature the same sealing principles apply.

D. Design for Pressure, Temperature, and Corrosion

Corrosive atmospheres in the system introduce a third variable in the design of appropriate gaskets. There are few problems that cannot be solved through proper use of the many materials available today. Examples are stainless steels, nickel, copper, aluminum, silver, Inconel steels, Hastelloys, titanium. Consult seal manufacturers; specific corrosion discussions exceed the scope of this book.

E. Final Conclusions

Gaskets can greatly influence the geometry of a joint or a closure configuration. Tightening diameter, tightening factor, and materials of construction are decisive factors for the geometry of the entire joint closure.

The tightening factor is highest when metal-to-metal contact of heavy-section metal gaskets is present. By using metal gaskets of the resilient type with springy components the tightening factor is essentially decreased. Sealing is further improved when gaskets are plated with soft metallic layers or coated with Teflon, Kel-F, or similar materials.

The tightening factor is lowest when elastomeric materials are used. Machining for large-scale vessels can be noticeably simplified by using an elastomeric O-ring with metal rings of the lens-ring type or similar configuration when temperatures permit.

Precise figures for tightening factors are presented in Section II.B.4 for high-pressure joints/closures with heavy-section gaskets for the determination of the required tightening force, thus providing a reliable basis for optimal design of high-pressure vessels at lowest cost.

Systems having vacuum and ultravacuum conditions are more difficult to seal than pressure systems because of outgassing of the elastomeric sealing components and of the systems as well.

In cryogenic systems bimetallic metal seals are preferable, because they use the differential of thermal expansion or contraction to improve sealing efficiency.

A few basic principles may be mentioned concerning the behavior of bolts in high-pressure systems. Bolts generally lose some of their preload as a function of time through material relaxation. This can result in leakage when gas is in the system. It is important, therefore, to know the relaxation behavior of the bolting material and to provide corresponding precautionary measures to prevent a loss of the contact force in the gasket interfaces.

A second factor concerning bolts is a general misunderstanding of the torquing procedure, which often causes considerable problems. Two facts are often overlooked. First, once the proper tightening factor has been chosen, the torque for establishing the contact pressure can be computed. By relying on the torque alone, the following mishaps may occur. If the bolt-nut combination consists of mild steel, it is possible—most of the time even likely—that the thread surfaces will roughen up after repeated opening and tightening. Rougher threads require increased friction; thus a higher torque will result. In other words, the torque calculated for sealing is reached earlier, the contact force will not be sufficient, and leakage may occur when assembly is controlled by initially calculated torque alone.

The second frequently overlooked fact is that if high-strength steels are used for bolt and nut, some work hardening will take place, the thread surfaces will become harder and smoother, and the bolts may already be overstretched so that the threads begin to yield before the computed torque moment is reached.

The only way to solve this problem is to calculate the torque required for proper sealing. Now the torque must be converted to the axial force imparted in the bolt shaft during tightening. The axial force in the bolt shaft produces an elastic elongation of the shaft that can be computed by the laws of elasticity. This axial lengthening of the bolt shaft is the safest criterion for the magnitude of the torque to be applied for tightening. By measuring the bolt elongation a precise control of the tightening torque is achieved. Dirt in the thread can further produce increased friction, which also results in torque increase without improving the tightening effect. Consequently, a given torque moment reflects in a clean thread a corresponding elongation of the bolt shaft, proportional to the torque. The measured elongation of the bolt shaft is, therefore, a function of the torque moment applied. This is the guarantee that the joint will seal with the necessary contact pressure as required by calculation and for safe operation.

F. Temperature Limitations for Gaskets

Gasket effectiveness is affected by upper temperature limitations. In the cryogenic range the gaskets tend to become brittle. At elevated temperatures they expand and and lose their resiliency. Rubbers become brittle and harden. They retain their sealing properties over a long period of exposure time at specified maximum recommended temperatures.

Rubbers are subject to thermal expansion, expanding to 10 times the values experienced for steel or aluminum. For this reason O-rings must have large grooves, requiring approximately 40% more space. Higher temperatures restrict the life expectancy of rubber. In seal practice life expectancy of rubber may be defined as the time required to reduce initial mechanical properties by one-half of their value in one year.

Asbestos, the most common gasket material, has extraordinary temperature resistance along with excellent resistance to chemical corrosion. White asbestos is used for sealing weak acids. For stronger acid crocidolite asbestos is an excellent material.

Pure asbestos is seldom applied, because of its low strength and high porosity. At temperatures in excess of 400°C it decomposes to a fine powder. When mixed with rubber or plastic it is designated as *compressed asbestos fiber* (CAF), and its temperature resistance increases to 450°C. Polymers commonly mixed with asbestos are neoprene, nitrile, butadiene, and styrene.

Asbestos and its combinations are very useful as materials for static sealing devices, better suited for moderate pressures and high temperatures than any other composite material. It is a rule of experience to assume the algebraic product of pressure p (psi) times temperature t (°C) as a constant value equal to 150,000. For all cases where the temperature exceeds 380°C product pt is reduced to a value of 120,000, considering a suitable safety margin. See G. J. Field (4).

In equipment handling corrosive chemicals and solvents or when hygienic conditions prevail, the asbestos is encased in pure or filled PTFE. For these materials the temperature must be lowered to 250°C at relatively low pressures.

For metal gaskets operated in the high-temperature range the upper temperature limits can be raised. Metal gaskets filled with asbestos or PTFE, such as spirotallic-wound or metal-jacketed ones, must be serviced at lower temperatures than the solid metal gaskets.

Temperature limits for spiral-wound gaskets are presented in Table 1.7. When these gaskets are filled with asbestos the upper temperature limit is approximately 500°C. For spirals filled with PTFE, the temperature maximum is 250°C.

For solid metal gaskets, such as lens, delta, octagonal, and other rings, the upper temperature limit can be raised well beyond 600°C. For plain metal O-rings the practical upper temperature limits are indicated in Table 1.8. For metal O-rings plated with indium, cadmium, silver, gold, or PTFE, the upper practical temperature limits are shown in Table 1.9. More specific details on this subject are presented in chap. 6 of the author's book on high-pressure technology (2).

Chapter 2 and others provide additional information on material characteristics in view of dynamic sealing concepts.

Table 1.7 Metals for Gaskets (Static) Spiral-Wound, and Jacketed

Material	Temperature max (°C)	Temperature max (°F)
Lead	100	212
Brass	250	482
Aluminum	400	752
St-steel	500–850	932–1562
Titanium	500	932
Nickel	750	1382
Inconel	1000	1832
Hastelloy	1000	1832

Table 1.8 Metal O-Rings

Material	Temperature max (°C)	Temperature max (°F)
Copper	400	752
Mild steel	550	1022
Cupro-nickel	600	1112
Monel	600	1112
Nickel	700	1292
Stainless steel	800	1472
Inconel	850	1562

Table 1.9 Plating Materials for Metal O-Rings

Material	Temperature max (°C)	Temperature max (°F)
Indium	140	284
Cadmium	200	392
Silver	800	1472
Gold	850	1562
PTFE	300	572

VI. Sealants

Sealants are essentially adhesives that are used to contain liquids and gases and to exclude dust, dirt, moisture, chemicals, and the like. Sealants are basically applied for operating conditions noticeably less severe in temperature and pressure than gaskets. There are, however, certain formulations that can withstand temperatures up to 5000°F for several minutes. There are also sealants that prevent leaking when pressures in hydraulic systems are as high as 3000 psi. Sealants can render a reasonably wide range of resistance to chemicals and are relatively inexpensive compared with gaskets.

A. Mechanism of Operation

The operating characteristics of sealants differ considerably from those of gaskets. Sealants are high-viscosity liquids that fill the surface asperities during clamping. Metal-to-metal contact takes place at the high spots whereas the sealant fills the void areas with a continuous film at low contact force. Gaskets require relatively high-contact clamping force to establish a compression seal. The thin film acts like a flexible gasket that perfectly matches the surface asperities of the contact areas of the joint.

The ability of sealants to fill voids in contact surfaces through gravity flow enables the sealants to tolerate surface irregularities that could never be accommodated by preformed gaskets. Sealants further permit intricate and irregular contact geometry of the seal areas where preformed gaskets could never be used.

B. Types of Sealants

Sealants are generally categorized as hardening and nonhardening. For simplification, certain types of tapes may also be included in the family of sealants. Sealants are marketed in nonsolid forms in a wide range of viscosities. Certain epoxy sealants are offered in powder form and have to be melted during application. Asphalt-based sealants and waxes are solid. For application they must be melted by a hot-melt system. Other sealants are available as thermosetting film adhesives. They also are marketed in tape form and must be heated and then pressurized for final curing.

Hardening sealants come either in rigid or in flexible form, depending on chemical composition. Nonhardening types are characterized by the nature of the plasticizers that come to the surface continually, producing the impression that they are "wet" after application.

Nonhardening sealants are characterized by the "mastics" that are applied to seams with a brush or trowel.

Tapes have a great variety of backings, including pressure-sensitive or solvent-activated adhesives.

C. Selection Factors

Sealant properties are sensitive to composition and even minor changes result in major property alterations. Other influencing factors to be considered are thermal sensitivity, chemical resistance, weatherability, mechanical properties, abrasion resistance, adhesion, electrical characteristics, color stability, toxicity, and production procedures. For details, consultation with a manufacturer is mandatory.

D. Typical Sealant Applications

Sealants are generally used for static pressure conditions, encountered in pipe thread, flange joints, tanks, concrete, potting, molding, encapsulating, coating, caulking, glazing, and numerous other operations. Metals, glass, masonry-type joints, or electrical components are involved. Wherever sealants are employed they are largely independent of mechanical conditions of the interface areas.

References

Publicly available literature on gaskets is scarce, consisting mostly of brochures, pamphlets, and papers, usually distributed by manufacturers. The lack of standards in a great many areas, discussed in this chapter, does not make the study easier. Because of the lack of reliable design analyses, emphasis has been given to practical applications that will help the man in the field more than vague theories that cannot be used to provide a reliable prediction of the actual behavior. Numerous brochures of all major seal and gasket companies, catalogs, papers, and fabricator publications in the United States and in Europe were evaluated. Considerable and highly valuable NASA information was also surveyed.

1. Bridgmen, P. W.
   The Physics of High Pressure
   G. Bell, London 1952.

2. Buchter, H. H.
   Apparate and Armaturen der Chemischen Hochdrucktechnik
   Springer-Verlag, New York, 1967.

3. Daniels, C. M.
   "Aerospace Cryogenic Static Seals"
   Paper presented at the Annual Meeting of ASLE, Houston, TX, May, 1972.

4. Field, G. J.
   "Seals that Survive Heat"
   Mach. Des. (May 1975).

5. Hull, J. W.
   "Ring Spring Design for High Performance Metal Static Seal"
   Hydraulics and Pneumatics (September 1960).

6. Logan, S. E.
   "Static Seal for Low-Temperature Fluids"
   Jet Propul., J. Am. Rocket Soc. (July 1955).

7. Prince, W. A.
   "Bimetallic Seal Solves Cryogenic Problems"
   Hydraulics and Pneumatics (November 1964).

8. Seufert, W.
   "Untersuchungen über das Dichtvermögen von Dichtleisten"
   J. Brennst. Wärme, Kraft, No. 3 (1951).

CHAPTER 2

Mechanical End Face Seals as Axial Sealing Devices

I. Introduction

The seal classification of Chapter 1 indicates two major categories of seals for rotating shafts: interfacial and interstitial seals. The interfacial seals again subdivide into axial seals and radial seals. Chapter 2 covers only axial seals. It provides numerous solutions for sealing problems associated with any type of machinery or equipment using rotating shafts that must be sealed when passing through any type of concealed equipment.

Compared with the gaskets, end face seals, which are also generally known as mechanical seals, employ a unique and different sealing principle. Mechanical seals were first massively applied in the automotive industry for sealing engine coolants and water systems. They are now used even more extensively in this application and have confirmed their enormous importance to an industry.

In chemical, petrochemical, utility, and related industries, the mechanical seal has gained considerable importance because of the constantly improving technology with respect to sealing techniques and construction materials used in sealing components. Ever-increasing temperature and pressure demands, in addition to increasing rotational shaft speeds, have created a climate that forces the progressive seal designer contantly to widen his horizon.

II. Fundamentals of Mechanical End Face Seals

Mechanical end face seals are now fabricated by the millions and by a considerable number of renowned sealing companies all over the world.

The state of the art for mechanical seals has developed to a degree that pressures from high vacuum of 10^{-5} torr up to 5000 psi ($\sim$350 atm) can be handled safely. New materials and particularly metal bellows have made it possible to handle temperatures up to 1000°C ($\sim$1830°F) and down to the cryogenic range with confidence. Rotational shaft speeds of up to 50,000 RPM are no longer impossible.

A. Sealing Mechanism

A mechanical seal represents a complex design that consists of a series of single design components. The sealing function is achieved by two primary seal rings with faces to prevent leakage. One of the primary seal rings is attached to the shaft and rotates with it; the other primary seal ring is stationary and is attached to the housing. To seal the shaft of a centrifugal pump the stationary primary seal ring is mounted to the gland plate ring. During rotation of the pump shaft, the primary seal ring attached to the shaft rubs with its seal face along the counterseal face of the stationary ring. Thus the two interface contact areas function like bearings and are subject to frictional wear. Any kind of leakage of the system fluid must pass across this interface.

The rubbing contact is continuously maintained by forces acting in axial direction. The origin of the axial pushing forces can be either mechanical or hydraulic. In many designs it is the sum of both. Pushing forces are required to establish and maintain a continuous contact between the components forming the interface. This steady contact either prevents or minimizes leakage across the rubbing areas.

Rubbing action between solid interface areas produces heat and wear even when there is good lubrication. This heat builds up and finally results in the destruction of the interface contact areas. In order to prevent this, lubrication is applied that has a twofold purpose. First, it removes the heat from the area of rubbing contact and thus reduces the detrimental temperature buildup. Second, the lubricant covers the interface with a minute film that reduces friction and simultaneously establishes a tight seal.

The lubricant fluid can be either the fluid of the pump system or a secondary fluid, which can be water or any other liquid compatible with the system fluid to be sealed.

The extremely thin lubricant film across the interface is the key to good sealing performance of mechanical end face seals, the function of which is still considered a mystery.

It is not a secret that a reliable and accurate design analysis for mechanical end face seals does not exist. Accordingly, all mechanical seals have thus far been developed on a purely empirical basis. Any new seal design obviously must be tested empirically, because a prediction of the eventual performance characteristic of the seal is not possible on any reliable theoretical basis.

B.　Basic Seal Requirements

The unique sealing principle where the mating seal contact faces are arranged in a plane perpendicular to the axis of the rotating shaft can best be explained through the description of a primitive seal device that functions like a mechanical end face seal. If we assume a pump shaft with a shoulder that faces a bearing support wall so that the shoulder face functions as a seal, we have practically the basic sealing principle of a mechanical end face seal.

This configuration readily demonstrates all the characteristic properties a mechanical end face seal should provide but cannot offer because of its primitive design and obvious shortcomings.

The actual seal function of the shoulder is established by the vertical face of the enlarged shaft, which rubs against a parallel counterface in the support wall.

This primitive seal arrangement has a number of deficiencies that can be readily observed even by those unfamiliar with the technology of mechanical end face seals. They are

1.　Both contact areas forming the seal interface are very rigid and have no relative flexibility to compensate for face wear or to counteract vibration, deflection caused by shaft whip, runout, wobble, or any other disturbance of the shaft rotation.
2.　The shoulder containing the wear face is an integral part of the shaft and is rigid, compact, and inflexible. It rotates with the same speed and follows every motion of the shaft, contrary to accepted practice of a component of a mechanical seal.
3.　The countercontact face is also a rigid component, integral with the housing wall and without specific alignment specification.
4.　It is very unlikely that a good shaft material that provides a high degree of stability for the shaft also satisfies seal requirements in the face of the shoulder. This is also true for the counterface in the support wall.
5.　Where there is wear the continuity of the rubbing contact is quickly destroyed and leakage occurs, since the parallel arrangement of the interface areas, which is mandatory for proper seal performance, is disturbed and then interrupted. No device is provided to supply automatic contact at the interface, a fundamental prerequisite for adequate

seal performance. Thus a compensation for face wear during operation does not exist.

6. A replacement of any of the two seal faces is not possible without taking the entire system apart. Refining the seal faces is practically impossible.

7. Even if truly flat seal faces could be provided they would be of little help, since the shaft has to obey rotational laws that do not coincide with those required of the seal faces.

The preceding disadvantages provide a basis for designing a functional mechanical end face seal. It is further obvious that a satisfactorily operating mechanical seal represents a relatively complex system of independent components that must be well coordinated to establish the desired overall function of a reliable seal. Flexibility of independent components requires good secondary sealing components with a wide range of flexibility, compatibility to pressure, temperature, corrosion, and shaft speed.

We find that a mechanical end face seal is composed of primary and secondary sealing components. The primary seal ring contains the rubbing seal face that is part of the final interface. The rings are designed as solid rings with one ring in rotation with the shaft while the other ring is mounted stationary with the housing. The secondary sealing components provide the seals between the primary rings and the shaft and housing.

To simplify the understanding of a mechanical end face seal and its operation in its entity, we show Fig. 2.1, which illustrates mechanical seals for a shaft of a horizontal centrifugal pump. The primary seal head is rotating with the pump shaft and is inserted in a cartridge that is slipped as a unit over the shaft and attached to the surface of the shaft by a simple set screw. The stationary seat ring is arranged in a corresponding cavity of the gland plate using an elastomeric O-ring as a secondary sealing component against the gland cavity. The seal action is accomplished in the interface, shown in the graph as line A-A, perpendicular to the shaft axis.

The primary seal head rings also use an elastomeric O-ring for secondary sealing component against the shaft. Other secondary sealing configurations are discussed in a later section.

In Fig. 2.1 the seal cartridge is surrounded by the system fluid, which tends to leak across interface A-A and then escape toward the external atmosphere. If leakage is to occur the system fluid has to overcome the interface contact pressure acting in plane A-A, produced by the multiple springs in Fig. 2.1b and by a single helical spring in Fig. 2.1a in addition

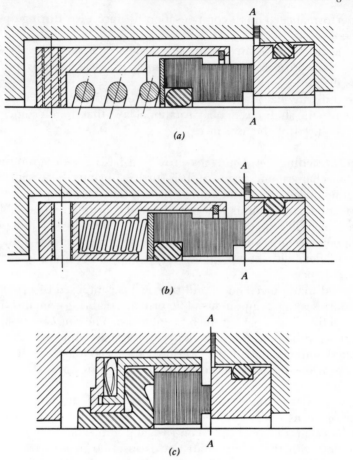

Fig. 2.1 (a) Seal unit with single helical spring. (b) Seal unit with multiple springs. (c) Seal unit with elastomeric device and wave spring.

to the pressure of the hydraulic system fluid. In Fig. 2.1c the contact force is supplied by an elastomeric bellows in combination with a wave spring and enhanced by the system fluid.

C. Primary Sealing Components

The primary sealing components, designated as the *seal seat* and *seal head*, characterize the entire design configuration in a mechanical end face seal. The seal head generally, although not always, rotates with the shaft. In most designs the seal seat is in a stationary position, but sometimes the seal seat is in a rotating position, rotating with the shaft, and the seal head is stationary.

The geometry of the primary seal components is governed by a multitude of factors. Both types of primary seal components have distinct design configurations. The design geometry of the rotary seal head is usually the result of the degree of dynamic balance to be utilized to control the hydraulic force activities. The secondary seal components greatly influence the shape of the gland. Last, but not least, the selection of the pusher springs is another influential factor in determining the configuration of the seal head ring.

The geometry of the seal seat ring is determined primarily by the environmental control methods chosen, to be discussed in detail in Section II.C.2.

1. Design of the Seal Head

The seal head is the predominant constituent of any mechanical end face seal. As Fig. 2.1 indicates, the seal head represents a unit consisting of several parts required to provide the interface function. The seal head must be designed to achieve optimal flexibility, adequate loading in conjunction with suitable pressure balance, uniformity in the circumferential distribution of the automatic pushing forces, supplied by compatible spring action, and a guarantee of proper positive drive facilities for the seal head ring when subject to rotation with the shaft.

As is shown in Section II.D, the secondary seal components play a significant role in head ring geometry and are next to the most decisive factor of hydraulic fluid balancing.

2. Design of the Seal Seat

The mating partner of the primary seal head, the seal seat, can be designed to be equipped with either one or two seal faces. The second seal face is usually machined to be used after the initial front face has been worn beyond repair. The seal seat generally functions in a stationary position.

The seat ring design is governed primarily by the configuration of the gland, which satisfies one or the other of several diversified environmental control methods, such as flushing, quenching, cooling, or a combination of several. The second significant factor determining the shape of the seal seat ring is the selection of the secondary seal components. These can be of many types as, for example, a cup ring or a V-ring or even metallic gaskets of the flexotallic type, as illustrated in Fig. 2.2.

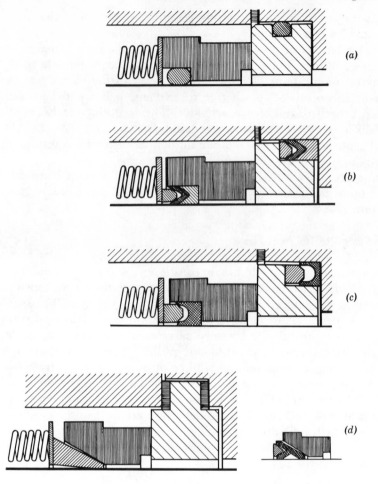

Fig. 2.2 Modifications of secondary sealing components.

The factors governing the design of primary seal seats by environmental control methods are discussed in detail in Section VI.

Seal seat design must incorporate simple components, simplicity of installation, ease of replacement and maintenance, and a secondary sealing component that permits flexibility of the seat ring, providing optimal seal effectiveness. A pressed-in design is not favorable, although this method is found in a variety of design configurations.

In seal design in which the seat rotates with the shaft, the seat arrangement with regard to the shaft can either be rigid or use elastomers in the form of highly elastic O-rings with drive pin attachment for positive drive. They provide some flexibility, which allows compensation for irregularities in the shaft motion. Some of the conventional drive methods are press-fit, pins, set screws, dents, and many others.

For rotating primary seat rings a fixed attachment to the shaft is sometimes preferred to assure positive drive action.

Seal seats frequently consist of brittle or fragile materials, such as carbon, ceramics, and the like. Such materials are very sensitive and, therefore, susceptible to stress, particularly tensile stresses. This must be considered when seal seats are attached solidly to the rotating shaft. If O-rings are used, the elastomer must be chosen from a group of materials that provide optimal resiliency and freedom from swelling. Swelling O-rings develop tensile stresses in the ring cavity against the mating ring wall, leading finally to destruction of the seat ring. Accurate information on the swelling characteristics of elastomer O-rings is mandatory if failures are to be avoided.

When using standard gland ring plates without built-in environmental control devices great care must be taken in attaching the gland ring to the housing. It is important that the gland ring be properly aligned with the housing to secure appropriate location of the seal seat ring in relation to the shaft. The use of adequate gland pilots provides a means for achieving reliable gland centering. Gland pilots can be designed in many different ways, particularly since they do not create any problem.

With the solution of alignment of the pump gland ring plates to the housing, other problems in seal assembly are not critical, except for uneven bolting in fastening.

As is discussed in a later section in connection with seal balancing, perfect alignment of the seal seat with the seal head on the shaft should be the aim of assembly. The pilot centering device of the gland ring plates facilitates this requirement; it also represents the simplest way at minimum cost.

It is common experience that the elastomeric O-ring as a secondary sealing component for the stationary seal seat represents an elegant possibility to compensate for rotational irregularities of the seal head on the shaft.

3. Springs for Face Loading

Steady contact between the rubbing faces of the primary seal rings to ensure proper seal performance is accomplished by using elastic springs, which permit a steady automatic pushing action. This activity is further assisted by the hydraulic fluid pressure provided by the pump. System fluid pressure and spring selection must be closely balanced against each other. A wide range of springs is available and in practical use.

a. Single-Coil Springs. A typical example of the application of a single-coil spring for the seal head is shown in Fig. 2.1*a*. Compared to the multiple-coil spring design the single-coil device has a relatively large wire cross section and thus provides more substance to combat corrosion from the system fluid. The performance characteristics of the spring should not in any way be affected by corrosive action by the system fluid.

Whether single-coil or multiple-coil springs are used is a matter of choice; there are advantages and disadvantages on both sides. A single-coil spring, for instance, is rigid, making it difficult to achieve a perfectly uniform load distribution across the entire shaft circumference, which may result in a distortion of the primary seal ring face. This is particularly critical at high rotational shaft speeds. Each seal design necessitates individual spring applications with careful evaluation of all possible design factors.

b. Multiple-Coil Springs. Instead of using one large single-coil spring many seal designs employ a series of several small coil springs uniformly distributed along the circumference of the seal cartridge. Distortion of the seal face of the primary seal head ring due to multiple springs is unlikely if it is properly assembled; this is particularly true at higher rotational shaft speeds. See Fig. 2.1*b*.

The application of multiple-coil springs makes seal design independent of seal diameter sizes. The achievement of higher forces for face loading by a multitude of small coil springs is no problem. A large-sized single helical spring could be too rigid to be flexible enough for a suitable seal performance.

With the use of stainless steel for the manufacture of springs the danger of deterioration of multiple-coil springs from chemical corrosion is greatly minimized.

c. Wave Springs. There are many seal designs for which only very small spring forces are required. In these cases a metallic wave spring can be used. Wave springs are thin washers that supply a springy action once they are under compression. Wave springs occupy the smallest space of all common spring devices, acting axially. They are simple in design and easy to install and maintain.

Wave springs cannot handle heavy loads. They are very susceptible to even minute changes in an axial direction, because of the relatively small amplitude of deflection when subjected to compressive load.

Figure 2.1 illustrates all three basic seal designs, comparing a single helical spring, multiple-coil springs, and a wave spring application in connection with elastomeric bellows.

There is no generally accepted rule other than personal experience regarding which of these arrangements should be given preference. Each case must be evaluated carefully, considering all aspects of service conditions, before a decision is made. All design configurations have demonstrated long periods of trouble-free service life in diversified industrial applications. Furthermore, the designs can be compared with each other. There will not be any two sets of service conditions alike.

In corrosive environments stainless steel spring materials generally solve most of the corrosion problems. Table 2.1 indicates materials that have successfully been used for temperatures ranging up to 850°F, as reported by NASA sources.

d. Metal Bellows. Metal bellows have been developed for a number of purposes; however, the primary one is to meet elevated temperature demands. Only metals can cope with high temperatures, far exceeding Teflon or Kel-F capacities, without losing much of their spring characteristics.

Metal bellows are not suited to withstand excessive pressures in a system. However, they exhibit a high degree of resistance to chemical corrosion if the suitable material is selected. Metal bellows provide

Table 2.1 Maximum Operating Temperature of Spring Materials

Material	Temperature °F
Inconel X	850
Inconel	760
Duranickel	550
Stainless steel 301 and (17-4 PH)	525
Stainless steel 302	500
K-Monel	450
Oil-tempered CR-V AMS 6450	350
NI-Span 'C'	300
Oil-tempered spring steel	300
Music wire	225
Hard-drawn spring steel	200
Beryllium copper	200
Phosphor bronze	190

Source: NASA publication.

sufficient stability to withstand drastic rubbing velocity at long service life. Most manufacturers claim that a maximal pressure differential of up to 1000 psi can be absorbed without detrimental consequences.

With regard to rotational velocities manufacturers claim that rubbing velocities on the order of 20,000 RPM are common for metal bellows. For speeds in excess of 20,000 RPM special materials should be chosen. Consultation with manufacturer is recommended.

Irregularities in shaft rotation in combination with excessive frictional forces as a result of rubbing wear in the interface of the primary seal components cause vibration in both axial and torsional directions. These vibrations impair the performance of the bellows, finally leading to seal failure. Manufacturers can generally provide useful and safe guidelines on resonance frequencies at critical rotational speeds.

In spite of the large quantities of bellows manufactured, it should be kept in mind that they are still in some kind of developmental phase and consultation with the fabricator is essential.

D. Secondary Sealing Components

Secondary sealing components are by definition designed to seal the primary seal rings in order to prevent leakage along the shaft and in the cavity of the gland ring plate to the atmosphere. They must allow axial motion of the primary seal ring to satisfy the requirement of compensation for rubbing wear in the interface contact areas as well as irregularities in the shaft motion during rotation. The term *secondary* is by no means used to downgrade this concept.

The most commonly used secondary sealing components in conventional design of mechanical end face seals are designated as O-ring, V-ring, U-cup ring, and wedge ring, as illustrated in Fig. 2.2. The illustrations are self-explanatory. Bellows of metal and elastomers are other important components and are presented in Fig. 2.3.

Elastomeric O-rings made of all kinds of highly elastic rubbers are desirable where temperature and chemical compatibility permit their application. Silicone rubbers are an excellent choice and generally perform with a considerable degree of memory, the most outstanding and desirable material characteristic a secondary seal can have. The elastic behavior compensates for minute axial motions, even in the presence of occasional high loads. In Fig. 2.2a O-rings utilize the self-energizing sealing principle under the action of the system pressure. The new Du Pont material Kalrez is even better than Viton and is perhaps the best elastomeric seal material on the market for use in O-rings.

The V-ring in Fig. 2.2b is a component of the pusher type of secondary seals, also designed to utilize the self-energizing sealing principle. When the metallic base ring pressurizes the V opening as a result of the hydraulic system pressure, the lips spread and enhance the seal effect. Customary materials for V-rings are Teflon in virgin form or reinforced by glass fiber or molybdenum-disulfide, carbon-graphite, asbestos, polyimids, or combinations of any of these materials.

The seal unit of Fig. 2.2c uses a U-cup ring for the seal seat and for the seal head as well. A metal base ring again is applied to open the lips of the U-ring to provide the seal once the system pressure is acting. Both U-cup and V-rings are fabricated from the same type of construction material. They can be interchanged arbitrarily. All secondary sealing components in Fig. 2.2a–c can also be used for seal heads and seats. Both secondary seal configurations need not be the same in the same seal unit.

The seal ring for Fig. 2.2d is designed as wedge ring, which is never found in the stationary seal. Wedge rings are generally made from Teflon, either of the virgin type or reinforced, asbestos, polyimid, or any conventional combination mentioned earlier.

The O-ring, V-ring, U-cup ring, and wedge ring represent components that are often referred to as pusher-type secondary seal components, because they require additional mechanical members to function and maintain continuous contact with the rubbing face elements. The component used is referred to as a *metal base ring*.

The secondary sealing components are not designed to provide the preloading force, as is the case with bellows-type secondary seal components, which can operate without the generally mandatory mechanical spring devices.

Along with the pusher-type secondary seal components the seal industry also employs nonpusher-type secondary seal elements that incorporate bellows. These utilize a high degree of built-in elasticity; their flexibility allows them to perform motions in an axial direction by supplying the primary seal rings with an automatic pushing force to compensate for face wear without using separate mechanical spring devices. Springs, however, may be employed in addition to the elastic built-in flexibility to combat fatigue of the bellows, which will occur once the bellows are overloaded or overcycled. Designs using elastomeric bellows require springs because the elastomer will fail by fatigue very rapidly as a result of overload by torsion without a spring backup device.

Bellows for secondary sealing components may be fabricated from metals, elastomeric polymers, Teflon, Kel-F, polyimid, and the like. Bellows made of metals use either the corrugated or the welded designs. The selection of the proper material and the suitable design configuration is governed exclusively by the severity of the process conditions that

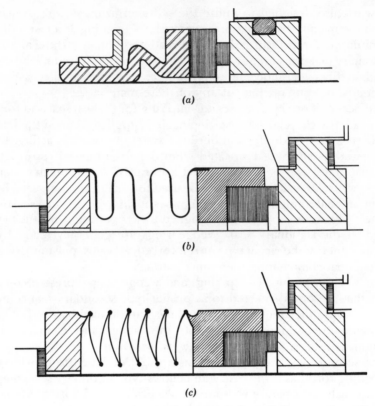

Fig. 2.3 Bellows as pusher devices. (*a*) Elastomeric bellows. (*b*) Corrugated metal bellows. (*c*) Welded metal bellows.

must be met. Metals generally withstand elevated temperatures and, depending on composition, a wide range of chemical resistance to corrosion.

A corrugated metal bellows design is illustrated in Fig. 2.3*b*. A seal unit using welded metal bellows is presented in Fig. 2.3*c*.

A seal configuration with an elastomeric bellows is presented in Fig. 2.3*a*. This particular design does not use any additional mechanical spring devices to enhance the face contact force. The bellows is attached to the shaft by a metal spanner ring. This design will handle only moderate service conditions at low hydraulic pressure.

Seals with elastomeric bellows generally use helical springs to supply the required face load. On the other hand, the elastomer would not be strong enough to provide satisfactory contact force for the interface.

The spring finally compensates for the loss of flexibility when the elastomer deteriorates because of hardening effects.

It may be surprising to learn that welded bellows are much less subject to vibrational fatigue failure than are corrugated metal bellows as long as they are serviced following manufacturers' specifications. Bellows are susceptible to axial deflection amplitudes. The fatigue is further accelerated by high-temperature exposures, which superimpose the fatigue stress caused by pressure pulsations. For this reason the fabricators' specifications should be carefully observed.

Finally, besides initial sealing, secondary sealing components have to accomplish the significant task of allowing the primary seal rings to maintain parallel positions of the interface contact areas, particularly since the shaft never runs true. Elastomeric components are best suited to cope with just this kind of problem.

E. Temperature Limitations of Secondary Sealing Components

As discussed in depth in Chapter 1, elastomers have severe temperature limitations for application as seal construction materials. They are similarly limited with regard to chemical compatibility. Temperature guidelines for materials conventionally used as secondary sealing components in mechanical end face seals are shown in Table 2.2. The temperature ranges indicated in this table are representative and generally accepted. The value of $-320°F$ for Teflon is commonly accepted in the cryogenic industry.

Table 2.2 Temperature Guidelines for Secondary Seal Components

Material		From	To
		($°F$)	
Synthetic rubber		−20	+ 200
Silicone rubber		−70	+ 400
Viton		−65	+ 300
Teflon, generally (−320°F)		−100	+ 450
Kalrez (Du Pont)		—	+ 550
Asbestos	up to	400	+ 650
Metals	in bellows	−400	+1200
	in gaskets	−400	+ 850 → 1000

Source: NASA publication.

Great care must be exercised when estimating the temperature conditions within a seal unit, since primary seal rings never show the same temperature as may be measured in the direct environment. Primary rings generate heat at the interface due to friction and wear. The actual face temperature is, therefore, definitely higher than the temperature of the environment.

In seal practice the temperature of the interface is considered to be the basis for determining whether an application will be successful. As a rule, if the rotational speed of the primary ring is high the allowance for maximum service temperature should be reduced in proportion.

NASA has published practical values of temperature limitations derived from missile and jet engine service. An excerpt is shown in Table 2.3, listing preferred materials in chemical and related industries, including Teflon, Kel-F, silicone rubber, Buna-N, butyl rubber, and thiokol.

Because of the widespread use of Teflon as a construction material in the chemical industry, some rules of caution may be presented: Teflon is a plastic, not an elastomeric material. Although it provides resiliency it has very little memory. This, however, can be imparted to special sections of relatively small parts, but this method is not common practice. Teflon deforms under compression resulting in permanent deformation. Teflon is incompressible and is subject to cold flow without limits, a significantly detrimental characteristic for a seal material. Cold flow cannot occur when the material is placed in a cavity with restraint volume where an expansion cannot take place. This method is frequently utilized in numerous applications in chemical and related industries taking advantage of the positive properties of Teflon and avoiding the negative ones. Repeated use of a Teflon O-ring generally results in flattening of the circular cross section, making the ring worthless for repeated application. If Teflon must be used, an elastomeric O-ring can be coated with a thin layer of Teflon, thus combining elasticity with chemical compatibility. Now Du Pont's Kalrez can be used instead.

Teflon is suitable for temperatures up to 450°F. In many a fabricator brochure the upper temperature limit for safe service is often given as 500°F. This is optimistic "sales talk." At 500°F Teflon definitely softens markedly and begins to decompose by vaporizing. These vapors are toxic.

At 450°F Teflon still renders reliable service, even under long periods of exposure. With steady-state service at 500°F or under temperature pulsation Teflon tends to become brittle and to shrink. The shrinking

Table 2.3 Temperature Limitations for Mechanical Seals

Fluid Environment	Teflon	Kel-F	Silicone Rubber	Buna-N	Butyl Rubber	Thiokol
			Temperature (°F)			
Air	450		500	250	160	
Water/steam	450			300		
Lub oil Mil-1-6082	450		350	300		
Lub oil Mil-016081	450		350	300		
Lub oil Mil-1-6086	450			300		
Syn hyd oil Mil-1-6387	450		350	300		
Syn lub oil Mil-1-7808	450		350	300		
JP-4,5, fuel	450			300		180
Gasoline fuel	450			300		180
Hydrolube Mil-1-7083	450			225		
Skydrol	450		300		160	
Liquid oxygen	450	−300				
90%Hydrogen peroxide	450					
RFNA/ nitric acid	450	160				
Ethylene nitrate					180	
Propyl nitrate					180	
Hydrogen/	450	−250				
Hydrazine	450	250				
50/50 UDMH/ hydrazine	450	250			180	
Pentaborane	450	250	350			
Tetrafluoro- hydrazine	450	250				
Nitrogen tetroxide	450	250	350			
UDMH	450	250				

Source: NASA Report.

effect becomes especially obvious under repeated temperature pulsations with high frequencies and differential amplitudes.

The limitations on application of contruction materials for secondary seal components in mechanical end face seals are generally dictated by the operating parameters pressure, temperature, and fluid medium. On the other hand, the availability of suitable construction materials limits or even dictates the design possibilities for seals.

F. Drive Methods for Rotating Seal Rings

One of the two primary seal rings must rotate with the shaft. This necessitates the application of a simple device to ensure positive rotation without losing relative flexibility. The seal industry offers quite a range of practical devices to provide mechanical engagement between the rotating shaft and the corresponding primary seal ring. Numerous devices are applied to ensure reliable drive of the rotating primary seal ring.

When using elastomeric bellows as a secondary sealing component a compulsory link must be added to establish positive drive engagement. As mentioned earlier, a metal ring is used to attach the bellows to the shaft while a helical spring provides the pushing force to accomplish rubbing contact in the interface at all times. Elastomers are sensitive to torsional load, twisting, and vibration. Thus if they were used, hardening with subsequent fissuring would develop, resulting in complete mechanical breakdown of the elastomer materials in a relatively short time.

Rotation by mechanical engagement is secured by dents, keys, set screws, pins, slot and ear, snap ring, elastomer bellows, and springs. Even press-fit and shrink methods are applied. The functions of these devices are simple, easy to understand, and straightforward.

G. Hydrostatic Fluid Force

Seals for rotating shaft equipment always operate in an environment characterized by a system fluid to be handled and subsequently sealed. This is particularly obvious in a system using a mechanical end face seal, mounted in a horizontal centrifugal pump. Depending on the specific design the system fluid may be acting internally or externally, considering the direction of the leakage when passing from OD to ID or from ID to OD of the interface seal area.

Once again, looking at a horizontal centrifugal pump, one notices that the system fluid pressurizes the seal interface either from the inside or from the outside, reflecting line A-A in Fig. 2.1a. In reactors with a vertical shaft arrangement the seal generally sees the pressure of the vapor phase in the vessel above the process fluid. As a result, the seal must be designed for the pressure acting in the seal environment inside the seal cavity, stuffing box chamber, or the like.

In addition to pressure, the shaft velocity is another parameter to be evaluated as significant in seal design. As is discussed in more detail in Section IV, seal balance is a predominant concept for adequate seal performance since seals function properly only if the requirements for hydraulic balance have been satisfied.

It has been found, however, that even unbalanced seals may operate properly at relatively high pressures as long as the shaft velocities remain correspondingly low. On the other hand, certain seals of the balanced design are used at high velocities at pressures that normally do not require balancing. Pressure-velocity concepts are discussed in Section VI.

III. General Modes of Arrangements for Mechanical Seals

To simplify discussion, seals are considered as they are applied in horizontal centrifugal pumps, because in this area most mechanical end face seals are used. Arrangements of seals for reactors, agitators, mixers, and others follow the same basic concepts and design philosophy.

With regard to seal arrangements distinctions are pointed out as they are used in everyday industrial application. Seals for large-scale agitators and their specific requirements are treated in a later section of this chapter.

A. Single Seals

Mechanical end face seals may be classified in a variety of ways. A distinction can be made as to whether the hydraulic fluid pressure is acting to pressurize the seal contact faces or whether this pressure tends to separate the faces, as is the case with outside mounted seals. They can further be classified with regard to whether they are balanced or unbalanced against the action of the hydraulic fluid pressure. Finally, a distinction can be made regarding use, such as single, double seals, and so on.

Single seals employ one set of primary seal rings to establish the seal of the system. They are designed to satisfy simple seal requirements where the process fluid provides the required lubrication for the interface.

140 Mechanical End Face Seals as Axial Sealing Devices

Single seals are also used in designs where the system fluid cannot be used to lubricate the interface. In these cases a secondary fluid must be chosen that has to be supplied by a separate and independent fluid system.

1. Single Inside Seal

A typical design of a single internally mounted seal is illustrated in Fig. 2.4, predominantly inserted in the stuffing box chamber of the pump where the hydraulic fluid pressure offers a force to close the interface. The seal head is attached to the shaft. Should leakage occur the leakage path would be from the external diameter across the interface toward the shaft and then on to the outer atmosphere. The hydraulic pressure assists sealing and promotes face closing. The net closing force is the sum of the hydraulic force plus the spring pressure, reduced by the separation force of the fluid film acting in the interface.

2. Single Outside Seal

In outside seals the hydraulic system pressure inside the stuffing box chamber penetrates to the seal head and tends to separate the primary seal rings, as may simply be derived from Fig. 2.5. In this design the net closing force is given by the spring contact force reduced by the opening force of the system fluid.

The leakage path in the seal of Fig. 2.5 is from the shaft surface across the seal interface to the OD. As is shown in Section IV, both inside and outside seals can be hydraulically balanced.

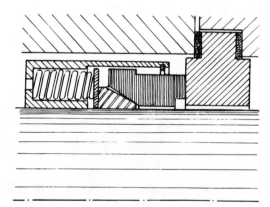

Fig. 2.4 Single seal, mounted internally.

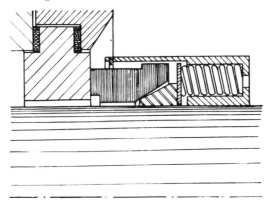

Fig. 2.5   Single seal, outside mounted.

There is no fixed rule or specific binding preference for either internal or external seal arrangement. Equipment and operating parameters generally dictate the principle on which such a selection is made. In most cases, however, the internal arrangement is preferred, because the shaft is held closer to tolerance conditions than is possible for housing and gland ring plate.

In high-speed applications where the masses of rotating parts must be kept at a minimum because dynamic balancing can become critical, it is convenient to design for stationary seal heads.

In Figs. 2.6 and 2.7 the primary seal ring arrangements of Figs. 2.4 and 2.5 are reversed. For both internal and external configuration the seal heads are mounted stationary while the seal seats are rotating with the shaft.

When a system is handling highly corrosive or even toxic fluid a seal with a rotating seat and the pressure acting on the inside of the face usually employed in externally mounted seals is most convenient. The spring and other significant parts of the seal unit are not in contact with the system fluid. Any amount of leakage can be readily detected. All parts are easily accessible for installation, maintenance or replacement, and visual control. Attachment of automatic leakage detection devices can be accomplished in a simple way. Rotating seal seat units are more difficult to balance dynamically and are, therefore, less suitable for high rotational velocities.

In the presence of contaminated fluids, particularly those containing solid and abrasive particles, the design of rotating heads mounted internally, with the system fluid acting on the OD of the interface, is of advantage. The centrifugal motion of the shaft tends to keep fluid away

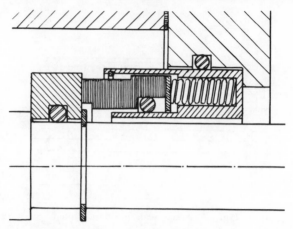

Fig. 2.6 Inside seal, head stationary.

from the critical area, freeing the spring section from depositing solid particles. This in turn contributes to facilitate flow of lubricant to the interface, where it is needed the most.

There is a definite tendency to use seal devices with fluid pressure on the external diameter of the seal contact areas, since theoretically at least leakage would have to overcome the centrifugal action on the outside in order to penetrate through the interface.

At the present time centrifugal action of the fluid is not fully understood. Seals with fluid acting toward the ID of the interface are exposed

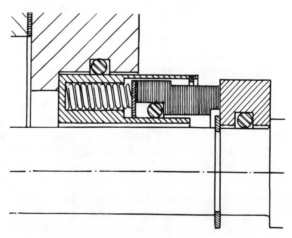

Fig. 2.7 Outside seal, head stationary.

to loads producing predominantly tensile stresses. They are, therefore, much more critical than compressive stresses, especially in rings made of brittle material, like carbon, ceramics, stellites, and even carbides of the silicon or tungsten variety.

As a rule, when the fluid pressure is acting on the external seal face diameter, the design is always oriented in such a way that the entire seal unit is mounted in the stuffing box chamber. When the fluid pressure is acting against the ID of the interface, the seal unit is usually mounted externally.

The internally mounted seal is less accessible and thus less convenient to handle and the interface is not available to visual inspection. However, in case of a sudden breakdown of the seal in the presence of valuable, toxic, flammable, or otherwise hazardous fluids the internally mounted seal offers a fundamental advantage of providing some sort of flow restriction in the passageway, functioning as a throttle bushing. In critical cases an additional auxiliary external packing may be installed in the gland ring. All these provisions are not possible in externally mounted seals offering no flow restriction of any kind at unexpected sudden major catastrophic seal failures.

B. Double Seals

In chemical, petroleum, and related industries numerous fluids are handled that can under no condition be considered as lubricants between the rubbing contact areas of the seal. A similar situation exists when the fluids contain abrasives or any foreign particles that could affect the rubbing faces. Pumps operating under such service conditions can be sealed only by providing special means to prevent these particles from reaching the seal interface environments. In the presence of few solid particles a cyclon filter will provide some relief. Abrasive fluids should preferably be handled by designing a second fluid system to supply the interface with a protective fluid, isolating it from the system fluid.

The new environment provided by the secondary independent system fluid must be controlled. This can be done by using a so-called secondary fluid flushing system, introducing a neutral fluid with lubricating properties, free from abrasives, into the stuffing box chamber and providing here a clean environment for the rubbing contact areas. This secondary fluid can be either water or a neutral lubricant that is circulated within a separate independent liquid cycle. The fluid will also serve as a coolant if this is required. Care must be taken to ensure that the secondary fluid is compatible with the process fluid so that any resulting dilution or contamination of the process fluid will not be harmful. Where the two fluids must be kept isolated from each other a double-seal system will perform suitable service.

1. Double-Seal: Internally Mounted

The arrangement of multiple mechanical end face seals can be achieved in a number of ways, with either both seals installed inside the stuffing box chamber, designated as double-inside mounted, or as double-tandem, or as double-inside and outside mounted.

An example for a double-seal arrangement with both units mounted inside is illustrated in Fig. 2.8, representing a design of the Crane Packing Company of Chicago. This design is not typical of the designs of this fabricator company. Figure 2.8 reflects a seal mounted back to back, filling the entire stuffing box space and providing an ideal opportunity to use a secondary fluid system independent of the process fluid with separate inlet and outlet openings. The secondary fluid can simply flush the seal interfaces along the OD circumference. When the secondary fluid is circulated its pressure should be slightly above the pressure of the process fluid to ensure penetration into the interface for the purpose of lubrication. If this fluid can be water, it will simultaneously be a good coolant, thus dissipating the rubbing heat away from the interfaces. Secondary fluids other than water also function as coolants. It is generally necessary, however, to employ a heat exchanger in the fluid cycle to prevent accumulation of heat.

In the design of Fig. 2.8 the outboard seal seat prevents the secondary fluid from leaking to the external atmosphere. The differences in design configurations in the individual devices are small and of no consequence, considering a variety of major fabrication companies.

The double-inside seal arrangement encounters two pressure differentials, one across the inboard interface and the other across the outboard interface. To ensure lubrication for the inboard rubbing areas at all times the system pressure for the flushing fluid must be higher than

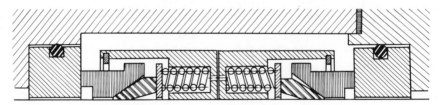

Fig. 2.8 Double seal arrangement, mounted internally.

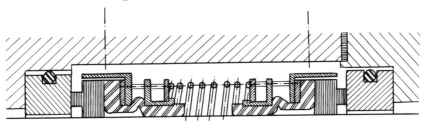

Fig. 2.9   Double seal with elastomeric bellows, internally mounted.

the operating pressure of the process fluid. The outboard interface sees a pressure differential, ranging between the pressure of the secondary fluid system and the external atmosphere. If these differential pressures exceed the limitation requirements of unbalanced seals, both or at least one of the seals should be balanced.

Double-inside mounted seals are not necessarily restricted to any specific seal design configuration. The selection of seals for a double-inside seal arrangement is a matter of personal choice and will preferably be derived from diversified evaluations of the process conditions. Figure 2.9 shows a design with double seals using elastomeric bellows and springs to supply the required face contact force.

2.   Double Seal: Tandem Arrangement

A popular conventional double-seal arrangement is the double-tandem seal configuration illustrated in Fig. 2.10. This seal also employs elastomeric bellows as secondary sealing components.

Tandem seals use two seals of the same design mounted in the same direction. The first seal is mounted in a stuffing box chamber just sufficiently long to incorporate the seal with a small gland ring plate for the seal seat. The second seal is added in the same direction in an additional seal cavity provided by a second gland ring plate, creating a so-called buffer zone. This arrangement makes it possible also to sepa-

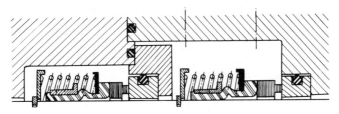

Fig. 2.10   Double seal, tandem arrangement.

rate the process fluid from the second seal, and leakage from the inside seal will enter the buffer zone and not reach the external atmosphere.

The inboard seal stops and confines the process fluid to the initial stuffing box chamber, functioning like any conventional inside-mounted seal. The cavity between the inboard and the outboard seal is flushed from a separate external fluid system, providing lubrication to the outboard seal. The interface of the inboard seal is lubricated by the process fluid inside the stuffing box chamber and not by the neutral fluid for the buffer zone. Should the inboard seal fail, the external seal will take over until measures can be taken to prevent further disaster. Sensing devices in the secondary fluid system can be arranged to trigger an alarm. Precautionary action can then be taken.

Since the secondary fluid is used at a lower pressure than that of the process fluid, the inboard seal can be balanced against high stuffing box pressures without necessitating a higher pressure in the secondary fluid system.

3. Double Inside-Outside Seal

Certain process requirements may necessitate a tandem double seal; however, the pump seal cavity may not provide sufficient space to install such a device. An alternate solution is to arrange one seal inboard and the second seal outboard, thus establishing a configuration termed an *inside-outside double seal*. A design of this type is illustrated in Fig. 2.11. This arrangement is customarily available using a balanced inboard seal with an unbalanced outboard seal. The inboard seal is arranged in a conventional way in the stuffing box chamber, and the outboard seal is mounted externally, using the same primary seat ring for both seal units.

If the secondary system fluid, serving two interfaces at the same time, is supplied at a pressure higher than the pressure of the process fluid in

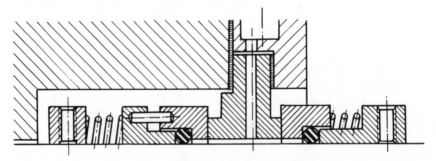

Fig. 2.11 Double inside-outside arrangement.

the stuffing box chamber, then the secondary fluid safely lubricates the interface of the inboard seal. Consequently, the system will function as a double seal with an artificial environment. But if the secondary fluid between the seals circulates at a pressure less than the pressure in the process fluid in the stuffing box chamber, then the inboard seal operates like any other single seal whereas the outboard seal simply functions as a backup seal, taking the role of a safety device as described earlier for the inside-outside seal configuration or the tandem seal.

C. Summary Considerations for Multiple Seals

Multiple mechanical seals generally are applied for a number of practical reasons:

1. Multiseal arrangements satisfactorily seal a gas or a vacuum under-conditions that could not be met by a single seal.
2. Multiseal devices are capable of sealing liquids containing slurries and abrasives in suspension, and liquids in concentrations higher than those safely handled by single seals.
3. When handling fluids with low or zero lubricity or high tendency of flashing at pumping pressure and temperature the multiseal design provides a better seal life expectancy compared to single-seal arrangements.
4. By circulating a secondary neutral fluid at lower temperature a pressurized barrier is produced with multiseal devices that will seal process liquids operated at high temperatures.
5. Multiseal arrangements offer safety features guarding against the escape of toxic and other hazardous fluids to the external atmosphere. For highly toxic fluids, using a design with the separation device of a buffer zone is often the only practical solution.
6. Multiseal devices represent the only design configuration capable of handling extreme pressure differentials. The establishment of separate pressurized fluid chambers provides a means of downgrading the pressure levels in stepwise fashion until tolerable levels for a single seal are achieved.

D. Multiseal Pressurizing Systems

Multisealing arrangements are generally equipped with a pressurized injection liquid supply system delivering the liquid from an external source. The secondary liquid has to meet two basic requirements. First, it must have satisfactory lubricity; second, it must be chemically compatible with the pumpage liquid.

Pressurized-liquid systems for multisealing devices are commercially available. They represent auxiliary equipment for circulating secondary sealing liquids—mostly lubricating oils—through the stuffing box chamber at controlled pressures and temperatures. Standard units—with and without heat exchangers—provide flow rates ranging from 1 to 7 gpm and pressures up to 1500 psi.

The pressurizing systems must meet the following requirements:

1. For satisfactory operation of pressurizing systems of multiseal arrangements a positive pressure must be maintained in the seal chamber at all times to provide for suitable interface lubrication. This pressure further prevents foreign particles from penetrating into the seal chamber. It is general practice to establish a seal chamber pressure 15 to 50 psig higher than the pressure in the pumpage system at the inboard stationary seal seat. In the presence of vacuum the pressure in the secondary liquid system should be in the range of 15 to 25 psig.

2. Multiseal arrangements generate heat that can be dissipated only by compulsory circulation of the secondary fluid. This system represents a closed system comprising pressure pump, filter, measuring devices, and so on, when lubricating fluids are used. The system further requires the incorporation of a heat exchanger, since cooling of the injection fluid is generally mandatory.

IV. Criterion of Seal Balance

System fluid pressurized by the pump impeller fills the stuffing box chamber, thus pressurizing the entire seal unit mounted in this closed environment. The hydrostatic fluid pressure is acting inside the seal cavity in all principal directions. For balance considerations, however, only the fluid pressure in an axial direction is of interest because this is the only pressure component that influences the contact behavior of the interface. The designer can, therefore, use this axial pressure component to compute the resulting face contact force. By modifying the geometry of the primary seal head ring in an appropriate manner the designer can utilize any proportion of the horizontal pushing force up to 100%, depending on the face area he provides for the pressure contact in an axial direction. This procedure has been referred to as *seal balancing*.

A. Balance Ratio

The concept of seal balancing can be clarified by describing the details of
Fig. 2.12, which represents seal units in unbalanced (Fig. 2.12*a*), partially
balanced (Fig. 2.12*b*), and fully balanced (Fig. 2.12*c*) conditions. A completely unbalanced seal unit is illustrated in Fig. 2.12*a*. The full fluid
pressure in an axial direction is acting against the primary seal head ring
comprising the area of front face *A'*. This pressure impulse is 100%
transmitted to face *A* on the opposite face of the seal ring, having the
same area as *A'*. Consequently, the hydraulic fluid force possible in face
A' is also present in face *A* without reduction. Increasing the hydraulic
fluid pressure in the system increases the face pressure in the interface
by the same amount.

By modifying the geometry of the primary seal head ring to establish a
new frontal area *A'* in Fig. 2.12*b*, however, smaller than area *A'* in Fig.
2.12*a* by providing a shoulder on the opposite side of the seal head ring,

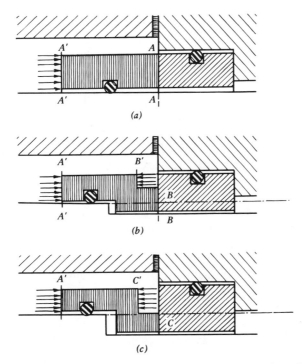

Fig. 2.12 Balance conditions. (*a*) Unbalanced. (*b*) Partially balanced. (*c*) Fully balanced.

a front face B' is formed where the hydraulic pressure can counteract a certain amount of the hydraulic pressure effect, which then is canceled out. Consequently, the remaining face pressure in the contact interface can be computed by the algebraic difference down to the balance line. The seal unit of Fig. 2.12b is termed *partially balanced*.

A further reduction of area A' and a further increase of area B' to the size shown in Fig. 2.12c results in the situation that A' equals C' and the hydraulic force effect is practically eliminated. This condition is termed *fully balanced*. In other words, interface C does not receive any hydraulic force component. The required seal contact force must be exclusively supplied by mechanical spring devices if interface contact is to be applied.

By staggering the geometry of the primary seal head ring any desirable degree of seal balancing can be achieved by pure geometric means. It may become necessary to provide a shoulder in the shaft or, where this is too costly, a sleeve over the shaft may be used to provide the intended degree of staggering. Sleeves are easy to fabricate, simple to replace, and convenient for selecting adequate construction material.

Seal balancing is a design manipulation to reduce the effect of the hydraulic fluid pressure against the primary seal head ring, diminishing the contact force in the interface area to a desirable and technically controllable level. If a seal design is designated as balanced to a ratio of 0.65 this value expresses a balance value of 65% of the total unbalanced condition of the existing design. Specific details on seal balancing are outlined in a paper by K. Schoenherr (49).

B. Pressure Distribution in the Seal Interface Due to the Hydraulic Fluid Pressure

In any standard seal design configuration the hydraulic pressure acts across the seal interface either from the OD to the ID or vice versa. In either case the pressure at the point of action is a maximum that is reduced across the interface to atmospheric at the opposite side of the contact area.

An extensive search of the literature has revealed that there is no valid analytical method for an accurate prediction of the pressure gradient across the interface, nor have reliable measurements ever been reported on this important subject. Several theories have been advanced but no particular one has gained general recognition.

Some theories state a linear pressure drop from ID to OD and vice versa, whereas other theories assume that the pressure distribution is nonlinear and even varies with the exposure time, attributed to topo-

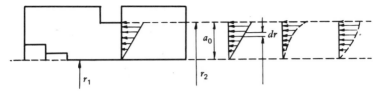

Fig. 2.13 Patterns of pressure distribution across seal interface.

graphical changes of the interface surface geometry and variations as a result of friction and wear.

Some theoretical methods tend to approximate the actual conditions more closely by considering such effects as elastic deformation of the contact faces, surface roughness, and axial movements on the interface surfaces.

It is most likely that the distribution of the pressure across the interface approaches the shape of a wedge. With this assumption the gradient could be expressed mathematically by the relation

$$\bar{p} = \frac{2\pi}{a_o} \int_{r_1}^{r_2} pr \, dr$$

where the designations of the various factors refer to the dimensions indicated in Fig. 2.13, where the hydraulic pressure acts externally against the interface.

With reference to Fig. 2.13 three concepts for pressure gradient are conceivable that may be designated as follows: linear, concave, convex. Whatever the true gradient may be, the pressure wedge tends to separate the contact faces of the primary seal rings, counteracting the contact forces. As a result the definition of the seal balance may be expressed by the relation

$$B = \frac{a_c}{a_o}.$$

Consequently, the seal balance reflects the ratio of a_c/a_o times 100 as a percentage value.

Figure 2.14a shows the pressure wedge of a seal unit where the hydraulic pressure is acting externally from OD to ID. In Fig. 2.14b a pressure wedge is given for a seal unit with the hydraulic force acting internally from the ID to the OD. Consequently, using the dimensions indicated in Fig. 2.14 the balance ratios then follow the relations:

$$\text{Bal}_1 = \frac{(D^2 - B^2)\pi/4}{(D^2 - d^2)\pi/4} = \frac{D^2 - B^2}{D^2 - d^2} \qquad \text{for externally acting hydraulic pressure}$$

$$\text{Bal}_2 = \frac{(B^2 - d^2)\pi/4}{(D^2 - d^2)\pi/4} = \frac{B^2 - d^2}{D^2 - d^2} \qquad \text{for internally acting hydraulic pressure}$$

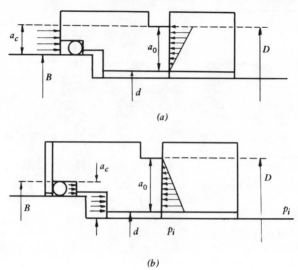

(a)

(b)

Fig. 2.14 (a) Pressure wedge at external hydraulic pressure. (b) Pressure wedge at internal hydraulic pressure.

Seals can either be balanced, unbalanced, or overbalanced; even the concept of underbalanced is used. Each of these concepts is applicable for seals with external and/or internal hydraulic fluid pressure.

C. Actual Face Pressure

Mechanical end face seals are influenced by three major phenomena known as hydrodynamic effects, mechanical effects, and thermal effects. Hydrodynamic effects involve fluid pressure, dimensional influences, surface conditions in the form of interface topography, and finally fluid properties, such as viscosity, density, lubricity, surface tension, and others. The influence of mechanical factors includes interface deformation as a result of centrifugal and hydraulic forces, shaft eccentricity, runout, misalignment, friction and wear, contact forces, and vibrational irregularities in the rotation. Thermal effects are the result of heat produced by friction, wear, deformation of the interface, and vaporization of the hydraulic fluid film that lubricates the interface. It is obvious that these factors affect the seal performance, making an analytical design theory of the seal face behavior very complex. Seal performance is basically evaluated on rate of leakage, frictional torque, and face wear. These three fundamental concepts must be accounted for.

For an analysis of the force relations in the interface we may consider an internally mounted seal unit provided with a fluid pressure of p_1 acting on the OD of the interface. The assumption is made that the pressure at the corresponding ID is p_2. It can also be assumed that p_1 is greater than p_2. Consequently,

$$\Delta p = p_1 - p_2.$$

With the assumption that pressure p_2 at the ID is atmospheric, the gauge will read zero pressure for p_2.

The face pressure acting in the seal contact areas represents the resultant of the closing force and the counteracting opening force, with reference to interface area a_o.

Face pressure p_F geometrically represents the algebraic sum of the hydraulic pressure p_h and spring pressure p_{spr}. Therefore,

$$p_F = p_h + p_{spr}.$$

For a closer mathematical approach the major seal fabricators use a gradient factor k on the order of 0.50 for a linear distribution. For a convex gradient k will be larger than 0.50.

With these assumptions the face pressure can be computed by the relation

$$p_F = \Delta p (b - k) + p_{spr}.$$

This equation states that the face pressure equals the pressure differential Δp multiplied by the algebraic difference between the balance ratio b and pressure gradient k, and increased by the spring pressure p_{spr}.

The following conclusions can be drawn:

1. Face pressure p_F is a quantity produced by the pressure of the process fluid. It is independent of the size of the interface area a_o. Modifying the face width does not change the contact pressure. Face width is designed for strength stability to resist deflection. A reduction of face width, however, results in a corresponding reduction in the drive power.
2. The portion of the face pressure attributed to the spring action is a function of the face area, computed by the relation

$$p_{spr} = \frac{F_{spr}}{a_c}.$$

Spring force F_{spr} is a purely mechanical characteristic that does not change within certain limits, ranging from about 10 to 40 psi, usually varying with the operating conditions. Under vacuum service the spring pressure may increase noticeably.
3. For a seal with a balance ratio of 0.50 and a linear gradient factor $k =$

0.50 the algebraic difference $(b - k)$ goes to zero. If such is the case the face pressure reaches the value of the spring pressure. This in turn indicates that the hydraulic force effect cancels out and the primary seal ring actually floats, pressurized by the spring only. If the spring forces are not carefully controlled the faces may separate in spite of positive face loading. Because of this a minimum balance ratio of at least 0.60 should be established. Accordingly, it has become common practice to choose a balance ratio ranging from 0.65 to 0.75 dictated by the operating conditions.

In seal units with hydraulic pressure acting on the interface OD it has been experienced that a minute deflection takes place that finally leads to a decrease of face area a_o. It is likely that this is one major reason why faces tend to separate in spite of having a balance ratio of 0.55 to 0.65.

In spite of all the practical experience and information available today, a fully satisfactory analysis method for accurate seal balancing has not been developed. There are too many factors that influence the complex function of seal interface behavior. Development of acceptable and satisfactory design calculations is hindered by many complexities. For these reasons a margin of safety is used. It has become general practice to select a minimum balance of 0.65 limiting the maximum sealed pressure.

By selecting a material for the primary seal faces that is capable of withstanding rubbing contact pressures of 200 psi the maximal fluid pressure that can safely be handled is about 5000 psi. Seals that operate at this pressure have become commonplace.

V. Friction and Wear

Friction and wear are the two major factors influencing seal performance and the life expectancy of a seal unit. They destroy the surface topography of the rubbing interface areas. These two phenomena are closely interrelated.

Wear is generally defined as the removal or transfer of solid material from one surface to another, while the two are in rubbing contact with each other. Friction can be described as the resistance that a material develops against sliding contact. This magnitude can be defined by a friction factor.

Wear phenomena are more complex, mainly attributed to four basic mechanisms generally known as adhesion, abrasion, corrosion, and surface fatigue.

A. Adhesion

Adhesion may be defined as a bonding between two solids in contact under given conditions usually with generation of heat. Chemical composition of the solids involved and the environment in which the bonding takes place are of prime importance. Adhesion may be described as the transfer of a particle from one position to another within the contact surface. Heat is generated particularly at localized points of contact when they are in motion, since cooling influences are not generally sufficient. The rubbing heat generated at these points becomes detrimental and a so-called surface welding may take place. This phenomenon is a particular problem in a vacuum environment where any lubricant film vaporizes between the contact surfaces.

B. Abrasion

Areas in rubbing contact can undergo abrasive wear by interaction of microscopic and macroscopic topographical asperities along the contact areas. This phenomenon is termed plowing one solid into another.

C. Corrosion

In the presence of a corrosive atmosphere in the vicinity of the interface the activity of adhesive and abrasive wear will facilitate interface surface reactions. Chemical corrosion can be either beneficial or detrimental. If soft oxides are formed, for instance, wear may be reduced, as long as the oxide layer is not disturbed. Free hard oxide particles, however, floating between the rubbing surfaces can act as grinders and increase wear.

D. Surface Fatigue

Rubbing contact generates heat, which facilitates and accelerates elastic deformation of the surface topography. This results in disturbances, superimposed by relaxation of stresses within the structure of the solids, which can result in surface fatigue.

Since friction and wear are closely interrelated, it is obvious that complex phenomena are involved. Despite many serious attempts there is no reliable analytical method available that permits a prediction of the service behavior of rubbing faces. The only reliable design criteria so far known are derived from empirical data.

The lack of a suitable design theory stems from three significant reasons:

1. The basic function of a mechanical end face seal is the restriction of the interchange of process fluid across the seal interface. This interchange does not take place along predictable leakage paths, but instead through interconnecting microscopic and macroscopic voids. This arbitrary leakage path unfortunately is the most dominant factor affecting the performance of a seal. In all rubbing processes the contact areas are continuously changing because of the arbitrary mechanism of wear. This forces the leakage flow to follow in random directions, depending on load, temperature, pressure, fluid composition, microstructure of the rubbing faces, and rubbing velocity.
2. Although considerable detailed information is available on both friction and wear, it has not been possible to combine these two factors into one appropriate deterioration theory applicable to seal interface wear.
3. Literature on friction is comprehensive and friction factors have been developed for numerous materials under many service conditions. However, there has never been a single case where the behavior of a new material under given conditions was predictable, the very least for rubbing seal contact areas.

E. Summary

Pressure, fluid composition, lubricity, rubbing velocity, temperature, density, surface tension of the fluids, and exposure time are all essential factors influencing the wear behavior of rubbing faces. Interface lubrication is also involved and may be of the mode of full film, mixed film, boundary film, or dry running. Depending on the system involved friction and wear characteristics of mating materials obviously affect the leakage behavior. No reliable means are available that accurately predict seal wear. Proper design still depends on empirical approaches, based on long-range experience.

VI. Pressure-Velocity Considerations

Seal manufacturers use the so-called pv factor to characterize the ability of a material to resist wear. This is the algebraic product of two factors p, the face contact pressure, and v, the rubbing velocity. Product pv is equal to the rate of heat energy generated, assuming the coefficient of friction is known.

Pv curves are characteristic for given seal materials, and seal manufacturers furnish such curves in connection with seal balance relationships. *Pv* curves are established on an experimental basis for practically
all materials used for primary seal rings in combination with known fluid
systems.

As indicated earlier the face pressure follows the relation

$$p_F = p_h + p_{spr}\,; \qquad d_m = \frac{D + d}{2}\,; \qquad v_m = \frac{d_m \pi n}{12}$$

and product

$$pv = p_F + v_m$$

From the equation

$$p_F = p_h + p_{spr}$$

results the relation

$$pv = (p_h + p_{spr})v_m$$

Thus

$$pv = [\Delta p(b - k) + p_{spr}]v_m$$

In this equation the wedge pressure is not included. The mathematical
relation

$$pv = \text{constant}$$

represents graphically a hyperbola, and product *pv* is bounded between
the hyperbola and the limiting values of pressure *p* and velocity *v*.

In industrial seal practice the upper limit of *pv* values is on the order
of 2,250,000 lb/in. ft/min. For further details see K. Schoenherr (49).

By establishing *pv* relationships on an experimental basis, it was found
that not only do wear differences exist for different face material combinations, but that wear also changes with the system fluid. Manufacturers
furnish *pv* curves for a variety of conditions, and consultation with the
manufacturers is recommended for new seal materials.

VII. Design of Pump Gland End Plates

When a horizontal centrifugal pump is to be sealed with a mechanical
end face seal instead of a packing, the gland end plate has to be
modified. In chemical, petrochemical, and related process industries
centrifugal pumps are used for an extremely wide range of process

media. For each of these diversified fluids under various pumping conditions a different mechanical end face seal may become necessary. Each seal may have to satisfy different environmental conditions and certain design changes may become necessary to accommodate them. Judicious design of the pump seal environment is many times just as significant as the design configurations of the particular seal unit itself. Seal manufacturers offer a great many possibilities for achieving optimal service life for end face seals.

The various environmental requirements can be met by modifying the gland end plate of the pump, which enables the designer to provide control devices for flushing, quenching, cooling, lubrication, and/or any combination of these.

In industrial seal terminology all modifications of the gland end plate permit the use of various environments, which are termed *environmental control systems*. The fundamental objective of environmental control is the continual provision of a clean liquid around the interface. The method selected depends on the nature of the process fluid to be sealed.

For volatile liquids, for instance, the lubricant film in the interface can be secured by lowering the process temperature in the seal environment or by increasing the pump suction pressure. When pumping fluids with a high melting point, it is important to operate at temperatures where solidification or crystallization of the flow medium is impossible.

When the system cannot tolerate changes in the process conditions a modification of the pump design is necessary to prevent any process disturbances that might affect the characteristics of the process fluid.

A. Plain Seat–Gland Combination

Plain glands refer to mechanical seal glands that are machined only to incorporate the stationary seat ring. No other control devices are employed. Such plain gland ring plates are either round or elliptical, with or without pilot. Flexotallic metallic gaskets or elastomeric O-rings are used to seal the gland ring plate against the pump housing.

The arrangement of the primary seat ring in the gland ring plate is a matter of choice as long as certain process requirements are not demanding. Conventional arrangements of the seat ring are possible because a specific dynamic seat ring balance is not required.

Some cases use the gland plate as the seal seat; other designs apply the shrink-fit betwen gland and seat ring. It is usually the method of sealing the seat ring with the gland plate that distinguishes the details for the specific design of the component combinations. A series of conventional

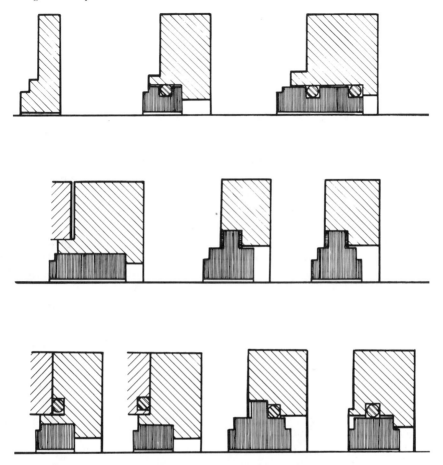

Fig. 2.15 Modifications for seat-gland combinations.

design configurations of seat-gland arrangements are presented in Fig.
2.15. As will be noticed, the elastomeric O-ring and the flexotallic metal
gasket play a dominating role. In conventional designs clamping and
bolting are preferably used.

B. Environmental Control by Flushing

Flushing as a method of environmental control is performed to secure
continuous flow of a liquid to flush the seal externally on the interface
OD, thus enhancing seal effectiveness and improving service life as a

result of lubrication and cooling. The steady fluid circulation supplies lubrication and prevents sedimentation of foreign particles around the critical interface area. Lubrication by direct injection prevents dry-run conditions in the rubbing areas at all times.

The liquid flushed to the interface environment can be supplied from the pump discharge or the fluid can be introduced from an external supply source. Continual flow of the flushing liquid is mandatory. When a bypass line from the pump discharge volute to the lantern ring connection of the stuffing box chamber is used, the operation is termed *internal flushing*. This method produces a flooded stuffing box chamber under positive pressure and a flow through the chamber to the stuffing box bottom with eventual penetration into the pump housing. The flushing action simultaneously provides cooling of the interface through removal of heat generated in the rubbing faces in the chamber cavity.

Internal flushing is required when petroleum products are processed. With a specific gravity less than one, these products vaporize at relatively low temperatures. They tend to flash at and within the interface. Flushing thus also provides cooling of the seal environment.

Flushing of the interface by direct jetting is mandatory for all fluids with a specific gravity of less than 0.63. In Fig. 2.16 various designs of gland ring plates indicate how glands are machined to accommodate flushing by directing the flushing fluid toward the seal interface.

In systems with flushing of externally mounted seals the liquid is generally supplied from a separate external source. This method is suited for pumps handling dirty and/or abrasive liquids with zero lubricity.

The secondary circulation system for flushing must be operated at a higher pressure than the liquid pressure of the pumpage fluid. This prevents passage of the pumpage fluid into the stuffing box chamber. Some of the secondary flushing liquid penetrates through the stuffing box throat into the pump volute, thus diluting the process fluid to some extent. Accordingly, the secondary flushing liquid must be compatible with the pumpage fluid. An accurate pressure control in the secondary

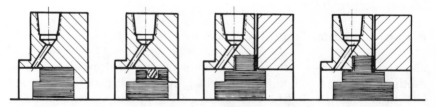

Fig. 2.16 Methods of flushing interface through gland plate.

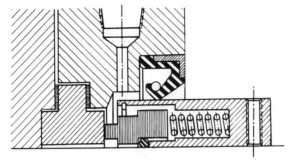

Fig. 2.17 External flushing for outboard seal.

circulation system is important. Leakage into the pump interior can be reduced to a minimum when a throttle bushing or a lip ring seal is mounted to the stuffing box bottom.

External flushing is applied to seals that are internally and externally mounted. Seals handling fluids that tend to crystallize at the interface must be flushed.

A design for a seal externally mounted using external flushing is illustrated in Fig. 2.17. The rotating seal head is mounted in a cartridge. The flushing fluid is sealed against the external atmosphere with a lip ring seal. External flushing of externally mounted seal units is frequently used to cool the interface because these units cannot be cooled by any other control over the temperature of the interface environment.

To enhance the cooling effect of the secondary fluid system circulation a heat exchanger is used as part of the fluid cycle. In all systems where a heat exchanger is used in the circulation, the flow resistance is correspondingly higher. To overcome this loss in pressure, an additional pumping ring is attached to the seal head cartridge, which acts like a small centrifugal pump impeller.

Boiler feed pumps for power stations, especially nuclear stations, operate under severe pressure and temperature service conditions. They employ floating bushing ring seals using several throttle rings and have to tolerate a certain amount of water leakage. By using mechanical seals this leakage is essentially minimized, in spite of considerable water circulation requirements. Instead of applying conventional-type pumping rings for circulating the buffer cooling fluid the Burgmann Seal Company of Germany and their branch company in Los Angeles, California, supply a device which is known as the Golubiev screw impeller. This device is an integral part of the seal and functions very effectively circulating the cooling water for the seal. It is capable of providing flow in the amount of several hundred liters per minute in practically all varieties of operating conditions at reasonable power consumption.

The purpose of internal and external flushing may be summarized as follows:

1. In the presence of highly volatile liquids, such as gasolines, liquefied petroleum gases, ammonia, or solutions of ammonia in water, the flushing operation reduces the tendency of vaporization in the vicinity of the seal interface.

2. Systems handling dirty liquids or containing solid and abrasive particles have a tendency to accumulate these particles in the region of slow and restricted flow. Flushing produces turbulent flow in the environment of the exterior of the interface, preventing these particles from making contact with the seal device or from accumulating and forming sediment.

3. When pumping liquids at temperatures close to their melting point, crystallization may occur, jeopardizing the effectiveness of the seal. Flushing counteracts solidification of these products.

4. Externally mounted seals, generating noticeable frictional heat, require external flushing to reduce the interface temperature and prevent flashing of the lubricant film in the rubbing faces.

5. The circulation of flushing fluid through the stuffing box chamber requires a pressure higher than the system pressure of the pumpage fluid. This prevents prenetration of pumpage fluid into the seal chamber. Flushing fluid must be compatible with the process fluid.

6. External flushing of externally mounted seals is frequently a very effective method of reducing seal temperatures, fire and explosion hazards, and certain types of chemical corrosion through the precipitation of corrosive salts.

7. The amount of deterioration of the pumpage liquid through dilution can be kept within reasonably controllable limits. Pressure differential, flow rate, and throttle bushing designs are conventional means of producing tolerable control conditions.

8. The major objective of flushing is to provide a means of isolating the seal face from undesirable liquids and generation of excessive heat. It is therefore necessary to adjust flow rate, pressure, and temperature of the flushing cycle to produce optimal operating conditions for maximum seal life and performance.

C. Environmental Control by Quenching

Certain chemical processes handle liquids that crystallize when in contact with the oxygen of the atmosphere. Other liquids may create problems

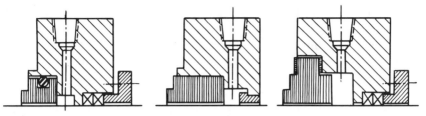

Fig. 2.18   Environmental control by quenching.

of toxicity or fire hazards when contacted by air. To prevent such hazards the quench-type gland plate design has been developed and used successfully. Three designs are shown in Fig. 2.18 that provide environmental control through quenching. The gland is termed a *quench gland* when it creates a neutral buffer zone behind the seal seat in the direction of the external atmosphere.

The fluid flushed through the buffer zone must be clean and should not contaminate the interface area. The flushing fluid in the buffer zone is sealed against the external atmosphere either by an auxiliary packing or by a throttle bushing.

Quenching for environmental control has the following advantages:

1. Quench glands provide protection against the oxygen of the external atmosphere. Toxic and other hazardous leakages through the interface are drained from the pump before they can reach the external atmosphere.
2. Quench glands are used to prevent sedimentation or precipitation of toxic and hazardous salts that may leak through the interface. The flushing fluid drains this leakage before contact with the oxygen of the atmosphere occurs.

The temperature of the quench liquid should be at least as high as the pumpage liquid. A vapor that leaks through the interface can crystallize in a cool quenching liquid. A cool interface can also cause the lubricant film to crystallize, resulting in scoring of the contact faces of the seal.
3. Other liquids, such as water glass (sodium silicate), may tend to harden or solidify when contacted with the atmosphere. The quench fluid smothers and excludes air, washes away minute leakage, and protects the seal.
4. Systems handling hazardous petroleum products require quench glands to provide a safety zone in case of a catastrophic seal failure. Likewise, throttle bushings or auxiliary small packings make a spray of large amounts of volatile liquid rather difficult.
5. Quenching is also a useful means of minimizing the transfer of excessive heat along the shaft to the pump bearings.

D. Environmental Control by Cooling

Many process pump systems operate under temperatures that in the long run are critical for seal performance. Where cooling of the seal environment is mandatory, the gland plate can be designed to allow cooling of the seal seat. The coolant can be either water or any other compatible liquid. Proper design of the gland end plate is important because it allows the fluid to reduce the seat temperature. Continual removal of heat prevents detrimental heat buildup. Cooling is suitable for systems handling liquids with a low boiling point or liquids that must be transferred at extreme temperatures.

Some of the customary designs using cooling of the seat ring are illustrated in Fig. 2.19. The first seat is mounted with press-fit whereas the second is arranged in a flexible fashion using two elastomeric O-rings. The third design clamps the seat and uses an auxiliary packing.

In the design with the seat mounted with press-fit to the gland ring the cooling water hits the seat ring and no secondary seal components are needed. In the design with flexible arrangement of the seat two elastomeric O-rings are used for sealing the coolant. In the design with a clamped seat ring, the O-ring seals the seat ring on the back of the seat. An external auxiliary packing prevents the water coolant from draining to the atmosphere.

Cooling reduces excessive interface temperatures, preventing vaporization of the lubricant film in the rubbing areas. If the temperatures of the process fluid are excessively high, it is recommended that a pump with a jacketed stuffing box chamber be used. By circulating cooling water through the jacket, the stuffing box chamber is preferably dead-ended. The pumpage fluid then enters the stuffing box chamber through the throat at the bottom.

Pumpage fluid temperatures in dead-ended seal chambers can be reduced with this method from 500 to 200°F with circulating cooling water of 70 to 80°F at a flow rate of 3 gpm through the pump jacket. The process fluid, however, must be clean.

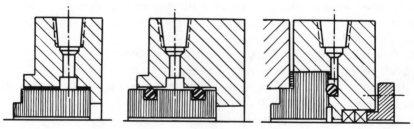

Fig. 2.19 Environmental control by cooling.

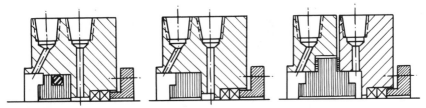

Fig. 2.20   Flushing-quenching combinations.

There are cases where an externally mounted seal should be provided with a cooled seat, but where the gland ring cannot be used to provide the coolant. The problem can be solved by using the stuffing box chamber to introduce the coolant. To reduce the stuffing box volume to a minimum a special sleeve with U-shaped cross section is used, sealing the coolant flow with two elastomeric O-rings. The design of the sleeve makes it possible to also minimize the leakage of the pumpage fluid into the stuffing box cavity from the pump housing. A proper clearance gap between sleeve and shaft should establish the effectiveness of this arrangement and create no other problems.

E.   Combinations of Environmental Control Methods

Environmental control systems are also used as combinations, such as flushing-cooling, flushing-quenching, or flushing-cooling-quenching. With these combinations the design of sealing units becomes quite versatile and flexible, making it easier to control contamination, corrosion, and temperature.

In Fig. 2.20 three double systems are shown to combine flushing-quenching operations using modified seat designs. Flushing-cooling combinations are illustrated in Fig. 2.21. A design that combines flushing-cooling and quenching is shown in Fig. 2.22.

These designs are only representative and seal manufacturers offer even more versatility. The design of the gland ring plate offers an almost unlimited number of possibilities for modifying environmental control devices, which in turn satisfy a wide range of diversified process requirements.

F.   External Supply of Interface Lubrication

In systems handling dry gases the interface must be lubricated by a separate external lub system. Figure 2.23 illustrates two possibilities for forced external lubrication of the interface environment through the

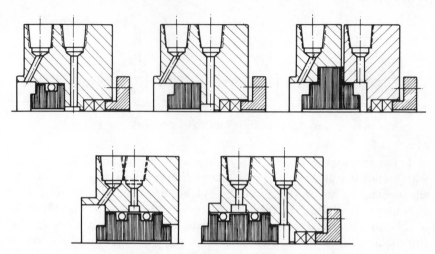

Fig. 2.21 Flushing-cooling combinations.

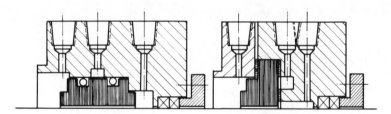

Fig. 2.22 Combinations of flushing, cooling, and quenching.

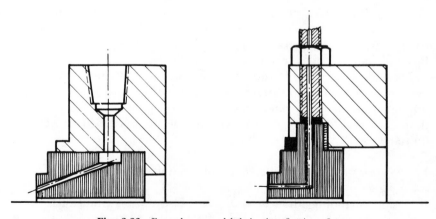

Fig. 2.23 Forced external lubrication for interface.

seal seat ring. The amount and pressure of the lub supply system depends on the relative process conditions of the pumpage fluid.

VIII. Design Requirements for the Mating Seal Faces

A major factor in establishing satisfactory seal performance is the interface topography of the two contact areas. The two mating faces should be smooth and reasonably flat, permitting minimal leakage across the interface during rubbing contact.

A. Requirements of Flatness

For end face seals it has become generally accepted practice to provide surfaces of two to three light bands of optical flatness with surfaces from 4 to 5 μin. finish. With this standard condition the seal will probably produce satisfactory performance for all liquids at varying temperatures and pressures.

By utilizing the interference effect of light it is possible to measure a few millionths of an inch in terms of a fraction of a wavelength. This can be done by using very high-quality reference standards known as optical flats together with supplementary test equipment now generally available from industry as shelf items.

B. Light Waves and Interference

To use an optical flat it is necessary to understand the mechanism on which it is based. Only the basics are shown here. For details one may consult one of many excellent textbooks on this subject or manufacturers.

Assuming that light travels in a wave motion at a constant velocity of 300,000 km/sec or 186,000 mi/sec, the wavelength is defined as reflecting the distance between two successive peaks. In the visible spectrum the units of wavelength are expressed as angstroms with 1 angstrom = 10^{-8} cm.

Each color corresponds to a different wavelength. As an example, typically green light has a wavelength of 5500 angstroms. The visible wavelengths range from violet light with 4000 angstroms to red light with a wavelength of approximately 6500 angstroms.

Monochromatic light is defined as light having a single wavelength, which is produced by isolating one wavelength from the many produced

Table 2.4   Standard Wavelengths

Source	Wavelength (Angstroms)	Wavelength	One Fringe
		(μin.)	
Helium	5876 (yellow)	23.1	11.5
	7065 (deep red)	27.8	13.9
Hydrogen	4861 (blue)	19.1	9.5
	6563 (red)	25.8	12.9
Mercury	4047 (violet)	15.9	8.0
	4358 (blue)	17.2	8.6
	5461 (green)	21.5	10.7
Cadmium	4800 (blue)	18.9	9.5
	6439 (red)	25.4	12.7
Sodium	5893 (yellow)	23.2	11.6
Helium-neon laser	6328 (red)	24.9	12.4

by a light source. For example, this can be achieved by placing an interference filter in front of a mercury vapor lamp.

Table 2.4 gives the various standard wavelengths emitted by mercury, hydrogen, sodium, and helium lamps. A monochromatic light source is required to produce the effects utilized for identifying the degree of flatness of a surface. The faces of primary seal rings require this precision.

For producing a particular wavelength a suitable light source and a corresponding filter must be selected. By superimposing two wavelengths three different effects can be observed, namely, troughs and peaks coincide, troughs coincide with peaks, and two waves are out of phase by a given amount.

When we assume one set of waves where the peaks coincide with the troughs of the other set at any particular point of equal wave amplitudes, it is said that the two sets cancel each other out. This is a phenomenon that is defined as interference and represents the basic principle on which the flatness testing techniques are based. When two wave sets of the same amplitude are out of phase, complete elimination of light results. When this condition is met regions of complete darkness will be formed. When two beams are not equal an interference pattern will be produced that becomes clearly visible.

When two amplitudes are different, no effects will be observed or distinguished. Another condition without fringes occurs when the path difference introduced between the two sets of waves is made too great.

For flatness measurement, an optical reference flat is placed in contact with a test specimen surface whose flatness is to be determined. Both surfaces must be perfectly clean so that an air film can be formed between the two surfaces. By subjecting the air film to monochromatic light, an interference pattern will be produced that is formed by eight waves reflected from the test sample and reference surfaces. Whenever twice the thickness of the air film is an integral number of wavelengths (such as 1, 2, 3, 4) a dark fringe is discerned. Such a fringe is a line that joins all points where the film thickness is the same. The fringe is, therefore, similar to the contour lines on a map and since the contact surface of the reference flat can be considered perfectly flat, the fringes form contour lines of the surface of the test sample. In this manner minute imperfections in the sample surface can be clearly identified and remachined by lapping if necessary to supply or meet the established specification requirements.

C. Typical Interference Patterns and Their Interpretation

The wedge method is applicable only to surfaces that are almost comparable in flatness to that of the reference flat. Both surfaces must have approximately the same diameter to minimize scratching both contact surfaces.

Irregular test sample surfaces show nonstraightness of the fringes. In the wedge technique the curvature of the fringes is a measure of the deviation from flatness, whereas the number of fringes is a property of the wedge having no relationship to flatness.

The unit of measurement is the wavelength of light, and one fringe is equivalent to a separation of one-half of a wavelength between the two surfaces. If the fringes are sufficiently far apart from each other, it is actually possible to estimate to an accuracy of one-tenth of a fringe, corresponding to a flatness of one-twentieth of a wavelength.

For assessment of the actual flatness using the air wedge configuration method refer to Fig. 2.24, in which an interference pattern is formed between two flats. At any particular zone the flatness is determined by joining the ends of a particular fringe either by using an imaginary line or by placing a straightedge across the rear surface of the reference flat. By using a photograph of the pattern, the fringes can be accurately evaluated.

Figure 2.25 indicates the curvature of one fringe corresponding to one-half wavelength and shows an air wedge of four fringes as two light waves. The actual dimensions in microinches are a direct function of the

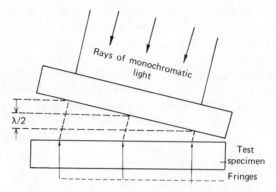

Fig. 2.24 Light interference produced by air wedge.

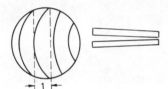

Fig. 2.25 One-fringe curvature.

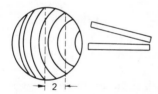

Fig. 2.26 Flatness error of two fringes.

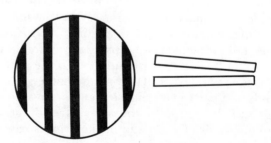

Fig. 2.27 Test piece perfectly flat.

wavelength of the light source to be taken from Table 2.4. For all practical purposes the helium light represents a wavelength of 23 μin. Consequently, the irregularity is 11.5 μin. and the overall air wedge is 46 μin. Figure 2.26 reflects a surface out of flat by slightly over two fringes. The air wedge represents eight fringes corresponding to four wavelengths. If the fringes are straight and parallel with equal spacing, the test specimen displays a perfectly flat surface. The resulting fringe pattern will be similar to that illustrated in Fig. 2.27.

D. Actual Flatness Measurements

Before a measurement can be made and light bands show up, reflection from a lapped surface must be obtained. Because ground surfaces are not smooth enough to be checked, it is usual practice to hand-polish the surface to be measured after it has been lapped. This can be done with paste and an alloy-polishing plate.

When placed under light with an optical flat straight, parallel, and equally spaced light bands are visible, indicating that the surface is equally flat within a millionth of an inch. The dark bands that can be seen underneath the optical flat do not represent light waves. They actually show where interference will occur because of reflection from two surfaces. These dark bands are used to measure the flatness of the polished surfaces to be checked.

When an optical flat is placed on a flat surface under monochromatic light the intensity or sharpness of the lines varies, depending on the degree of polish or the color of the surface to be measured since both polish and color affect the degree of deflection obtained. Neither of these factors, however, has any influence on the degree of flatness.

Variations in the spacing of bands or lines from one piece to any other test sample are not indications of variations in surface flatness, but merely express a difference in the thickness of the air wedge. For practical purposes the band spacing should not be too close together. It can be separated by putting pressure on the top surface of the flat to reduce the air wedge. Conversely, the pattern of bands can also be set closer together by inserting a sliver of tissue underneath the edge of the flat, which will increase the thickness of the air wedge.

E. Measurement of Surface Finish

A significant factor in the discussion of surface topography for the rubbing surfaces of mechanical end face seals is the concept of *surface finish*, also frequently termed *surface roughness*. For the measurement of the degree of surface roughness many different methods are in use. One method employs a specially prepared diamond needle, connected with a piezoelectric crystal. When the needle is moved across the surface to be measured, it detects relative differences in elevation and registers them through its action on the piezoelectric crystal. When moved across the surface the needle senses the topographical differences, which produce deflections against the instrument. The deflections are then translated into voltage, from which they are converted into electrical plots.

In the general machine industry two basic types of systems are in use to specify the degree of surface finish. One method is termed the RMS method, using the first letters of the words *root-mean-square*. The second method is known as the CLA, or AA method, where CLA stands for *center line average* and AA is the abbreviation for arithmetical average.

Mathematically the definitions are based on the following relations:

$$\int_0^L y\, dx = 0$$

$$\text{CLA or AA} = t \int_0^L |y|\, dx$$

$$\text{RMS}_{av} = \left[t \int_0^L y^2\, dx \right]^{1/2}$$

The designations of these equations are taken from Fig. 2.28, the curve actually reflecting a surface measurement, determining the topography.

Surface roughness alone does not accurately describe a surface. Consequently, a second concept must be introduced termed *surface waviness*. In surface topography both terms are used and are known as patterns in the rubbing surfaces. Figures 2.29 and 2.30 show these concepts. Their interpretation is self-explanatory.

Waviness is the designation for surface irregularities and asperities of

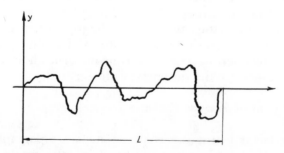

Fig. 2.28 Basic pattern of surface topography.

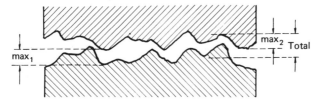

Fig. 2.29   Pattern of surface roughness.

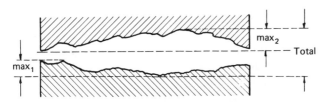

Fig. 2.30   Pattern of surface waviness.

a spacing greater than that of normal roughness. Both the height and the width of waviness are pertinent.

The depth of scratches is measured in microinches, or millionths of an inch. Readings are generally referred to as 5 μin. rms, indicating that the finish is within 5 μin., or microinches, meaning that tiny irregularities on the surface are 5 millionths of an inch deep.

Flatness, however, should not be confused with surface finish. Polishing removes minute peaks of a surface for the purpose of obtaining a shiny and reflective area without changing flatness.

As far as roughness is concerned a surface finish for interfaces of mechanical seals of the quality of 10 to 5 μin. rms is satisfactory. Finish conditions of better than 5 μin. are too smooth and do not offer sufficient traction to the film to establish suitable adhesion. Surfaces with a roughness of better than 5 μin. represent waste of effort and money and may facilitate the separation of the faces because of the lack of necessary friction for the lubricant film.

In conclusion it can be said that the phenomena of surface roughness and waviness are far from being completely understood. The surface topography still involves too many unknown variables to allow reliable correlations for predicting seal performance. It also must be stated that surface inspection techniques and methods used for defining surface topography are still inadequate and need noticeable improvements.

As far as inspection of surface patterns after operation of the seal in the field is concerned, it is very difficult if not impossible to interpret the interference band pattern adequately.

The correlation of one single profile trace with the interference band pattern is also difficult. The interference pattern does show that the single profile trace will not yield an adequate dimension for surface identification.

IX. Requirements for Shaft Design to Accommodate Mechanical Seals

In the design of mechanical end face sealing devices for rotating shafts, the actual geometry of the shaft plays a significant role. The most economical shape for the shaft is a straight-through cylinder, as this requires a minimum of material and machining. Even this configuration is not possible in all cases. There are a number of important reasons that force the designer to deviate from the ideal geometry and provide steps, shoulders, recesses, and the like, to accommodate the introduction of sleeves, bushings, bearings, and other mandatory components.

Perhaps the dominating reason for staggering the shaft is for hydraulic seal balancing. In many cases the seal can be properly balanced only when there is a shoulder in the shaft.

For a number of seal configurations sleeves provide great versatility. A shoulder in the shaft facilitates fastening of the sleeve to the shaft and makes it easier to seal the sleeve against the shaft. It is often simpler and less expensive to use a sleeve and provide the staggering requirements in the sleeve, which can be modified or replaced if this should become necessary. Material selection is simpler if a sleeve is chosen.

A small recessed shoulder in the sleeve locates the seal head and a throat bushing minimizes the leakage of the pumpage fluid. A metal bellows permits high-temperature service. When a flushing cycle is used, the pump may handle abrasive liquids. Quenching of the seal seat will prevent overheating of the seal unit.

A sleeve over the shaft also offers a good opportunity for attaching the rotating seal seat without damage to the shaft. Staggered sleeves are further well suited for attaching and mounting cartridge seals.

X. Design of Seal Systems Handling Abrasive Liquids

Abrasives in the pumpage fluid are detrimental to the performance of any design of mechanical end face seals. Rubbing contact in seals can be successfully achieved by providing adequate lubrication. Abrasive solid particles in the pumpage fluid will tend to penetrate into the interface, break down the lubricant film, and bring the seal to complete failure. Abrasive particles can originate from a number of supply sources, such as:

1. The particles may be an inherent characteristic of the process fluid.
2. The process fluid itself produces abrasives through crystallization or precipitation due to thermal influences.
3. The fluid reverts to a crystalline structure when contacted by air from the external atmosphere.

For each of these possibilities a thorough evaluation must be made. To minimize the detrimental influence of abrasives in the process fluid the following general design rules may be considered:

The destructive action of abrasives in the process fluid can be fought by injecting a clean secondary liquid compatible with the process fluid into the interface environment. The secondary fluid is preferably injected around the interface circumference, thus keeping the solid particles from penetrating into the interface. Even though the secondary fluid is chemically compatible with the process liquid, it must be determined whether any dilution of the primary liquid is tolerable.

Although basic principles of this problem have already been discussed together with environmental control devices, a few more details may be added to underline the significance of these factors in practical seal design.

Pumpage fluid and operating conditions must be analyzed to define the nature of the abrasive and susceptibility to dilution. Composition of the secondary fluid, flow rate, pressure, and sensitivity to dilution are all factors to consider in establishing adequate prerequisites to combat the deteriorating conditions and produce optimal service life of the seal.

One of the major concerns in dealing with abrasive liquids is their susceptibility to dilution by the secondary flushing fluid. Some liquids are not affected by dilution; others will tolerate no or only a minute amount of dilution.

In cases where minor dilution is of no consequence to the process, the application of a throttle bushing in the stuffing box throat as a restricting flow device represents minimal cost at reasonable effectiveness.

Many methods have been developed to reduce and minimize leakage flow of the secondary system fluid into the pump housing to dilute the process fluid. Diversified designs of restrictive bushings in the stuffing box throat are available and have been widely applied to prevent the flow of abrasive system fluid into the stuffing box chamber, which could jeopardize the function of the seal interface.

Among many configurations it was found that the bushing principle gave the most enduring success. The throat bushing can be of the lip-seal-type in a wide range of design modifications usually arranged and mounted in metal cartridges. Some devices have been provided with multiple V-ring seals mounted between carbon-graphite front and back-up rings. These devices, inserted in the vicinity of the stuffing box throat, use multiple or single helical springs to establish compulsory contact with the throat face and practically automate the seal contact.

XI. The Design of Throttle Bushings in the Stuffing Box Throat

Throttle bushings in the throat of the stuffing box bottom are designed for two major purposes: First, they prevent abrasive fluid from the pump volute from leaking excessively into the stuffing box chamber where it could contaminate the interface environment of the mechanical seal. Second, they prevent the secondary flushing fluid from leaking into the pump housing and contaminating or diluting the process liquid beyond a tolerable amount. The throttle bushing accordingly becomes a component of the environmental control system.

The key for full utilization of the flushing action in the chamber lies in the proper control of the flow of the secondary flushing fluid. The throttle bushing, therefore, plays a critical role in this control mechanism. The conventional throttle bushing is designed with a clearance gap toward the shaft that throttles the flow of liquid but does not fully stop it, as can be seen in Fig. 2.31.

The size of the clearance gap c varies with the construction material for the bushing and with the length of the gap. When made of metal the clearance gap may amount to 0.030 in. and when made of carbon, Teflon, Kel-F, or polyimid, the gap can be 0.005 to 0.010 in. The size of the clearance gap for the bushing is closely associated with the accuracy of the shaft movement. Small shafts readily deflect and whip and thus

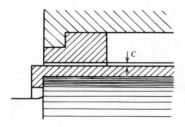

Fig. 2.31 Conventional throttle bushing [From (37), excerpted by special permission from *Chemical Engineering*, September 1956, Copyright © 1956 by McGraw-Hill, Inc., New York, N.Y. 10020].

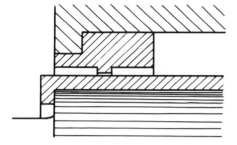

Fig. 2.32 Modified throttle bushing [From (37), excerpted by special permission from *Chemical Engineering*, September 1956, Copyright © 1956 by McGraw-Hill, Inc., New York, N.Y. 10020].

require larger clearance gaps than is customarily experienced with larger shafts.

Pumps handling abrasive fluids should be fully drained when they are not in operation. This precaution prevents blocking of the clearance gap caused by solidification and sedimentation.

Figure 2.32 illustrates a bushing design that eliminates most of the disadvantages of the straight-through bushing. The length of the bushing is reduced to a minimum and reflects only a small projection with a closer clearance. The effective length of the projection ranges approximately from 0.125 to 0.250 in. For the size of the clearance gap the same rules apply as mentioned for the bushing of Fig. 2.31.

Such bushings are conventionally made of metal, carbon, Teflon, Kel-F, and polyimid where the clearance can be kept at 0.005 in. radially.

A different approach is shown in Fig. 2.33. A long throttle bushing is replaced by a short bushing section. A stripper lip is added, mounted between two short sleeves of carbon or metal, held in place by a spanner ring. The material of the lip ring is generally Teflon, Kel-F, or polyimid mounted with interference fit to the shaft. The lip is always directed toward the stuffing box throat from where the leakage flow is expected.

Under suitable pressure differentials the lip permits limited leakage of

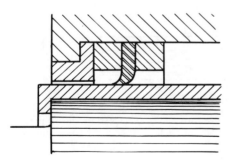

Fig. 2.33 Bushing with additional stripper lip [From (37), excerpted by special permission from *Chemical Engineering*, September 1956, copyright © 1956 by McGraw-Hill, Inc., New York, N.Y. 10020].

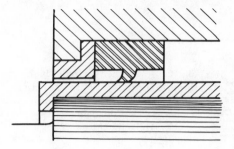

Fig. 2.34 Stripper lip integral part of sleeve [From (37), excerpted by special permission from *Chemical Engineering*, September 1956, copyright © 1956 by McGraw-Hill, Inc., New York, N.Y. 10020].

the flushing fluid through the throat. But the lip always acts as a check valve to prevent backflow from the pumping housing. The plastic lip ring is relatively thin and very flexible. Its thickness may range from 0.020 to 0.040 in. with an average radial interference of 0.002 in. The lip component is also designed as an integral part of a sleeve, mounted in the same way as shown in Fig. 2.34. The device may basically consist of Teflon, Kel-F, or polyimid always mounted closely adjacent to a metal throat bushing. The diametral clearance of the lip to the shaft is 0.002 in. Reports confirm that this design has performed successfully for operating periods of up to 25 months of service under pressures of up to 75 psi differential.

Another approach is illustrated in Fig. 2.35. This design incorporates a metal bushing of short length in the throat of the chamber and a plastic sleeve with serrations toward the shaft surface. The short metal bushing in the throat provides the major portion of the flow restriction. The adjacent lips of the plastic sleeve further reduce the leakage like the chambers of a labyrinth seal. If cooling by the flushing fluid is provided, the clearance gap for diametral clearance is kept at 0.010 in.

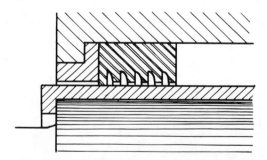

Fig. 2.35 Bushing and serrated Teflon sleeve [From (37), excerpted by special permission from *Chemical Engineering*, September 1956, copyright © 1956 by McGraw-Hill, Inc., New York, N.Y. 10020].

XII. Flow Relationships in Clearance Gaps of Throttle Bushings

Clearance gaps in restrictive bushings allow certain amounts of flow that can be computed when the geometry and the process conditions are known. For a given clearance gap the pressure of the fluid imparts a velocity to the fluid that becomes the basis of the computation. For a given geometry of the gap and a known velocity the flow rate can be calculated. A paper of Tao and Donovan (56) gives the basic principles of such a computation. The paper discusses flow through fine annular clearances with and without relative motion of the participating boundary surfaces. The diagram of Fig. 2.36 permits the determination of laminar flow velocity through annular close clearance gaps with water as a flow medium at room temperature (68°F). The basic assumption is that the gap is concentric. The graph is established on a friction factor of 96/Re, where Re represents the Reynolds number.

The Tao-Donovan curves should be considered as guidelines because

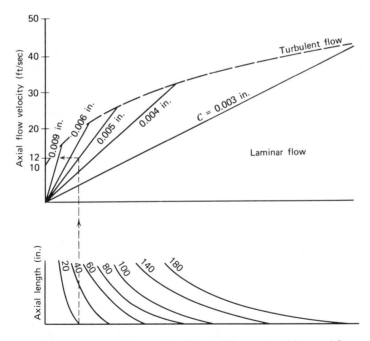

Fig. 2.36 Flow in annular clearance gap [From (37), excerpted by special permission from *Chemical Engineering*, September 1956, copyright © 1956 by McGraw-Hill, Inc., New York, N.Y. 10020].

the prerequisites of their test devices deviated from the bushings conventionally used in seal units. They give, however, a good indication of the effects of clearance gaps on flow under given pressure differentials and fluid velocities. Further details are presented in Chapter 7 in the section on bushing seals.

An example shows how to use the diagram of Fig. 2.36. With a sleeve of 2 in. length and a pressure differential of 20 psi the flow velocity in an axial direction through a concentric axial clearance gap of 0.005 in. amounts to approximately 12 fps, following the dotted line in the diagram.

XIII. Selection of Materials for Primary Seal Rings

The selection of suitable materials for the construction of primary sealing components is complex and should not be done by merely taking them out of catalogs and tables. The solution of this problem necessitates the evaluation of a number of factors to be discussed in some detail. These factors are face pressure, rubbing velocity, system temperature with reference to the interface, friction, wear, thermal effects, compatibility of seal materials, process fluid, and flushing liquid, to name just the most important ones. Unfortunately, the ability of materials to meet these requirements in their relation to each other has not been identified or related to the rate of leakage across an interface.

As stated earlier, comprehensive information on friction and wear for many suitable seal materials is available. This information, however, is not directly applicable for solving seal problems because important correlations are missing. Therefore, the selection of appropriate seal face materials cannot be assured with absolute certainty, but promising candidate materials can at least be identified, as is discussed below.

A. Factors that Influence Seal Life

Failure or malfunction of primary seal components may be the result of many detrimental factors. It is usually not the effect of a single property, but of a combination of complex interrelations, that leads to a breakdown of an interface. The failure can be caused by long-time exposure in the form of fatigue or it can be a sudden breakdown involving a combination of forces, such as mechanical, thermal, chemical, or radiation. Most of these energies react in combination with each other, producing sudden or eventual failure. This is why the interpretation of a failure or an accurate prediction for a new design is so complex.

1. The Effect of Mechanical Forces

Hydraulic system pressure, spring action, and rubbing velocity in the contact faces produce mechanical forces that control the formation of face topography and the seal performance. As stated earlier, the operating pressure of the process liquid acts against the seal components in two principal directions, axially and radially. In the axial direction the system pressure effect can be controlled within a desirable range by modification of the geometry of the primary seal ring. The result is a range of degrees of hydraulic balance designs. In a balanced seal the design configuration varies slightly, showing a total seal face contact area greater than the area exposed to the hydraulic fluid in an axial direction and tending to close the mating faces.

Balance is expressed as a percentage of the exposed area in relation to the interface contact area. The fluid-exposed area is usually greater than the shielded rubbing face area. In seal practice mechanical seal balance is designed to incorporate balance values ranging from 60 to 75%, indicating that the static face pressure against the interface is only 60 to 75% of the value of its unbalanced face configuration. The values of balance become apparent when one considers a seal with 65% balance operating at a stuffing box pressure of 400 psi. Consequently, the seal face unit load is reduced to 260 psi increased by the spring force in pounds, divided by the area of the seal face, expressed in square inches. In industrial practice no mechanical end face seal in service is ever completely balanced. The principle of pressure gradients and the formation of the lubricating film across the interface applies to both balanced and unbalanced seal designs. In a seal unit that is completely balanced the only force acting against the interface for contact is the force generated by the springs.

The pressure gradient acting in the interface tends to open and thus separate the contact faces. The gradient increases proportionally with the pressure of the process fluid when acting against the seal faces. The spring force performs in favor of the closing action. It is, therefore, practical to reduce the face load through hydraulic balance to allow sufficient overbalance. This in turn means that the exposed area is greater than the shielded area of contact, which now imparts sufficient face load to eliminate the separation tendency through the wedge of the pressure gradient. This arrangement allows for an automatic adjustment to compensate for variations in stuffing box pressure, keeping the seal face loads at a possible minimal value. The radial components of the fluid pressure in the stuffing box chamber tend to distort complex sections of the primary seal ring in a nonuniform pattern. Any distortion in the seal ring must be detrimental to the degree of flatness in the contact areas, which are machined to a flatness of two to three light bands of accuracy. It is, therefore, mandatory to secure seal faces with cross sections commensurate with the modulus of elasticity of the construction material to diminish or eliminate face distortions.

Excessive face loads of primary seal rings, especially at higher pressure, create noticeable disturbances. The fluid film may be squeezed out, producing burnup of the faces. Power consumption increases, excessive heat is generated, and the seal finally breaks down.

2. The Effect of Velocity

Velocity of the rubbing faces cannot be balanced, as is possible with the hydraulic fluid pressure. Both, velocity and fluid pressure, result in generation of frictional heat in the seal faces. The actual amount of heat generated by rubbing at the interfaces can be computed by the relation:

$$\text{heat (Btu/min)} = \frac{\text{torque (ft-lb)} \times \text{rpm}}{123.6}$$

In this function the torque reflects the work required to move the rotating primary sealing component relative to the seal face. Velocity is generally expressed by revolutions per minute (rpm) and is computed from the relation:

$$\text{rpm} = \frac{\text{velocity}}{\pi \times d_m}$$

where d_m represents the mean diameter of the interface area.

The rate of heat input increases with velocity and the temperature rises correspondingly at the contact face until a heat balance is established. With the variation in temperature the mechanical and physical properties of the face materials change. The lubricant film between the faces is also subject to physical changes. Rising face temperatures, however, result in a loss of face loading. A rubbing velocity of 15,000 fpm is considered normal seal practice and is far from the maximum capability of manufacturers.

3. The Effect of Temperature

Heat can have very detrimental effects on seal performance if it is not properly dissipated. Temperature at the seal faces is in most cases higher than suspected. Often the temperature between the faces is several degrees higher than the temperature of the immediate face environment. This is the reason for selecting a material for the primary seal rings that has good thermal conductivity. It should be recognized that the viscosities of liquids decrease with rising temperatures, whereas the viscosities of gases increase with the temperature.

Materials with a low value of thermal conductivity, low tensile strength, high coefficient of thermal expansion, and high modulus of elasticity experience thermal shock when subjected to large temperature differentials over relatively short and abrupt time intervals.

Parts with relative motion under poor lubrication conditions between the rubbing faces may be subject to heat checking. Thermal heat checking is a surface phenomenon, characterized by a multitude of fine cracks and fissures, often visible only under high magnification. These cracks are formed suddenly by thermal gradients at the rubbing faces in the presence of certain material combinations. Ceramics are particularly subject to severe heat checking. Among the many materials used in the seal industry, the tungsten-carbides offer the highest resistance to the phenomenon of heat checking.

Where limited lubrication prevails, heat checking may be mitigated by selecting materials with high rates of thermal conductivity. It should be kept in mind, however, that friction and wear are of equal importance.

Stellite is a typical material that is subject to spontaneous heat checking. Careful consideration should be given to a thorough evaluation of the process conditions, specifically when stellite is to be selected for seals operating with extremely light liquids or fluids providing poor or no lubricity.

Stellite is an excellent material for seal faces where fluids with good lubricity and lower temperatures are used and steady operating conditions are guaranteed. Stellite provides extremely hard surfaces and gives increased corrosion resistance compared to most conventional stainless steels.

When the seal unit has to operate in a liquid near boiling point at low pressure, the lubricant film will flash and break down, and typical heat checking will occur as a result of possible dry-run. For this type of operating conditions tungsten-carbides are the only reliable materials, as they have the highest resistance to heat checking disturbance. Their excellent thermal conductivity dissipates the generated heat and prevents the formation of "hot points."

Temperature may also tend to break down the process fluid. As an example, coking may develop when oils are pumped at high temperatures. Resulting deposits across the interface may in turn accumulate, giving rise to separation of the contact faces.

In shrink-fit parts the difference in coefficients of thermal expansion can be critical, often resulting in loosening or even separation at the shrink contact. On the other hand, shrinkage may produce irregular residual stresses that lead to face distortion and final seal failure.

Cast iron and nickel-resist represent excellent mating materials for primary seal rings, although their thermal conductivity is far below that of tungsten-carbides. They are used for a wide range of applications, especially for noncorrosive liquids.

Systems handling fluids at cryogenic temperatures likewise face temperature shock effects. Systems should be brought to the desired temperature level gradually, avoiding all sudden and excessive temperature differentials.

From a chemical standpoint water appears to be a harmless liquid. However, when water is pumped at a temperature of 140°F or over, the corrosion rate of steel can be accelerated by a factor of 4 times the normal rate at room temperature. After exceeding the 140°F limit the corrosion rate has been shown to double with every 20°F increment. Oxides are formed and break loose, floating around the interface, and they tend to precipitate in this environment. Suitable cooling is, therefore, most desirable.

Rust inhibitors should be considered and evaluated very carefully. Good rust inhibitors should produce clean fluids and all piping must be free of any mill scale or other foreign particles that may have accumulated during assembly of the system.

4. Chemical Effects

Chemical reactions become possible when materials contact the pumpage fluid. They can form destructive by-products on the exposed surfaces. Data from corrosion handbooks may be used as guidelines only. Unfortunately, they have been obtained in most cases from test conditions that do not necessarily correlate with conditions existing in mechanical seal systems. Pressure, temperature, and velocity can accelerate chemical corrosion with temperature, generally increasing it most dramatically. A rule of thumb indicates that a 30 to 35°F temperature increase will double the normal corrosion rate.

Although acids and alkalies are both strongly corrosive, in general acidic solutions are more corrosive. However, in high alkaline concentrations the amphoteric metals and some nonmetals may corrode more readily than in acids. Even carbon with its high degree of chemical resistance should be chosen with great precaution. Solution -pH is thus an important factor. Consultation with seal manufacturers is strongly recommended.

Oxidizing agents can be powerful corrosion promoters. Oxygen from the atmosphere or in solution can produce troublesome corrosion problems. Corrosion inhibitors, such as sodium dichromates, can be seriously detrimental to seal faces, particularly at concentrations of 250 ppm.

Galvanic corrosion also can be detrimental to seal performance. It can be minimized by selecting material combinations that are within 25 mV in the galvanic series.

Distilled water is a poor electrical conductor and hence does not generally create galvanic corrosion problems. Strong salt solutions, such as brine, sea water, and others are good conductors and therefore accelerate galvanic corrosion processes.

Chemical and galvanic corrosion are thus a serious threat to the integrity of machine parts. When these parts are subjected to additional forces, such as tension, compression, temperature, and rubbing velocity, the destructive influence becomes even greater, resulting in stress corrosion effects.

Materials like carbon and ceramics are generally acceptable for most acidic environments. Alkalies, however, will attack ceramics and strong nitric acid can attack carbon.

For highly corrosive service, filled Teflon, tungsten-carbides, boron-carbides, and the like, have given the most satisfactory seal performance.

Chemical corrosion, friction, wear and rubbing velocity must be evaluated together. In chemical environments many metals form protective oxide films on the contact surfaces. Inside the interface zone this protective film coating can break down, producing a threat of seal failure. All oxides considered beneficial for corrosion service are not necessarily suited for service with mechanical seals.

5. Radiation Effects

Radiation can deteriorate a wide range of materials and it can deprive them of a number of desirable properties that otherwise would make them useful for seal ring components of mechanical end face seals. The changes in materials through radiation are produced by the presence of gamma (γ) rays emitted from radioactive substances. The deterioration of materials exposed to alpha (α) and beta (β) rays is insignificant. Of primary concern, however, is the possible damage resulting from exposure to gamma (γ) rays and neutron penetration. R. E. Bowman (8) gives specific details on this subject.

The changes produced in materials resulting from exposure to gamma radiation may be either temporary or permanent. Radiation damages can persist even though the source of emission has long been removed.

The best resistance to radiation can be expected from carbon steels, stainless steels, and alloys of aluminum, nickel, and copper. Inorganic materials such as graphite, carbides, glass, and ceramics are strongly influenced by radiation and can change in basic material properties to a noticeable extent.

Elastomeric materials vary widely in their resistance to radiation emission. Plastics, with the exception of Teflon, have equal or better resistance to radiation than the conventionally used elastomeric materials.

B. Desirable Material Properties for Primary Seal Components

Although satisfactory analytical design theories for end face seals do not exist, certain guidelines are available that facilitate the selection of appropriate mating materials for primary seal rings. They are summarized as follows:

Mechanical Properties. Materials to be selected for primary seal rings must have a high modulus of elasticity, high tensile strength, good antifrictional behavior, and/or a low coefficient of friction. Good wear characteristics, optimal hardness, and, if possible, self-lubricating properties are also highly desirable.

Thermal Properties. Materials should have a low coefficient of thermal expansion, good thermal conductivity, high resistance to thermal shock, and a negligible dimensional change in extreme temperature pulsations.

Chemical Properties. Materials should be sufficiently inert and fully resistant to a wide range of corrosive chemical substances. Good wearability and solid adhesive characteristics are other important properties when used in the environment with chemicals.

Radiation Properties. Resistance to emission of gamma, beta, and alpha rays is mandatory.

Miscellaneous Properties. A major requirement is dimensional stability under pressure, temperature, and corrosion. Materials should have good machinability at low cost and should be readily available. Seal ring materials should have all the protective properties required for good bearing materials.

C. Suitable Mating Combinations

Primary seal rings in mechanical end face seals always perform in pairs. Thus the selection of adequate mating materials becomes even more complex, with countless possible combinations. A fundamental law in mechanics states that similar materials should not be mated when relative motion is involved. This law is highly applicable to primary seal ring components. In the case of carbide versus carbide, for example, one material must be a nickel-base and the other a cobalt-base configuration. Materials for primary seal rings in centrifugal pumps should be impervious to all chemical liquids involved.

D. General Considerations Concerning
Seal Materials

The result of an extensive literature search showed that carbon-graphite meets most of the requirements expected from a seal ring. It is very extensively used; perhaps it is the most frequently applied face material in the chemical and related industries.

Carbon-graphite is inert to most common chemicals and has an inherent quality of self-lubrication not readily found in other seal ring materials. Carbon is black, has a high emissivity factor, and, therefore, dissipates frictional heat generated in the interface. Carbon has excellent resistance to thermal shock. It provides very good wear characteristics desirable for seal performance. Many grades and combinations of carbon-graphite materials are available. For a suitable selection consultation with the manufacturer is recommended.

In actual seal practice it is immaterial whether carbon is ued as the rotating or stationary ring component. Carbon is generally used for the geometrically simpler design configuration, and because of its grain structure carbon provides favorable traction for fluid films of light liquids such as alcohols, kerosene, and even water. It offers excellent bearing faces in the contact areas, even under heavy loads.

Carbon-graphite, either plain or impregnated with resins, inorganic salts, polymers, or metals, such as lead, Babbit, antimony, or silver, is used in large quantities for primary rings in combination with a variety of metal alloys, alumina, ceramics, and tungsten-carbides. The unique properties of carbon-graphite account for approximately 95% of all sealing devices used for mechanical end face seals.

Carbon can be finish machined to tolerances of 0.0005 in. In oxidizing atmospheres and at a temperature of 700°F carbon begins to lose weight.

Carbon retains hardness at high speed, high pressure, and elevated temperatures. Its coefficient of friction varies from 0.04 to 0.25, depending on grade and type of practical application.

Carbon has very low porosity; however, some grades are capable of absorbing fluids in volumes up to 13%. These data must accordingly be thoroughly investigated with the manufacturer before selection decisions are made.

It is common practice to use dissimilar materials for mating seal faces. Best results are usually obtained with tungsten-carbide combinations with dissimilar binders. By combining carbon-graphite with a variety of materials and/or fillers its versatility can be greatly improved. For example, its range of applications is improved when it is treated with the following:

Parafin.   Provides a protective coating at noticeable reduced porosity and enhances lubricity characteristics.

Phenolic.   Additions produce an increase in hardness, accompanied by a reduction of the initial porosity.

Wax.   A high-temperature type can markedly lower porosity, affecting resistance to most corrosive chemicals.

Metals.   Various types, such as Babbit, cadmium, copper, copper-loaded alloys, lead, and silver diminish porosity and impart strength.

Metals can be introduced in carbon-graphites by impregnation between grains. The molten metal enters spaces in the carbon structure to fill the pores after the air has been removed by heat and vacuum application. The metals impart additional strength to the structure, higher resistance to general face wear, and higher resistance to increased hydrostatic pressure. Thermal conductivity can be doubled by this method. The increase in wear resistance produced by treating carbon with metals is attributed to an increase in conduction and heat dissipation. Cast iron, nickel-resist, and certain types of tool steels have been used successfully for primary seal rings under low-pressure conditions at low velocities when corrosion is not a factor.

In the presence of mild corrosive atmospheres hardenable stainless steels and/or stellites (cobalt-nickel-base alloys) can be used.

In operations handling stronger corrosive fluids at moderate pressures and velocities alumina ceramics are employed either as solid rings or in the form of thin coatings. Carbon rings with a layer of silicon-carbide impart excellent service qualities. For very strongly corrosive liquids a high purity ceramic alumina is available with 99.5% aluminum oxide. Same results are obtained with solid silicone carbide rings.

Where high pressures and high rubbing velocities are encountered carbon-graphite, treated or nontreated, can be mated with tungsten-carbides.

Tungsten-carbide has a thermal conductivity 8.3 times that of alumina ceramics and 7.5 times that of Ni-resist. This high thermal conductivity coupled with high tensile strength and low coefficient of thermal expansion is the basis for its high resistance to thermal shock. It is often used in corrosive service where alumina ceramics have failed.

Alumina ceramics have relatively low tensile strength, low thermal conductivity, poor resistance to thermal shock, but a high modulus of elasticity.

Bearing and leaded bronzes are used in certain applications where the low tensile strength and low modulus of elasticity of carbon-graphite are disadvantageous because of deflection problems.

Nickel-silvers exhibiting higher strength characteristics than the bronzes show some promise in high system pressures and where rubbing velocities are low.

Plastics can be improved for moderate service conditions with graphite.

Glass-filled Teflon, applied to virgin Teflon for bellows, has been used for severe corrosion service.

E. Practical Mating Combinations for Primary Seal Rings

In many years of industrial practice a wide range of mating face combinations have evolved for various environments. Only a few of the more important ones can be discussed since a complete listing could fill books.

1. Cold Water as Pumpage Fluid

Water is the most common pumpage fluid for which a wide range of mating combinations is available from all major seal manufacturers.

When steels are used for mating rings the operating temperature is a controlling factor because corrosion is proportional to the service temperature, as stated earlier. The following materials are generally used in service with water:

		Bronze
Carbon-graphite		Ni-resist
		Nickel, cast iron (for constant operation only)
(also carbon-graphite		Ceramic
containing various	vs.	Stellite (hard faced on 316 stainless steel or other 300 series stainless steel or any stabilized stainless steel)
metals—copper, lead,		Tungsten-carbide
Babbit, etc.)		Malcomized 316 stainless steel
		Carbon-filled Teflon
		Glass-filled Teflon
		Chrome plate on various parent materials (must be thick enough)
		Ceramic facing on stainless steel

Tungsten-carbide   vs.   Tungsten-carbide

Carbon, containing   vs.   Stainless steel (series 400, hardened
various metals   to $R_c \sim 50$ and higher)

Case-hardened steels, however, are not suitable, since they are subject to rusting. In addition, the hardness is never uniformly distributed across the surface, resulting in detrimental effects in the mating faces. Nitrided steels offer a moderate resistance to atmospheric corrosion. They provide a uniform distribution of surface hardness. Ni-resist and cast iron are only conditionally suitable because of their susceptibility to rusting. Various types of bronzes are often recommended for this type of service, although they seem to be somewhat soft. A major prerequisite for water service is a low solids content. Where rust cannot be prevented, carbon-graphite can be mated with ceramics. Some cases with ceramics versus ceramics have been reported as good and durable. Ceramics versus carbon is generally regarded as a good combination for mating face materials because of excellent corrosion resistance and negligible wear. For fluids containing abrasives, ceramics offer excellent mating materials because of their hard faces. Ceramics are also available as coatings on metal backings. They are produced by welding techniques or by powder spraying methods. The backup metal provides the stability.

In a water system pressures to 40 psi are considered low. Pressures to 200 psi are designated as medium range. All pressures exceeding the 200 psi limit are referred to as high pressures, using seal terminology.

2. *Acids as Pumpage Liquids*

The term *acid* should only be used in a broad sense. The following chart may be used as a guideline:

Carbon-graphite vs.
- Hard-faced 316 stainless steel
- Carpenter 20 stainless steel
- Stellite
- Chromium boride
- Ceramic
- Hasteloy A, B, C
- Carbon-filled Teflon (for nonoxidizing acids)

Ceramic vs.
- Stellite (attacked by many mineral acids)
- Class-filled Teflon (oxidizing acids)
- Teflon

Strong oxidizing acids attack carbon-graphite, the action increasing with temperature. For example, strong oxidizing agents, such as concentrated hydrogen peroxide or fuming nitric or sulfuric acid, rapidly attack carbon-graphite. Even 316 stainless steel has marked limitations. Ceramics have a wider application for acids. Ceramics versus ceramics or ceramics versus glass-filled PTFE often offer practical solutions. In the presence of hydrofluoric acid glass-filled PTFE and ceramics are not suited, but combinations of the boron-carbide family have been successfully applied. In concentrated form, 75% and higher, nitric acid and anhydrous sulfuric acid attack carbon, acting similarly to oxidizing agents. Further strong chromic acids are also detrimental to carbon-graphites. Hydrogen fluoride and other fluorine-containing compounds attack ceramics as well as glass.

3. *Mating Materials for Caustics*

Carbon-graphites provide better service in caustic environments than in acidic atmospheres. However, carbon must be of high purity and should not contain any resin fillers. The following chart indicates that carbon-graphite successfully mates with carbon-filled PTFE or stellite-faced stainless steels.

Carbon-graphite vs.
- Carbon-filled Teflon
- Stellite-faced stainless steel

Carbon-graphite
(nonmetallic)   vs. Hard-faced 316 stainless steel

Carbon-graphite
(metallic for dilute solutions)   vs. Stellite-faced stainless steel

Good performance has been experienced with nonmetallic carbon versus hard-faced stainless steel. For diluted caustic solutions metallic
carbon-graphite has been successfully mated with stellite-faced stainless
steel. For certain concentrations copper-free Ni-resist has been used,
depending on temperature and degree of concentration.

4. Mating Materials for Oils

Face combinations for oils are recruited from a wide range of well-suited
seal ring materials.

Carbon-graphite	vs.	Bronze (for some applications)
		Ni-resist
		Cast iron
		Ceramic
		Stellite (hard-faced on 316 stainless steel especially for high pressures and high velocities)
		Tungsten-carbide
		Malcomized 316 stainless steel
		Carbon-filled Teflon
		Glass-filled Teflon
		Sintered iron or bronze
		Nitralloy, hardened
		Tool steel, hardened
		SAE-1040 steel
		Stainless steel (400 series $R_c \sim 50$. This reflects general recommendation as 316 stainless steel is not hardened).

Cast iron   vs. Bronze

Graphite-molybdenum   vs. Bronze

The chart reflects the many possibilities for matching materials with
carbon-graphite. Metals of all types are possible, since corrosion is not a
factor. Carbon mated with bronze has only limited application in oil
environments. The steels most preferably used are hardened tool steels,
cast iron, Nitralloy, and the hardenable stainless steels of the 400 series.

Minimum hardness should not be less than $R_c \sim 50$. The most economical mating combination for oils is carbon-graphite versus cast iron or various sintered metals. For crude oils some manufacturers recommend the use of stellite as a layer on stainless steels versus leaded bronze.

5. Face Combinations for Gasolines

In systems pumping various types of gasolines, carbon-graphite has become an almost universal mating partner. It mates equally well with metals of the types cast iron, nickel-resist, Nitralloy, and hard-faced stainlesses of the hardenable 400 series and with glass-filled Teflon, ceramics, or stellite, usually faced on stainless steels. From the standpoint of low friction the combination of carbon-graphite versus cast iron has given good service as long as the environment is not corrosive.

Carbon-graphite	Cast iron
	Carbon-filled Teflon
	Glass-filled Teflon
	Ni-resist
	Nitralloy
	Ceramics
	Stellite faced on stainless steel
	400 series stainless steel

When corrosion is a factor more resistant mating partners must be selected.

6. Mating Materials for Salt Solutions and Seawater

The term *salt solutions* comprises a wide range of liquids, including those with possible crystallization under certain defined conditions.

Carbon-graphite	vs.	Stainless steel Ceramic Monel	Carbon-Babbit	vs.	Aluminum-bronze
Ceramic	vs.	Ceramic-faced stainless steel Ceramic	Stellite on stainless steel	vs.	Aluminum-bronze
			Tungsten-carbide	vs.	Tungsten-carbide
Carbon-Babbit	vs.	Phosphor bronze	Bronze	vs.	Laminated plastic

As the chart indicates, carbon-graphite, ceramic, and carbon-Babbit are mated with a variety of materials. Stainlesses, Monel, ceramics, and phosphor-bronze in various combinations have given satisfactory service. Plain carbons have not been suited for seawater environments. Carbon-graphite is not recommended in connection with stainlesses, since electrolytic corrosion of the stainless can be expected. Seawater does not deteriorate plain carbon. However, when it is mixed or filled with metals the possibility of electrolysis must be considered.

7. Considerations for Hot Water as Environment

Systems operating with hot water require special consideration. For temperatures in the order of 450°F the pressure ranges in excess of 400 psi. Seals for these pumps are designed to incorporate a device for circulating a small amount of water through a heat exchanger that maintains a temperature in the seal chamber at a level slightly higher than the temperature of the cooling water passing through the heat exchanger. In these installations the seals are actually sealing clean water at temperatures below 160°F, although the water going through the pump may be close to 500°F.

Water, by nature already a poor lubricant at room temperature, fully loses its lubricity at higher temperatures, thereby becoming a sealing problem. Specific gravity, surface tension, and viscosity decrease with rising temperature, but these are not the dominating factors that destroy the lubricity of the liquid film between the contact faces. Although higher temperature tends to deteriorate lubricity, increasing pressure tends to reverse the action because specific gravity, surface tension, and viscosity increase with pressure. It was this premise that led to the theory that water would not lose its lubricity characteristics if it was sealed at pressures above the flash point. It is known, however, that the pressure of the interface film decreases with penetration into the faces. The pressure level at which surface tension prevents further loss into the atmosphere is considerably lower than in the system handling the hot water. As pressure decreases within the liquid film while the temperature may even go higher, the advantage of pressure disappears, the film begins to flash, and the seal finally fails. Suitable cooling eliminates flashing and maintains at least some lubricity. It is important that the water stay clean and be filtered properly.

As a conclusion Table 2.5 presents a series of mating face combinations for a variety of process liquid systems as reported by NASA and Table 2.6 gives a variety of face combinations commonly used in chemical and related industries.

8. Advantages When Using Tungsten-Carbides as a Mating Material

Tungsten-carbide is an excellent mating material for seal faces, but it is not the answer to all seal problems. Considerable improvements may be achieved when tungsten-carbide is one component and any other suitable material the other.

The wear rates of carbon against nickel-iron materials are much higher than against tungsten-carbide. With tungsten-carbides the system pressures can be much higher without increasing wear on the carbon-graphite rings.

Numerous field tests and literature reports show that the following advantages are possible:

1. No heat checking occurs regardless of severity of operating conditions.
2. Negligible wear is observed when mated with carbon-graphite counterparts. All the wear is generally found on the carbon rings.
3. The reduction of wear resulting from tungsten-carbide use permits higher pv values, accommodating more severe operating conditions.

XIV. Summary Conclusions for the Design of Mating Faces

This section compiles and evaluates information taken from papers, brochures, books, numerous NASA publications, government brochures, and literature from all major seal manufacturers in this country and Europe. The evaluation further reflects the author's experience in over 40 years in industry in Europe and in this country.

The salient points may be summarized as follows: The mechanism of sealing interfacial seals is not yet fully understood. A reliable design analysis is not available that will accurately predict the behavior of a new design configuration.

1. Design criteria for end seals embody information related to the structure and character of the material of the rubbing faces. This information should be relevant to seal performance for a developed design. This latter criterion can obviously not yet be fulfilled.
2. Most of the theories developed thus far are predominantly concerned with forces produced by the hydraulic fluids that provide contact forces for the interface. None of the theories is associated with the effects of

Table 2.5 Material Combinations Providing Satisfactory Sealing

Face Material Combinations		Fluids Contained
Carbon-graphite	/Stainless steels	Pentaborane, 50/50 UDMH/ hydrazine
	/Stainless steel, 300 series	Hydrazine
	/Stainless steel, 400 series	Nitrogen tetroxide, water
	/440 Stainless, hardened to RC55-60	LOX
	/Hard-faced 316 stainless	Acids
	/Carpenter 20 stainless	Acids
	/Tool steel, hardened	Oils, hydraulic fluids
	/SAE 1040	Oils, hydraulic fluids
	/Chrome-plated steel	LOX, 50/50 UDMH/ hydrazine, hydrazine, hot combustion products (not extremely oxidizing), nitrogen tetroxide, water
	/Malcomized 316 stainless	Water, oils, hydraulic fluids
	/Stellite 25	Acids, 50/50 UDMH/ hydrazine
	/Stellite facing on stainless	Acids, gasoline, JP-4, kerosene
	/Stellite facing on 316 stainless	Oils, hydraulic fluids, acids, caustics, water
	/Cast iron	Gasoline, JP-4, kerosene, oils, hydraulic fluids
	/Nickel cast iron	Water
	/Bronze	Water, oils, hydraulic fluids
	/Sintered iron or bronze	Oils, hydraulic fluids
	/Hastelloy	Pentaborane, acids
	/Nickel-resist	Gasoline, JP-4, kerosene, water, oils, hydraulic fluids
	/Nitralloy	Gasoline, JP-4, kerosene
	/Nitralloy, hardened tungsten carbide	Oils, hydraulic fluids, LOX, hot products of combustion (not extremely oxidizing), oils, hydraulic fluids
	/Titanium carbide	50/50 UDMH/hydrazine
	/Chromium boride	Acids
	/Glass-filled Teflon	Oils, hydraulic fluids, water, gasoline, JP-4, kerosene
	/Carbon-filled Teflon	Oils, hydraulic fluids, kerosene, nonoxidizing acids, gasoline, JP-4, water, caustics
	/Ceramic	Acids, water, gasoline, kerosene, JP-4
	/Ceramic coating on stainless	Water

Table 2.5 *(Continued)*

Face Material Combinations		Fluids Contained
Graphite-molyb-denum	/Bronze	Oils, hydraulic fluids
Cast iron	/Bronze	Oils, hydraulic fluids
Kentanium	/Kentanium	Fluorine
Tungsten carbide	/Tungsten carbide	Water
25% Glass-filled Teflon	/Hard (RC40) Armco 17-7 PH stainless	90% hydrogen peroxide
Ceramic	/Boron carbide	90% hydrogen peroxide
	/Stellite	Nonmineral acids
	/Teflon	Oxidizing acids
	/Glass-filled Teflon	Acids
	/Ceramics	Acids

Table 2.6 Temperature limits for Seal Materials

Material	Temperature Limit (°F)
Aluminum oxide	2000
Bronze	350
Carbon	950
Cast iron (gray)	650
Cast iron (malleable)	650
Cast iron (ductile)	650
Cast iron (alloyed)	650
Hastelloy	1000
Inconel X	1200
Kel-F	300
Ni-Resist	800
Nitralloy	800
Nylon	290
S-Monel	950
Stainless (410, 17-4 PH)	900
Stellite	1200
Steel	800
Steel (high alloy)	1000
Teflon (glass-filled)	450
Titanium carbide	1800
Tungsten carbide	1500
Cermets (Haynes LT-1, LT-1B)	2000
Flame-plated tungsten carbide	1200
Flame-plated aluminum oxide	1600
Chrome-plated steel	600

wear, which establishes a predictable leakage pattern across the interface.

3. Most of the existing data on friction and wear are predominantly
related to bearing and have little application to rubbing face wear.

4. When two surfaces are in plain contact and in relative motion to each
other, a leakage path may be formed by the interconnection of voids
between the two surfaces. The possible leakage rate then depends on the
size and geometry of these voids. For minimal leakage the mating surfaces must have minimal topographical roughness and waviness. At the
present time a surface flatness should be within two to three light bands
with a roughness of 5 to 8 μin. rms finish for both mating faces. One face
should be noticeably harder than the other.

5. Leakage performance of interfacial seals is exclusively governed by
the degree of initial interface mating and the subsequent effects on the
rubbing faces. Optimal degrees of flatness and surface hardness establish minimum changes in the surface topography.

6. Both rubbing contact and topographical face changes produce forces
that influence the behavior in the interface. These forces are thermally
induced and are capable of altering the leakage rate. Vaporization in the
contact faces interrupts lubrication and the pressure wedge tends to
separate the contact faces. Vaporizing fluids lose their lubricity, create
increased wear, and lead to seal failure.

7. When rubbing contact loads can be kept within controllable limits
through lubrication and satisfactory heat dissipation, the changes in
surface topography as a function of rubbing velocity and exposure time
can be minimized.

8. The information on the selection of a seal design and the corresponding combination of mating face components desirable for a
minimum leakage rate are still empirical.

9. Industrial experience has shown that best seal performance is obtained when one face is harder than the other. Tungsten-carbides have
produced best results among all prospective mating materials selected
for optimal service life.

10. Environmental control devices contribute noticeably to improved
seal life.

XV. Special Design Configurations of
Mechanical End Face Seals

Extensive developments in chemical, petroleum, refining, utility, automotive, and many other related industries have stimulated the con-

tinual search for improved and more sophisticated sealing devices. This search for new technology has resulted in a wealth of new seal configurations and industrial products from which selections can be made.

Developments discussed in the following section are designated by their trade names in connection with their manufacturer. Learning about them will widen the reader's horizon and aid him in making difficult decisions.

A. Circo-Flex Seal

The Circo-Flex seal was designed and developed by Dunley Product, Inc., Cleveland, Ohio. It is designated as a dry-face seal and was originally applied to the Vanton pumps. The manufacturer claims that the seal can handle liquids and slurries. The seal is driven by frictional contact, enabling the floating element to operate at less than shaft speed. The floating arrangement permits speeds of the central floating element to range from zero to shaft speed.

The basic design principles of the Circo-Flex seal are presented in the cross-sectional view shown in Fig. 2.37. The seal consists of three principal parts: a drive element, attached to shaft, combined with a rotating primary seal ring; a floating spring-type element containing two equally arranged primary seal rings, arranged back-to-back; and a stationary element containing a primary seal ring.

The Circo-Flex design has the simple form of a double-seal arrangement. Each single composite part is of unit-type construction and cannot be disassembled any further once the primary seal rings are in space. The "heart" of the design is the floating center element with two parallel mating faces operating free of the shaft and housing bore. The seal is internally mounted and driven by rubbing surface friction on the contacting faces. The frictional torque permits variable speed that allows self-alignment and produces considerably reduced heat.

The stationary element is mounted externally to the housing, thereby protecting the stuffing box chamber from external contamination. The rotating drive element is attached to the shaft using a shoulder for gasket sealing.

The seal ring retainers are very flexible and compensate for contraction or expansion differentials from temperature variations or shaft motions without the need of any elastomers. Face distortions do not occur because of a very resilient sticking taper contact surface between the retainers and the primary seal ring components. This feature simply eliminates mechanical interference caused by shaft torque, wobble, or even end play respective to shaft runout.

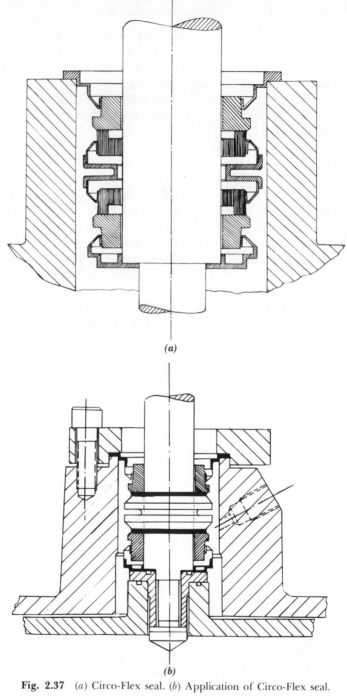

Fig. 2.37 (*a*) Circo-Flex seal. (*b*) Application of Circo-Flex seal.

The annular leaf spring of the central floating element imparts a constant but elastic spring force axially, thus automatically compensating for face wear.

The external surfaces of the primary seal rings are provided with a taper of maximal 7.5% slope to facilitate a press-fit between retainer and seal ring.

The seal unit is not affected by elevated temperatures because there are no elastomeric secondary seal components used.

The manufacturer claims that liquid pressures up to 100 psi can be handled. There is easy access to all seal parts. The application of environmental control devices in the form of cooling, flushing, or quenching is not a problem.

The Circo-Flex seal is ideally suited for reactor vessels with agitator shafts that use seals against the gas phase at moderate process pressures. Lubrication is not critical. Industrial design applications of this seal configuration are given in Section XVI.C of this chapter.

B. Syntron Seal

The Syntron Company of Homer City, Pennsylvania, has developed a mechanical end face seal designated as the Syntron seal. This design also represents a double-seal arrangement incorporating elastomeric secondary sealing components. The seal has gained its industrial recognition through its successful use on boating shafts with propeller attachments. A cross-sectional view of the seal unit is given in Fig. 2.38.

The design represents a double seal and incorporates two rotary seal heads and two stationary seal seats. The seat rings are integral parts of a cartridge that can be attached to the stuffing box of a pump, agitator, or any other device to be sealed. The entire seal unit can be arranged as a complete package, mounted to a sleeve, and then inserted into the stuffing box chamber. The location of the stationary seat rings is fixed; they are embedded in antifrictional angle rings.

The Syntron seal can be used as an internally or externally mounted unit. In either case the system fluid is acting at the ID of the interfaces. Attachment of the seal cartridge to the housing is possible.

The elastomeric secondary sealing components are U-shaped and are mounted to the shaft with an interference fit. This provides a satisfactory seal along the shaft surface against the external atmosphere. A helical spring provides the axial contact force for the interfaces, thus preventing fatigue of the elastomeric U-shaped elements. The elasticity of the U-ring components in combination with the helical spring imparts a high degree of resiliency to the seal unit.

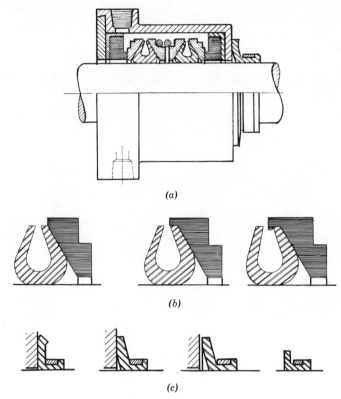

Fig. 2.38 (*a*) Syntron seal assembled. (*b*) Modifications of seat ring attachments. (*c*) Arrangement of flingers.

The elastomeric seal component governs the temperature limitation of this unit with Viton, Kel-F-Elastomer, polyimid, or silicone rubbers. The upper temperature limit ranges from 300 to 450°F.

The rotating seal head is modified in various ways as shown in Fig. 2.38*b*.

When the seal is used in systems handling abrasive fluids, it is feasible to use flingers as exclusion devices that protect the seal chamber from foreign particle penetration. The attachment of flingers is shown in Fig. 2.38*c*. Materials for flingers may be chosen from PTEF, Kel-F, polyimids, and the like.

Environmental control methods are easily applied if required for cooling of the interface.

The cartridge seals in either direction and is independent of the direction of the pumpage flow. The seal is applicable to a wide range of moderate process conditions.

C. Johns-Manville Seal

The Johns-Manville seal is a product of the Johns-Manville Company of Denver, Colorado. This design is generally referred to as the J-M seal. Its basic design principle is illustrated in Fig. 2.39. A gland sleeve is mounted to the stuffing box chamber and is sealed statically against the housing with an elastomeric O-ring. The external end of the sleeve is provided with a shoulder-type flange ring, threaded on the OD to accommodate a gland nut. This nut is designed to incorporate the stationary seal head. An antirotation stud, with the assistance of two guide lugs, prevents the sleeve from rotating.

The sleeve unit has freedom to move axially, forced by an adjustable helical spring that supplies the required contact force for the seal face. The spring is mounted externally, acting through a yoke plate against the back side of the gland nut. The seal seat ring is split to facilitate replacement without removing other parts of the unit.

The mating seal seat is provided with a conical face for contact with the seal head to form the interface. The seat is sealed against the shaft by an elastomeric O-ring.

The J-M seal represents a unique design deviating noticeably from other conventional end face seal configurations. The interface is formed by an edge contact with a corner of the stationary seal head rubbing against a plain designed as a slope in the seal seat.

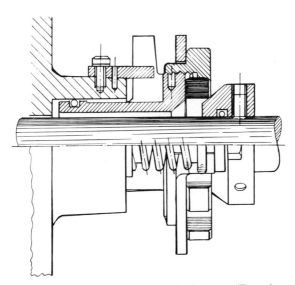

Fig. 2.39 Single-spring unit of Johns-Manville seal

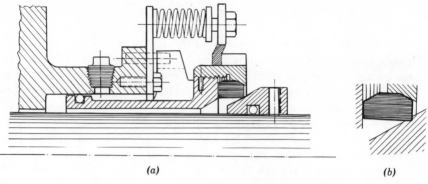

(a) *(b)*

Fig. 2.40   *(a)* Multiple-spring unit of Johns-Manville seal. *(b)* Interface detail.

For larger seal units the use of multiple springs is preferable, as shown in Fig. 2.40. This arrangement does not differ essentially from the design of Fig. 2.39.

These seals are designed to operate satisfactorily at pressures up to 150 psi. Environmental control devices can also be used. The manufacturer claims a wide range of applications in power plants, sewage installations, and the paper, brewing and distilling, gas, marine, chemical, and petrochemical industries.

D.   Magnetic Seals

A new concept of face loading is employed in the design of the magnetic seal. The magnetic seal configuration uses a mechanical end face seal by utilizing a magnet to provide the contact force for the interface. The unit consists of two primary seal rings with the same geometry found in conventional end face seals. One of the primary seal rings is magnetized in such a way that it can act as a magnet despite the rubbing rotation of the unit. In most magnetic seal designs the stationary seat is chosen as the magnet and generally is made of a high-grade Alnico-magnet alloy. If the housing is constructed of magnetic material, it is necessary to provide a nonmagnetic shield ring around the magnet ring seat to prevent a leakage of magnetic flux into the housing; otherwise the basic mechanism of the seal unit is destroyed.

Simple magnetic end face units for a centrifugal pump are illustrated in Fig. 2.41. They are manufactured by the Magnetic Seal Corporation, West Barrington, Rhode Island.

Four basic seal models are shown in Fig. 2.41 representing simple designs of magnetic end face seals. Figure 2.41*a* and *b* show two types.

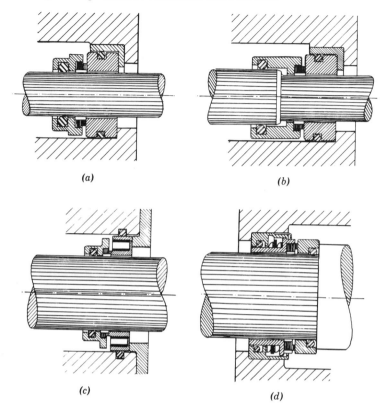

(a)   *(b)*

(c)   *(d)*

Fig. 2.41 Magnetic seal configurations.

The lower half illustrates the seal without a protective metal ring, whereas both upper halves are provided with a metal ring, press-fit to the housing, representing a shield to prevent loss of magnetic flux into the housing. The protective rings must be of nonmagnetic metals (usually stainless steel).

The design of Fig. 2.41c replaces the usual magnetized Alnico seat ring by a steel ring containing a multitude of small magnets inserted into the base metal ring. Size and number of the single small magnets depend on the shaft diameter in conjunction with the process conditions. Whatever the service requirements are, the unit must provide the necessary face contact pressure. The model shown in Fig. 2.41d incorporates a cartridge-type seal with additional wave springs to increase face contact pressure.

Magnetic seals have a number of shortcomings that can significantly limit their range of applications. They are as follows:

1. The magnetic seal cannot absorb any shaft end play exceeding 0.015 in. Beyond this limit separation of the faces occurs because magnetic force is not sufficient to maintain contact. The initial gap is a serious obstacle and any increase cannot be tolerated. The maximum values for shaft runout must be smaller than the flexibility the elastomeric O-ring can provide.
2. Magnetic attraction is a maximum with direct contact. By separating the surfaces the magnetic attraction decreases hyperbolically.
3. Alnico and all other magnetic materials have limited corrosion resistance and protective measures of questionable value may have to be taken.
4. All magnetic materials lose 3 to 5% of the magnetic field intensity within 100 hours as a result of metallic environmental influences.
5. The magnetic field intensity decreases with rising temperature. A thorough investigation of this temperature influence is necessary.
6. The magnetic field in the environment of the seal interface may attract magnetic particles that can contaminate the interfaces.

In summary, magnetic seals exhibit a simple design but have only a limited range of application because of their pressure, temperature, and corrosion sensibility. They can be recommended for simple agitator shaft applications under moderate process conditions.

E. Teflon Bellows Seals

For service with highly corrosive liquids a special seal with nonmetallic PTFE bellows has been developed. Such seals are made by a variety of major seal manufacturers. A typical example of this special configuration is shown in Fig. 2.42. The primary seal ring is designed to snap into the front face of the PTFE bellows. Snap-in rotary face rings can easily be interchanged. They may be made of either carbon-graphite, PTFE reinforced with glass fiber, or molybdenum disulfide.

The seal is externally mounted. The highly flexible PTFE bellows eliminates the seal problem that generally occurs when abrasive fluids are handled. The bellows act as spring to compensate for face wear, shaft runout, and a certain amount of shaft deflection.

Manufacturers claim that the seal is capable of handling pressures up to 100 psi at 75°F temperature or 75 psi pressure at 100°F without

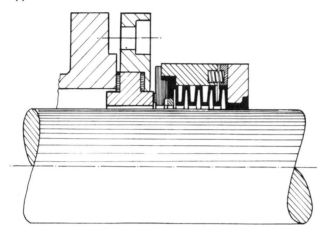

Fig. 2.42   Bellows seal with Teflon bellows, externally mounted.

additional coolants for the mating faces. When it is properly flushed, the seal can be used for handling abrasive liquids. Shaft runouts of up to 0.002 in. can be tolerated.

XVI.   Industrial Applications for End Face Seals

All discussions on mechanical end face seals in this chapter have been based on the prerequisite that the seals are applied to shafts of horizontally mounted centrifugal pumps. However, centrifugal pumps use by far the largest number of mechanical end face seals. The application of these seals need not be limited to centrifugal pumps only. Large numbers of mechanical seals are also used in all types of equipment in chemical, petrochemical, general process, and utility industries for reactors, columns, mixers, dryers, and any other equipment where rotating shafts are employed.

Centrifugal pumps primarily use conventional mechanical end face seals from the shelf. Larger sizes must be fabricated to meet specific requirements. Likewise, special features may be required for reactors, columns with rotating shafts, mixers, dryers, and so on, and, particularly, for reactors operating under severe or corrosive process specifications. This section describes a number of field applications involving special seal designs to familiarize the reader with the wide range of possibilities available.

A. Conventional End Face Seals
for Centrifugal Pumps

Centrifugal pumps are used extensively in the chemical industry, where a considerable number of widely diversified process liquids involving pressure, elevated temperatures, and corrosive atmospheres are encountered. Most sealing requirements can be handled with conventionally available seals. Where simple seal designs do not suffice, double seals with or without environmental control devices are used to provide continuous leak-free service.

Each pump, vertical or horizontal, represents a pressure vessel that is subject to the ASME code. Even though the mechanical end face seal is applied to seal a pressure vessel the seal itself is not covered by the code. Its design is, therefore, left to sound engineering practice of seal manufacturers and plant personnel.

Centrigugal pumps are generally operated at much higher rotating speeds than are used for the shafts in most agitated reactor vessels. Mechanical end face seals are equally suitable for horizontal and vertical centrifugal pumps. The presence of contaminated, abrasive, and corrosive liquids has been discussed in depth earlier and suitable solutions were presented there.

B. Mechanical End Face Seals for Equipment
Other than Centrifugal Pumps

Chemical, petroleum, utility, and many other industries use equipment where countless mechanical seals are employed. Here mechanical seals serve as closures between shaft and vessel to prevent leakage of the vessel contents to the environment. The vessel is usually a unit employing an agitator, a turbine, or other kinds of impeller blades. The vessels are designed with shafts for vertical and/or horizontal entry. Each category requires special design features based on the individual operating conditions that the seal must satisfy. Shafts mounted in vertically arranged vessels are termed *top entering* or *bottom entering* or *top and bottom entering* depending on the entry location. There are also vertical vessels using a shaft that enters the vessel from the side in a horizontal direction. This design is termed *side entering*. Finally, there is a category of vessels mounted in a horizontal position in which the shafts enter either on one side or on two sides.

Figure 2.43 illustrates four basic types of vessels with a rotating shaft used as mixer, agitator, and so on. Figure 2.43*a* shows a vessel with a top entry shaft, which is by far the dominating category of process vessels

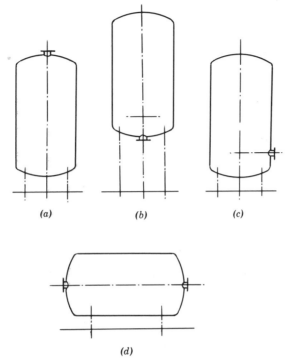

Fig. 2.43 Basic categories of reactor vessels with shaft entries. (*a*) Top entry. (*b*) Bottom entry. (*c*) Side entry. (*d*) Horizontal double-side entry.

used in practically all process industries. It requires one seal as a shaft closure, which seals the gas phase above the liquid against the external atmosphere. Dry-run can be prevented by adequate environmental control measures. The seal's accessibility makes maintenance easy, since the vessel contents do not have to be removed.

Vessels with bottom entry, Fig. 2.43*b*, are more difficult to handle. Their seal is always under fluid pressure and the vessel contents must be removed when repair work is required or the seal has to be replaced. The seal is located where the bottom outlet should be, making it difficult to empty the vessel completely, especially when highly viscous polymers are involved.

Vertical vessels are also used with side entry of the shaft, as in Fig. 2.43*c*. This design has many disadvantages from the mechanical and process standpoint. In chemical and related process industries they are used only where other designs are not feasible. For mechanical end face seals this configuration is impractical because agitators with horizontal shaft arrangement are subject to vibration due to shaft overhang, and mechanical seals are very susceptible to these shaft movements. In this design the seal is always exposed to liquid pressure.

The horizontal vessel with double side entry, Fig. 2.43d, offers many machine design difficulties. Vessels of this nature usually employ a long, large, and heavy shaft with all kinds of mixing devices. When handling very viscous polymers, the shaft may have to move several tons of load. The balance between shaft and vessel is very intricate, particularly with respect to the weight of the container provided with a heat jacket. Either the shaft is considered as the central part and the vessel must be aligned against it, or the vessel is the centering part and the shaft must be aligned against it. In both cases the design is intricate and requires considerable experience for proper alignment and shaft balance.

Mechanical end face seals are usually not applied to this type of vessel. Where cooling of the shaft is applicable, elastomeric seals are preferable. The visco pump seal is another possibility for solving the intricate seal problem. The relatively slow rotation of these shafts does fortunately facilitate the seal problem.

A seal for heavy-duty service with a shaft 6 in. in diameter mounted in vertical position with top entry is shown in Fig. 2.44. The shaft designed for a length of 20 ft is consequently subject to noticeable deflection. The vessel handles toxic liquids. In the original design the vessel was sealed with packings that failed frequently. After changing to a mechanical end

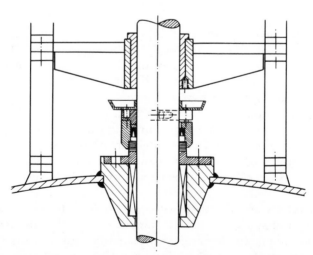

Fig. 2.44   Vertical agitator shaft—top entry for heavy duty [From (14), courtesy of *Chemie Ingenieur-Technik*].

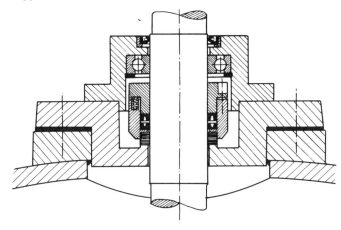

Fig. 2.45 Vertical agitator vessel—top entry [From (14), courtesy of *Chemie Ingenieur-Technik*].

face seal, the operation was successfully serviced. The major contribution to the success was the addition of a large bearing in close proximity to the seal. The rigid bearing provided a relatively deflection-free environment for the seal, which could now operate under normal conditions. The rigidity of the heavy shaft made it possible to extend the critical length for the shaft deflection away from the seal environment. The multiple-spring arrangement in connection with the elastic lips of the secondary seal components imparts sufficient flexibility to the primary seal head. The original packing is used for safety purposes.

Figure 2.45 illustrates another agitator vessel with top-entry seal closure. This vessel (14) was also initially sealed by conventional packing and frequent failures were experienced, practically every second day of service. After replacing the packing by a mechanical end face seal and mounting a ball bearing in close vicinity to the seal, the system operated normally without interruption. The U-cup rings as secondary components for the mechanical seal were chosen to prevent the formation of a vacuum on top of the vessel. The lip seal mounted above the ball bearing protects the bearing and seal arrangement from dust penetration. The vessel was serviced at a pressure of 4 atm and a shaft rotation of 50 RPM.

An interesting approach for sealing a gas atmosphere above the liquid level of a vertical vessel with a top-entry agitator shaft is shown in Fig. 2.46, as reported by G. Diefenbach (14). The design uses two primary seal head rings (*A, B*), both rotating with the shaft and mounted in concentric arrangement with the shaft. Both contact faces of rings *A* and *B* mate with the face of the stationary seal ring seat *C*. Lubricant is

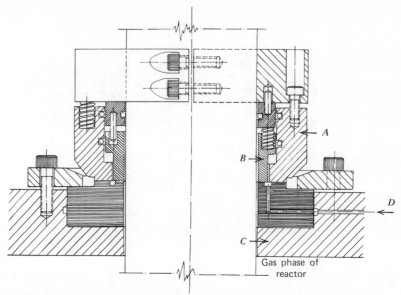

Fig. 2.46 Vertical agitator vessel—seal for special gas phase [From (14), courtesy of *Chemie Ingenieur-Technik*].

introduced through bore D at a pressure at least equal to the pressure inside the vessel to prevent possible dry-run. Simultaneously it provides cooling of the seat ring. The annular ring space between rings A and B is open to the coolant lubricant, which also cools rings A and B. The interface between rings B and C seals against the internal gas pressure, and the interface between A and C seals against the external atmosphere. The device is designed to seal a gas phase above the liquid level of the reactor vessel under a pressure of 150 psi and at a temperature of 200°F.

The design of Fig. 2.47 illustrates an internally mounted seal for a vertical vessel with top entry of the agitator shaft, with a shaft length of 15 ft and a shaft diameter of 6 in. Service conditions were relatively moderate at a rotation of 50 to 80 RPM. During initial operations the seal leaked badly before the ball bearing was placed in close proximity to it. After modifying the bearing location and placing it directly above the seal, the shaft deflection was brought under control and measured well below 0.030 in. This amount was simply absorbed by the seal without leakage. Reports indicate that after the displacement of the bearing the seal was in service for several thousand hours without developing leakage difficulties, an excellent performance considering badge-type operations in a very toxic atmosphere with excessive shaft length.

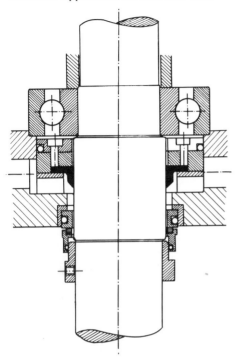

Fig. 2.47 Seal for vertical shaft with top entry [From (14), courtesy of *Chemie Ingenieur-Technik*].

An example of a vertically arranged top-entry agitator shaft seal handling intricate process conditions is illustrated in Fig. 2.48. This configuration uses a unitized double mechanical seal assembled along a sleeve that is then slipped over the shaft as a complete unit. The separate outside assembly as a unit guarantees that all parts are mounted properly without any damage to the mating faces. Relatively little heat is transmitted from the shaft to the sleeve since the annular air gap is a good insulator. Cooling and lubrication is supplied in sufficient quantities, keeping the unit from overheating.

The primary seal rings, both stationary and rotating, are sealed by elastomeric O-rings serving as floating bearings. This provides a high degree of flexibility, greatly dampening vibrational disturbances. Consequently, true face alignment is not critical. As reported by Diefenbach (14), this device is capable of taking radial side deviations up to 0.040 in.

For a series of large autoclaves a seal configuration was used, as illustrated in Fig. 2.49. A double-seal arrangement was chosen and mounted back to back. The agitator shaft had a diameter of 6 in. and a length of 10 ft rotating at a speed of 120 to 160 RPM under a pressure of

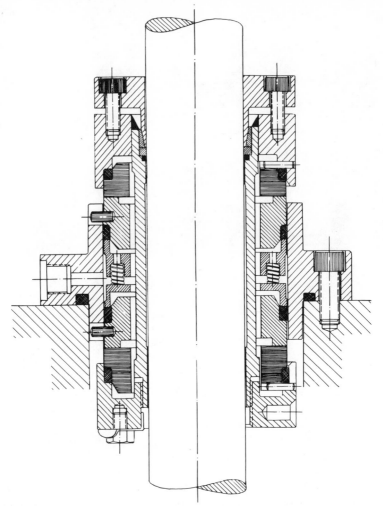

Fig. 2.48 Double-seal arrangement—top entry [From (14), courtesy of *Chemie Ingenieur-Technik*].

300 psi at 375°F temperature conditions. The autoclaves were operated as top-entry vessels.

The seals were assembled on a sleeve together with the ball bearing in one cartridge to provide accurate assembly with minimum deflection. Cooling and lubrication were supplied to keep the unit at temperatures tolerable by the secondary sealing components. The autoclaves have given excellent service without disturbances over a period of daily batch operations for several years.

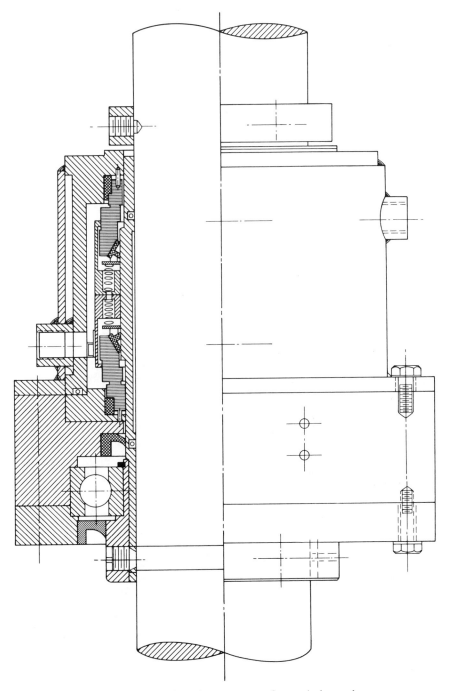

Fig. 2.49 Double-seal arrangement for vertical autoclave.

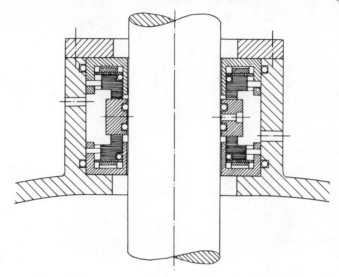

Fig. 2.50 Double seal for horizontal high-speed centrifuge [From (51), courtesy of *Chemie Ingenieur-Technik*].

A double-seal arrangement for a high-speed centrifuge is shown in Fig. 2.50 using a mutual seal seat as rotating component. Special wave-type springs impart the face contact pressure. This arrangement provides a short seal unit despite double-seal design.

An interesting design configuration is illustrated in Fig. 2.51, which was developed by NASA for a highly toxic and explosion hazardous environment. One seal is mounted internally and the other externally, with a double ball bearing between the two seals. Each of the seal units is mounted on a sleeve and can be assembled separately with a high degree of accuracy without damaging the interface contact areas. High accuracy of shaft alignment and true rotation was the basic prerequisite for trouble-free service of thousands of hours. The environmental control devices supplied sufficient fluid for cooling and lubrication.

The vertical double-seal configuration in Fig. 2.52 seals a vertical agitator shaft in top-entry position; it was developed by NASA to seal a very critical and toxic environment. A ball bearing is located between the two seal units to stabilize the shaft. Rigid flanges are employed to secure stability. The seals are assembled on a sleeve, facilitating assembly and alignment. Sufficient space is provided for employing environmental control device systems, controlling interface environment and temperature. NASA reports that seals of this type of design configuration have been used successfully in rocket service facilities.

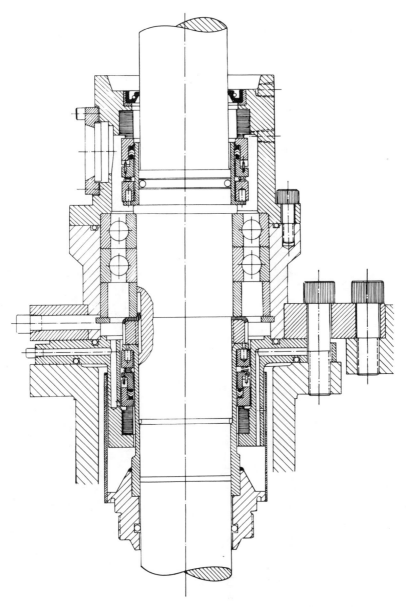

Fig. 2.51 Special NASA seal for vertical shaft.

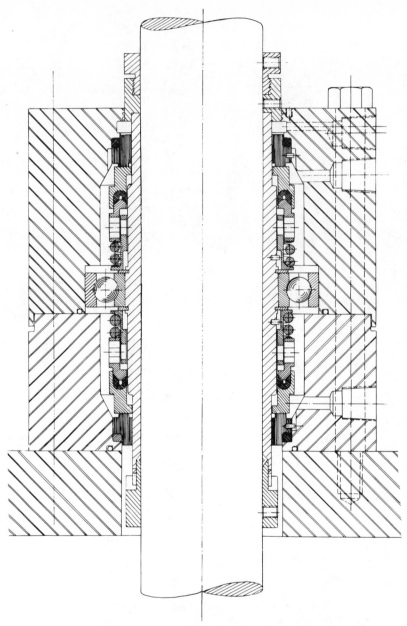

Fig. 2.52 Vertical agitator shaft (NASA design).

C. Applications for Magnetic Face Seals

Two examples of magnetic end face seals are illustrated in Figs. 2.53 and 2.54, sealing vertical shafts with top entry for relatively moderate vacuum service conditions. In both cases an oil sump is used to arrange the seal head units. O-rings provide relatively good flexibility. The unit of Fig, 2.54 is used to seal a system handling liquid sodium for nuclear reactor operations. The oil barrier prevents atmospheric air from penetrating the system. The design of Fig. 2.55 illustrates two magnetic seals mounted back to back in a double-seal arrangement. Cooling and secondary system fluid cycle are not indicated.

XVII. Summary of Concepts To Be Considered in Selecting Mechanical End Face Seals

Extensive information has been presented in this chapter offering a substantial basis for the proper selection of mechanical end face seals for a wide range of process conditions. Major seal mechanisms, operating principles and factors influencing seal performance are summarized in the following pages.

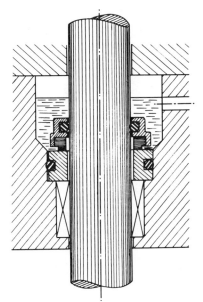

Fig. 2.53 Magnetic seal for top entry shaft.

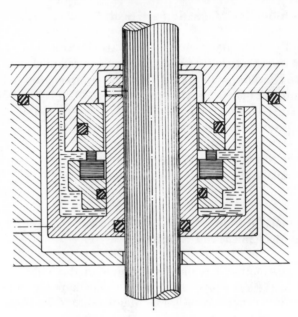

Fig. 2.54   Magnetic seal for liquid sodium (NASA).

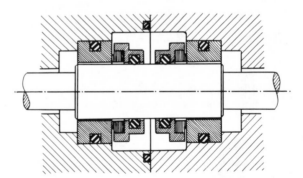

Fig. 2.55   Double-seal arrangement with magnetic seals.

A.   General Considerations

The selection of mechanical end face seals requires thorough evaluation of a number of significant factors, such as shaft diameter, rotational rubbing speed in the contact faces, pressure of the process fluid, service temperature, chemical and physical properties of the liquid constituents involved, environmental control systems, and mechanical and structural properties of the primary seal components. Finally, the properties and resistance characteristics of the secondary sealing elements must be considered.

Experience confirms that the size of the seal has a significant influence on seal performance. Under a set of given operating conditions a small seal may perform satisfactorily whereas the same seal configuration in a large size seal may fail.

The pressure of the process fluid determines the seal design, influences the selection of a balanced or unbalanced configuration, and helps define the degree of balance of the hydraulic force.

Velocity of the rubbing faces results in generation of heat, the amount of which will increase with rising hydraulic system pressure. Consequently, the algebraic product pv reflects a range of pronounced limitations for pressure and/or velocity for any given seal configuration at known process conditions.

Temperature influences seal performance in a number of ways. It affects the stability of the materials and elastic properties of the secondary sealing components. Temperature creates expansion problems of the seal components and is critical to viscosity and lubricity of the system fluids. It also has a negative influence on corrosion resistance. Temperature may finally vaporize the lubricant film in the interfaces. Chemical and physical properties of the process liquids are of primary concern to the stability and performance characteristics of the secondary sealing components, including the primary mating rings. These properties often force a compromise in the selection of the seal materials.

Finally, the best combination of primary mating components will not give satisfactory seal performance if the friction and wear characteristics of the mating materials are not compatible with each other.

1. Characteristics of the Process Liquids

The chemical characteristics of the process liquids play an important role in the selection of an adequate seal combination. Light hydrocarbons, such as liquefied petroleum gases and fluids with high vapor pressure, require liquid pressures in the stuffing box chamber that will ensure a level of 30 to 40 psi above the vapor pressure of the liquid at pumping temperature to prevent vaporization of the film. The pressure-temperature relationship dictates the use of environmental control methods, such as flushing, quenching, cooling, and lubrication because of viscosity variations with varying temperature-pressure conditions.

The volatility aspects of the process liquids may also become significant from the safety standpoint. The application of vents, drain, throttle bushings, auxiliary packings, or other important features may be necessary. In addition, environmental control methods may be required.

Corrosive environments dictate the selection of suitable material combinations for mating faces for two reasons: (1) resistance to friction and wear and (2) face metallurgy for extreme service conditions.

The selection of secondary sealing components is also determined by corrosive characteristics of the process fluids and their prospective temperature.

2. Suitable Combinations of Mating Materials

The carbon-graphites represent an almost universal type of mating material for rubbing conditions. They are available in a wide variety of compositions with a noticeable range of versatility for seal applications. They are widely mated with metals, ceramics, and various kinds of carbides.

Ceramics for seal rings are available in solids and facings and provide excellent resistance to wear, friction, abrasion, and corrosion. However, they are brittle and very susceptible to thermal shock.

Stellites offer excellent service and seal performance. The cast stellite materials are generally better than the face-welded materials.

In the category of steels for primary seal rings surface hardness and resistance to temperature, chemicals, and corrosion are the decisive factors. All steels that exhibit extreme surface hardness conditions can be used. Candidates are tool steels, hardenable stainless steels, nitriding steels, and certain case-hardening steels.

Within the nonferrous metals group a distinction may be made between antifrictional materials, such as bronzes and alloy combinations with bronzes and cast iron, Ni-resist, and carbon-graphites filled with bronzes, Babbit, and many others.

Tungsten-carbides, or carbides of titanium, tantalum, silicon, and other formulated carbides, represent excellent mating materials providing high mechanical strength and extreme surface hardness. They are available in many grades and under a variety of trade names. Silicon-carbide is used either as a layer or in solid form for the entire ring.

When mated with carbon-graphites the tungsten-carbides have extensive applications that provide good seal performance characteristics. Carbides are not recognized as materials that can be run against each other, even under extreme temperature and pressure conditions. When they are chosen, their binders should be dissimilar.

3. Process Liquids and Their Prospective Mating Combinations

The following guidelines for the selection of suitable mating combinations have been established with two basic facts in mind:

1. There are only a few face combinations for a given set of service conditions to which all major seal manufacturers could simultaneously agree.
2. The best seal combination is subject to failure if improperly handled and/or installed. The human factor is thus decisive.

a. Cold Water and Watery Solutions.   Industrial water is very seldom pure and clean. It is generally contaminated by salts, minerals that contaminate the interface. Further, they are corrosive, becoming more active with rising temperature. The following mating combinations are recommended:

ceramic, solid or faced		
carbon-graphites		carbon-graphite
carbon, faced with silicon-carbide	vs.	also filled with lead,
stellites, solid and faced		copper, silver, Babbit, etc.
Ni-Resist, conditionally		
bronzes		

      tungsten-carbides, solid and faced   vs.   tungsten-carbides
      Ni-Resist and cast iron only,
        when rust is not a problem

b. Hot Water.   Hot water produces accelerated corrosion, a deterioration of lubricity, and because of its temperature a danger of vaporization. The interface environment must be kept below 170°F temperature with flushing and cooling mandatory.
    Best mating combinations are

ceramic, solid	vs.	carbon-graphites
ceramic, faced	vs.	carbon-graphites
tungsten-carbides	vs.	carbon-graphites
silicon-carbide	vs.	carbon-graphites

c. Gasolines and Heavy Hydrocarbons.   Conditions are similar to those existing for water or water solutions. Gasolines may contain moisture, which creates a corrosion problem for steels, cast iron, and rust-sensitive metals. Suitable mating combinations are stellites and ceramics versus carbon-graphites.

d. Light Hydrocarbons.   Light hydrocarbons and liquefied petroleums are difficult to handle in that they tend to flash. Pressure must be applied to maintain liquid phase conditions.

For liquids with a specific gravity of around 0.50 and up the system pressure should be 30 to 50 psi higher than the vapor pressure of the liquid pumped. Suitable mating combinations are

stellites   vs.   carbon-graphites

For fluids with a specific gravity less than 0.50 and where the vapor pressure is at equilibrium or where the vapor pressure reaches the pressure of the process fluid the selection of the proper mating materials is critical. Under these conditions tungsten-carbides versus carbon-graphites have given satisfactory service. Carbon-graphite materials, on the other hand, have been found acceptable for both heavy and light hydrocarbons.

e. Oils and Oily Solutions.   Oils basically create no problem for the selection of suitable mating partners. As a general rule, all materials good for water service are also satisfactory for oils. It should be kept in mind, however, that materials such as stellite versus bronze, carbon graphite versus Ni-resist, and cast iron versus bronze are good for oils but not recommended for water service.

Oils with high viscosity offer good lubricity and are not a problem for seals. Major mating combinations are

cast iron   vs.   carbon-graphite, Ni-resist, ceramics, stellites

f. Service with Acids.   Inorganic acids usually have fair lubricating properties resulting from viscosity. Caution is required, since lubricity and corrosiveness vary greatly with temperature concentration and impurity levels. Recommended mating combinations are

ceramics, solid, faced		
stellite, solid, faced	vs.	carbon-graphites
ceramic, solid		
stellite, solid	vs.	glass-filled Teflon
stainless steels		
tungsten-carbides	vs.	carbon-graphites
tungsten-carbides		tungsten-carbides

Glass-filled Teflon should be used with caution. Glass as filler enhances mechanical properties but does not necessarily improve resistance to chemical corrosion.

B. Summary Conclusions

The selection of mechanical end face seals involves numerous steps and requires careful evaluation of all the process conditions involved.

1. Pressure of Process Fluid

The pressure of the process fluid is a criterion for selecting balanced or unbalanced seal designs for both single-seal and multiple-seal configurations. The pressure that can be handled by a single stage of a mechanical seal unit decreases with the diameter of the shaft, along with the size of the seal.

Figure 2.56 gives an approximate guideline for pressure versus seal size relationships generally used by seal manufacturers. Experience indicates that the optimal balance ranges from 60 to 75%, depending on pressure level. With increasing system pressure the balance ratio goes toward fully balanced condition.

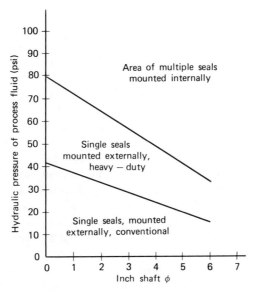

Fig. 2.56 Relationship pressure versus size for mechanical seals.

2. Rotating versus Stationary Seal Ring

In most conventional designs the seal head rotates with the shaft. Balance and rotating velocity decide which of the ring components is to be stationary. The rule is to rotate the head for speeds below 5000 fpm; above this limit the head should be stationary. Furthermore, by simply holding close tolerances on the rotating seat member a more perfect dynamic balance can be attained than is possible through control of all the tolerances of the seal head component parts, including clearance factors, spring effects, flexibility characteristics of elastomers, and others. Rotary seal seats have a simpler geometric configuration of space available for the seal unit.

3. Inboard versus Outboard Mounting

Where service conditions do not require special design configurations, convenience determines whether a seal should be mounted inboard or outboard. If a seal unit is mounted inboard, the hydraulic pressure can be utilized as a force component. All components of the seal unit are then under compressive stress rather than the undesirable tensile stress. Inboard mounting produces an externally acting hydraulic force against the interface. Outboard mounted seals with rotating seal head or inboard mounted seals with stationary seal head produce an internal hydraulic pressure against the interface.

In outboard mounted seals a minimum number of parts are in contact with the process fluid, an important factor in handling highly corrosive environments. This is particularly true when gumming up, precipitation, accumulation, or solidification in the clearance gap may occur. Outboard seals provide accessibility and easy observation of the interface. Leakage alarm devices are readily installed. In external mounting the stuffing box is empty and can, therefore, readily be used for flushing and the like. For equipment with insufficient seal mounting space, external mounting permits freedom from space limitations. Glands and seal seats can always be piloted for installation of outboard seals, regardless of space available in the stuffing box chamber.

Inboard seals are used for all design configurations and experience indicates that pressures up to 5000 psi can safely be handled. Outboard seals, on the other hand, are customarily used for pressures ranging from vacuum to an upper limit of 500 psi. For catastrophic seal failures the outboard mounted seal has a definite disadvantage in that there is no way of preventing the escape of fluid into the atmosphere. Cartridge seals with external elastomeric lip seal devices offer some protection until the pump has come to a stop.

4. Design of Primary Seal Rings

Whether the design configuration of the primary seal rings is made stationary or rotational is principally governed by the mode of application. The specific geometry of the rings is then dictated by the choice of the secondary seal components. For stationary mounting the seat is usually flexibly attached to the gland ring plate.

5. Secondary Sealing Devices

Secondary sealing devices are either of the pushing or of the nonpushing type. The pusher-type components require additional mechanical devices to maintain steady contact of the interfaces in order to compensate for face wear. Typical representatives are O-, V-, and U-rings and wedge rings. The wedge ring generally permits optimal axial motion and flexibility without being restricted by close tolerance requirements.

The nonpusher types of devices comprise the bellows, which are made from elastomers, plastics, or metals. Metal bellows impart their own elastic flexibility and seldom require additional spring devices. Plastic bellows are flexible; however, they have to be designed with additional springs to accomplish mechanical preloading and to prevent fatigue failure caused by excessive torque.

Secondary sealing components have upper temperature limitations beyond which either metal bellows, gaskets, or asbestos combinations must be used. Special graphoil rings provide the same service.

6. Face Loading

Face loading is the result of a partial utilization of the hydraulic pressure and the action of the springs. The selection of the mode of spring loading is a matter of personal preference and essentially depends on the size of the seal, since for large seals single helical springs are no longer applicable. In magnetic seals the only face loading force is provided by the magnetic field intensity.

C. Final Rules for Selection

The final selection of seals requires an intensive evaluation of all available process conditions, including information on single versus double seal, inboard versus outboard mounting, environmental control devices, balance ratios, secondary sealing components, and the like.

As far as mating materials are concerned, the following must be decided on:

1. Materials for primary seal rings as a function of pressure, rubbing velocity, chemical compatibility, and temperature
2. The effect of service temperature on
   a. Structural stability of the mating materials from a metallurgical standpoint.
   b. Resistance against thermal shock when temperature is pulsating.
   c. Secondary sealing elements, which may be limited by upper temperature restrictions.
   d. Lubricity and viscosity, which decrease with rising temperature.
   e. Flashing of the lubricant film at the interface.
   f. Pv values, which impart further pressure-velocity limitations.
3. Environmental control devices to combat overheating, and precipitation of solid and abrasive particles in the interface environment.
4. The use of water, light and heavy hydrocarbons as process fluids, which may create lubrication and flashing problems.

1. Maximum Temperatures of Seal faces

Rubbing faces generate heat that amounts to approximately 50 W/sq in. This is the main reason why the interface contact areas show a higher temperature than the ambient in the fluid environment. Table 2.7 reflects seal face temperatures as reported by G. J. Field (16).

Table 2.7 Actual Seal Face Temperatures

Material	Face Temperature (°C)	(°F)
Stainless steel	300	572
Ni-resist	180	356
Leaded bronze	180	356
Aluminum oxide	180	356
Carbon-graphites	275	527
Glass-filled PTFE	180	356
Stellite, solid	200	392
Stellite on metal ring	180	356
Cast stellite ring	230	446
Solid tungsten-carbide	400	752

2. Mechanical Seal Troubleshooting

Misalignment, static, and dynamic leakage occur if one of the primary mating rings has been mechanically distorted, which may be the result of nonuniform tightening of the bolts of the gland end plate. During misalignment the gland is distorted and the distortion is then transmitted to the seal seat ring. The interface does not mate as two parallel planes but at an angle, producing leakage. High spots are formed, resulting in a wavy wear pattern. Leakage occurs both statically and dynamically.

Distortion by misalignment can be prevented by checking the bolt length exceeding the nut for tightening. A second method is to check that the end face of the gland ring is absolutely perpendicular to the shaft axis over the entire circumference of the shaft.

Mechanical seals behave very much like bearings. Conditions that produce bearing failure also create seal failure. The rotary ring must run parallel with the shaft, whereas the stationary seat ring must be perpendicular to the shaft axis. The gland end face must check perpendicular to the shaft at all times. Gaskets should be used that are practically incompressible. Hand-tight tightening conditions will basically suffice.

Other mechanical forces are hydraulic fluid forces, axially and radially, thermal distortions, axial and radial temperature gradients, and superpositions of these various influences.

Lack of Lubrication. An uninterrupted liquid film in the entire interface is mandatory. Lack of lubrication may occur at the start of pump operation and/or when light hydrocarbons are handled; consequently flushing will be required to supply lubricant.

Overheating. Lack of lubrication and overheating usually occur simultaneously, particularly when the hard face is subject to heat checking. Surfaces then buckle upward and crack. The cracks open up and the edges scratch and destroy the mating rubbing faces. Overheating further affects secondary elastomeric sealing components, resulting in surface hardening, embrittlement, and cracking. The temperature limit of the elastomers must be carefully controlled.

Abrasion Damage. Minute abrasive particles eventually penetrate into the interface and groove the mirror finish of the contact areas, intercept the lubricant film continuity, develop hot spots, evaporate the film by partial overheating, and cause the seal to fail.

These abrasive particles may also accumulate and deposit at various spots along the shaft surface, thus damaging the secondary sealing components and affecting their flexibility. This condition particularly occurs when sealing wedges of Teflon (PTFE) are used because they also embed in the Teflon matrix and produce shaft damage.

Corrosion. The occurrence of corrosive action due to the system fluids can be simply recognized by signs of pitting on metal parts or even destruction of the entire metal surface. Seal parts are smaller than pump components and, therefore, corrode more rapidly than the pump parts. In case of doubt, a metal with higher corrosion resistance should be chosen.

References

1. Barunke, R. D.
   Heated Shaft Seal for a Liquid Nitrogen Centrifugal Pump
   Linde Intercompany Report 14, Wiesbaden, Germany, 1969.
2. Bialkowski, L. S.
   "B. F. Goodrich Metallic Seal"
   B. F. Goodrich Aviation Products Experimental Report 10086, November 1960.
3. Bialkowski, L. S. and Stachiw, J. D.
   "Types and Characteristics of High Pressure Seals"
   Mach. Des. (March 1965).
4. George, R. L. and Elwell, R. C.
   Bibliography of ASTIA Literature on Seals
   Report 63 GL 101, Technical Information Series, G. E., July 1963.
5. Allen, G. P.
   Bibliography—Rotating Seals 1968–72
   National Aeronautics and Space Adminstration, Lewis Research Center, Cleveland, Ohio.
6. Berkich, A.
   "Mechanical Seals Theory and Criteria for Their Design"
   Prod. Eng. (April 1950).
7. Boon, E. F.
   "Enige obmerkingen over abdichting van Chemische Pumpen"
   Chem. Ing. Tech., 4 (1955).
8. Bowman, R. E.
   "How Radiation Affects Engineering Materials"
   Mater. Des. Eng., Manual 173 (July 1960).
9. Buchter, H. H.
   Apparate und Armaturen der Chemischen Hochdrucktechnik
   Springer-Verlag, New York, 1967.

10. Craden. W. H.
    "The Mechanical Face Seal"
    University of Tennessee, Knoxville, TN., U.S. Dept. of Commerce, TID-15988, 1965.

11. Coopey, W. H.
    "New Seal for Superpressure Shafts"
    Chem. Eng. (September 1965).

12. Davis, M. G.
    "The Generation of Lift by Surface Roughness in a Radial Face Seal"
    Paper E4 Int. Conf. Fluid Sealing, British Hydromechanic Research Association, April 1961.

13. "Design Data on O-Rings and Similar Elastic Seals"
    WADC-TR-56-272, Parts I–IV, November 1956–March 1960.

14. Diefenbach, D.
    "Gleitringdichtungen für Drehende Wellen"
    Chem. Ing. Tech., **7** (1954).

15. Fekete, K.
    "Similarity Consideration for Radial Face Seals"
    Escher Wyss News, **34,** No. 2/3 (1961).

16. Field, G. J.
    "Seals that Survive Heat"
    Mach. Des. (May 1975).

17. Gardner, J. F.
    "Recent Developments on Non-Contacting Face Seals"
    ASLE Paper 73 AM-88-3, 1973.

18. George, R. L. and Elwell, R. C.
    "Study of Dynamic and Static Seals for Liquid Rocket Engines"
    G. E. Comp. Advance Technology Laboratories, Contract NAS 7-102, February 1963.

19. Giles, O.
    "New Design Concepts and Materials for Shaft Seals"
    Prod. Eng. (February 1956).

20. Greiner, H. F.
    "Rotating Seals for High Pressure"
    Prod. Eng. (February 1956).

21. Hull, J. W.
    "Development of Metal Seals for Hydraulic Actuators"
    Boeing Airplane Company, Document Na D 2-2554 Seattle, WA, March 1958.

22. Hummer, H. B.
    "Requirements for Sealing Liquefied Petroleum Gases"
    ASLE, Paper 73-AM-38-1, 1973.

23. Iny, E. H. and Cameron, A.
    "Load Carrying Capacity of Rotary Shaft Seals"
    Paper A2, Int. Conf. Fluid Sealing, British Hydromechanic Research Association, April 1961.

24. Ishiwata, H. and Hirabayashi, H.
    "Friction and Sealing Characteristics of Mechanical Seals"
    Paper D5, Int. Conf. Fluid Sealing, British Hydromechanic Research Association, April 1961.

25. Iwanami, S. and Tikamori, N.
    "Oil Leakage from an O-Ring Packing"
    Paper B2, Int. Conf. Fluid Sealing, British Hydromechanic Research Association,
    April 1961.

26. Keller, G. R. and Staffors, P. H.
    "Metal Dynamic Hydraulic Seals"
    Paper presented at SAE Annual Meeting, January 1957.

27. Langhaar, H. L.
    Dimensional Analysis and Theory of Models
    Wiley, New York, 1961.

28. Lindeboom, H.
    "Hydrostatic Design Concepts Applied to Dynamic Sealing Devices"
    ASME Paper 67-Pet-31

29. Mayer, E.
    "Unbalanced Mechanical Seals for Liquids"
    Paper E2, Int. Conf. Fluid Sealing, British Hydromechanic Research Association,
    April 1961.

30. Mayer, E.
    Mechanical Seals
    VDI-Verlag, Düsseldorf, Germany, 1966.

30a. Mayer, E.
    Mechanical Seals
    Newness-Butterworths, London, England, 1977.

31. Morris, T. L.
    "Bellows Mechanical Seals"
    Chem. Process Eng. (March 1972).

32. George, R. L. and Elwell, R. C.
    NASA N 63-19597,
    Bibliography of open literature on seals in general.

33. Elwell, R. C., Bernd, L. H., Fleming, R. B. et al.
    "Studies of Special Topics in Seals"
    NASA N 63-19598.

34. NASA
    Final Report, **2** (February 1962–February 1963).

35. NASA
    Numerous reports on basic seal technology

36. Nau, B. S. and Turnbull, D. E.
    "Some Effects of Elastic Deformation on the Characteristics of Balanced Radial Seals"
    Paper D3, Int. Conf. Fluid Sealing, British Hydromechanic Research Association
    (April 1961).

37. Norton, R. D.
    "Mechanical Seals for Handling Abrasive Liquids"
    Chem. Eng. (September 1956).

38. Paxton, R. R. and Strugala, E. W.
    "Wear Rates of Impregnated Carbons, Sealing Warm Water"
    Paper presented at 28th Annual Meeting of ASLE, Chicago, May 1973.

39. Paxton, R. R., Massaro, A. J., Strugala, E. W.
"Performance of Siliconized Graphite as a Mating Face in Mechanical Seals"
ASLE Paper No. 76-AM-68-3, 1976.

40. Pien, L. D.
"The Application of Tungsten-Carbide for Mechanical Face Seals"
Paper presented at 20th Annual Meeting of ASLE, Detroit, May 1965.

41. Pinkus, O. and Sternlicht, B.
Theory of Hydrodynamic Lubrication
McGraw-Hill, New York 1961.

42. Rankin, R. D.
"Give Mechanical Seals a Chance"
Chem. Eng. No. 1 (1952).

43. Ramsey, W. D. and Zoller, G. C.
"How the Design of Shafts, Seals and Impellers Affects Agitator Performance"
Chem. Eng. (August 1976).

44. Romine, C. F. and Morley, J. P.
"Thermal Distortion, the Frequently Overlooked Factor in Balancing Mechanical Seals"
Mach. Des. (January 1964).

45. Schaffer, R.
"Gleitringdichtungen für Kreiselpumpen der Chemischen Industrie"
Chem. Ing. Tech. No. 4 (1957).

46. Schoenherr, K.
"Materials in End Face Mechanical Seals"
Mach. Des. (January 1964).

47. "Profiting by Mechanical Seal Advances"
Chem. Eng. (March 1964).

48. "Materials in End Face Seals"
ASME, Paper 63-WA-254 (September 1964).

49. "Design Terminology for Mechanical End Face Seals"
SAE Conference Proceedings, Los Angeles, CA, May 1965.

50. "Fundamentals of Mechanical End Face Seals"
Iron Steel Eng. (November 1965).

51. Schwab, A. and Linck, E.
"Gestaltung von Rührern und ihren Antrieben sowie Abdichtung der Triebwerke"
Chem. Ing. Tech. No. 11 (1956).

52. Shaws, M. C. and Mocks, E. F.
Analysis and Lubrication of Bearings
McGraw-Hill, New York, 1949.

53. Spores, A. G.
"Controlling Pollution with Mechanical Seals"
ASLE-J., **31**

54. Summers, D. and Smith, W.
"Laboratory Investigation of the Performance of a Radial Face Seal"
Paper D1, Int. Conf. on Fluid Sealing, British Hydromechanic Research Association, 1961.

55. Tankus, H.
"End Face Seals in Abrasive Service"
Paper presented at ASLE 18th Annual Meeting, New York, May 1963.

56. Tao, L. N. and Donovan, W. F.
"Through Flow in Concentric and Eccentric Annuli of Fine Clearances with and
without Relative Motion of Boundaries"
Trans. Am. Soc. Mech. Eng. (1955).

57. Thayer, J. H.
"Construction of Mechanical Seals"
Nat. Assoc. Corros. Eng., Houston, **5** (3), 1966.

58. Tracy, H. E.
"Select Best Pump Seal"
Chem. Eng. (April 1957).

59. Tyrek, J. J.
"The Application of Mechanical End Face Seals for Hot Water Service"
ASLE-J., **29** (1961).

60. Wilkes, III, G.
"Hidden Sources of Seal Trouble"
Lubr. Eng. (September 1973).

61. Wilkinson, S. C. W.
"Rotary Shaft Sealing under Extremes of Temperature and Pressure"
Process Eng. (February 1970).

62. Wood, H.
"Mechanical Seals in the Chemical Industry"
Trans. Inst. Chem. Eng., No. 32 (1954).

63. Several Reference Issues on Seals
Mach. Des., (1973, 1974, 1975).

CHAPTER 3

Radial Seals for Rotating Shaft

The classification chart for seals, presented in Fig. 1.2, lists three basic types of radial seals for rotating shafts, such as lip ring seals, circumferential split ring seals, and packings. These three categories of potential seals are also applied to seal reciprocating shafts, as is discussed in Chapter 5.

PART 1. Lip Ring Seals for Radial Sealing

In modern industrial equipment lip ring seals have reached a high degree of technical perfection and have conquered an unusually wide field of applications in practically all sectors of industry.

I. Introduction

Circumferential lip ring seals represent a type of seal configuration for rotating shafts characterized by a component that provides a highly flexible leg as typical identification that is urged in a radial direction against the surface of both the rotating and reciprocating shaft.

Circumferential lip ring seals are typical representatives of elastomeric seal devices within the category of rotary shaft seals encountering interfacial flow. The high state of the art is attributed basically to low cost, excellent performance characteristics, persistently durable materials, simple design, high degree of reliability at minimal nearly predictable leakage rate, and easy replaceability.

As a result of the unusual volume of production and industrial demand there are numerous manufacturers providing the market with

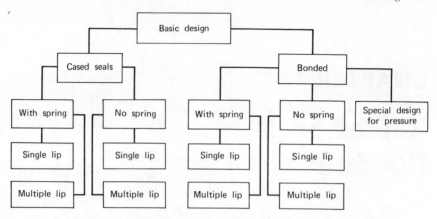

Fig. 3.1 Classification chart for circumferential lip ring seals.

countless seal design configurations of all kinds readily seen in the classification chart for lip seals, illustrated in Fig. 3.1.

II. Classificaton of Circumferential Lip Seals

The multitude of circumferential lip ring seals, also often referred to as oil seals or shaft seals, may be classified into two basic major categories, the cased seals and the bonded seals, as illustrated in Fig. 3.1. Each of these categories is again subdivided into groups with and without springs and groups with one-lip or multiple-lip configuration.

III. Principles of Operation

The initial purpose of any lip seal design is primarily to retain oil and other lubricants in equipment operating with rotating shafts. This definition also includes reciprocating and/or oscillating shafts, which are discussed in Chapter 5.

Circumferential lip seals are suitable for a wide variety of sealing requirements. They can be operated with almost all the technically used oils and hydraulic fluids from atmospheric to medium pressures across a considerable range of temperature. They are not sensitive to moderate shaft misalignments, dynamic shaft runout, and even variations in rotational shaft speeds.

Radial lip seals are further used for excluding dirt, abrasive particles, or liquids from the container to be sealed. Their design is also suited for sealing off vapors at low system pressure in agitated reactor vessels. Sealing function is the result of an interference fit between a flexible sealing element (leather, rubber, PTFE, Kel-F, or polyimid) and the rotating shaft. Since rubber generally interacts with lubricants, which it causes to lose stability and elasticity, a spring is often added to maintain and enhance the contact pressure of the lip. Fluid and vapor retention require a precise amount of lip contact pressure in conjunction with a minimum interference fit. Satisfactory interference must be provided for the lip, rubber tension and compression set, lip swell caused by interaction with hot oils, vapor, and other deteriorating factors. In addition, interference is mandatory to compensate for shaft eccentricity and shaft-to-bore misalignment.

The interference fit forces the lip to exert a mechanical force on the shaft surface, which is used to control the thickness of the film of the lubricating fluid that must exist between the lip and the shaft. Proper seal function and long seal life expectancy are obtained if the lubricant film is continuously maintained. Experience proves that the best film thickness is found around 0.0001 in. Leakage results from a greater film thickness. If the film becomes thinner than 0.0001 in., the lip is subject to wear caused by increased friction, which in turn generates heat, producing the so-called stick-slip oscillation, as described by L. A. Horve (8).

Stick-slip is a phenomenon created by friction. It represents a jerky motion that occurs when one surface is dragged across another. The mechanism of stick-slip is still not fully understood, but it appears that it is caused by dry contact, produced in local spots and dispersed along the surface, occurring occasionally and then sheared. An alteration of gripping and releasing may take place, producing a gradual surface wave that finally results in failure and leakage.

An adequate film thickness is critical. When the contact pressure of the lip increases, the film thickness decreases. With rising shaft speed the fluid temperature increases and the viscosity decreases, which in turn causes the film to become thinner. As the film thickness decreases, lubrication weakens and friction increases, causing the temperature to rise. This cycle continues until the film thickness is insufficient and seal failure occurs.

For a given set of operating conditions the elastomeric lip exerts a defined pressure load on the shaft surface. With the alteration of temperature and shaft speed this contact load also varies. To prevent premature failure a spring is used to force the lip to maintain continuous contact with the shaft and to compensate for motion irregularities.

Lubrication of the lip is mandatory. Start-up conditions, therefore, are critical, because there is always a short time before the lubricant reaches the lip contact area. This phenomenon of minute intervals of possible dry-run must be considered by the seal designer to provide seals that can withstand this critical period without damage or failure of the seal edge.

The average oil seal overcomes this critical phase and performs without leakage. However, some tolerable leakage always occurs. L. A. Horve reports that approximately 80% of all synthetic seals leak about 0.002 g/hr, which translates to about one drop every 11 hr. Since this is hardly measurable in the field, this rate of leakage does not create any problem.

Some 12 to 15% of all lip seals show a leak rate of 0.002 to 0.1 g/hr, which is considered as borderline in most applications. Very few synthetic seals leak in excess of 0.1 g/hr. If they leak they are either misaligned, defective, or not properly specified.

IV. Design Features for Conventional Radial Lip Seals

Lip ring seals for rotating shafts come in such a wealth of configurations and varieties of field applications that it is important to discuss in more detail their basic design features. In this chapter emphasis is given to sealing functions for rotating shafts. Details on requirements of seals for reciprocating shafts are presented in Chapter 5.

A. Design Parameters

The radial lip seal is characterized by a flexible lip, urged radially against the rotating shaft, designed for a certain defined amount of dimensional interference. There are three basic parameters that change during the life of a lip seal. With the details of Fig. 3.2 in mind, these parameters are

1. Radial load, imposed on the shaft surface by the flexible lip through interference fit. This load must prevent leakage without preventing the formation of the lubricant film.
2. Interference. The seal requires sufficient interference in order to follow the shaft eccentricities without influencing the film.
3. A sharp, nick-free lip at the area of contact with the shaft is essential.

These three parameters are essentially influenced by the characteristics of the materials used for the lip rings. Once the characteristics of the materials are subject to changes, the parameters themselves undergo changes, which is the case with alteration of temperature as a function of the time, as reported by O. Ostmo (13).

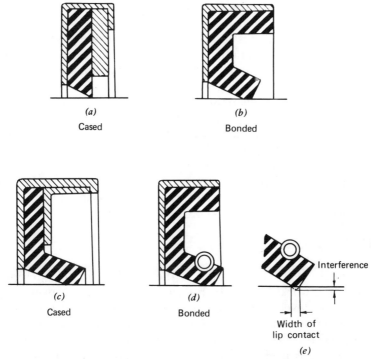

(a)
Cased

(b)
Bonded

(c)
Cased

(d)
Bonded

Interference

Width of
lip contact

(e)

Fig. 3.2 Basic concepts of lip seal designs (Courtesy of Crane Packing Company, Morton Grove, Illinois).

B. Design Considerations

As indicated in Fig. 3.1, lip seals are categorized into two groups, cased and bonded seals. There are many design modifications that may apply to type and shape of the sealing lip, the design of the body, and the arrangement of the lip. Several of the basic design configurations are available with the sealing lip on the OD of the ring body; these are used when the shaft is stationary and the housing is rotating.

Cased seals are those in which the sealing element is retained in a precision-manufactured metal casing by means of the case construction. This type of seal is available with either a leather or a synthetic elastomeric sealing element.

Bonded seals are generally made only with synthetic sealing elements permanently bonded to the metal case structure.

Both categories are offered in either single- or multiple-seal design with a wide range of design modifications.

Leather seals are very effective and less sensitive to shaft surface finish variations and can function satisfactorily on rougher shaft surfaces than synthetic seals. Because it is absorbent, leather provides a self-lubricated seal that is capable of operating with minimal or no additional lubrication for a given length of time. Both the allowable rotational shaft speed and the appropriate operating temperature are lower for leather seals than is tolerable for synthetic seals.

Synthetic seals must be specified where shaft speeds exceed the 2000 fpm limit. Since they are more resilient than leather they further allow more shaft runout, whip, deflection, and eccentricity. For synthetics the degree of surface finish must be higher than that for leather seals.

Certain synthetic rubber compounds permit higher operating temperatures than leather. Manufacturers provide specific information on this subject.

1. Single-Lip Seals

The single-lip seal configuration is the simplest lip seal design and represents the workhorse of the lip seal industry. The unsprung single-lip design was developed to contain viscous fluids or grease at relatively low to moderate shaft speed. When fluids with low viscosity must be handled, which is necessary for many industrial lubricant oils at higher shaft speeds, it is preferable to add a garter spring to compensate for wear of the elastomeric sealing lip in connection with gyration irregularities of the shaft. Unsprung lip seals are the least expensive of all basic lip seal configurations.

Figure 3.2 reflects a series of typical designs of single lip seals of both cased and bonded categories. Figures 3.2a, b, and c are devices that operate without springs. Figure 3.2d shows a lip seal provided with a garter spring to enhance the contact force of the lip with the surface of the shaft.

The spring-loaded configurations are the most commonly used lip seal devices. Their conventional purpose is the retention of oils or grease at all kinds of rotational shaft speeds.

Most conventional lip seal designs allow operating pressures of 10 to 15 psi. Some seal configurations are suited to handle pressures up to 25 psi. Special designs may go as high as 500 psi.

A variety of common lip seals are provided with an extra lip directed toward the external atmosphere to prevent penetration of dirt or dust into the system. Figure 3.3 illustrates a single-lip device with an additional dust prevention lip. Figure 3.4 represents a seal capable of operating at 500 psi as claimed by the Crane Packing Company of Chicago, Illinois.

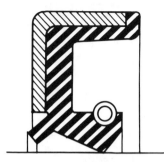

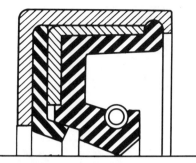

Fig. 3.3 Lip seal with additional lip for dust prevention (Courtesy of Crane Packing Company. Morton Grove, Illinois).

Fig. 3.4 Lip seal for 500 psi, Crane Packing Company (Courtesy of Crane Packing Company. Morton Grove, Illinois).

2. Double-Lip Seals

Double-lip seals come in a wide range of design configurations in either category, cased or bonded. They can be installed in tandem as parallel form or dual opposed, or they can be arranged in a combination of a series of sealing elements, as illustrated in Fig. 3.5. These designs represent basic concepts for a double-lip seal that can easily be modified in a variety of ways. Whatever the specific design features may be, the concept remains always the same.

V. Modifications of Lip Seal Configurations

Despite the relatively advanced development of lip seal technology, continuous attempts are made to improve seal efficiency further. Helixseals, hydroseals, bidirectional seals, and wave seals are typical examples of progressive design developments in recent years.

A. Helixseals

All lip seals are interference seals and are generally designed as unidirectional seals. They are characterized by a resilient member that also seals in static condition (i.e., before the shaft rotation is started). The sealing member is made of resilient material that can follow the irregular shaft movements, such as deflection, runout, tolerance variations, and the like.

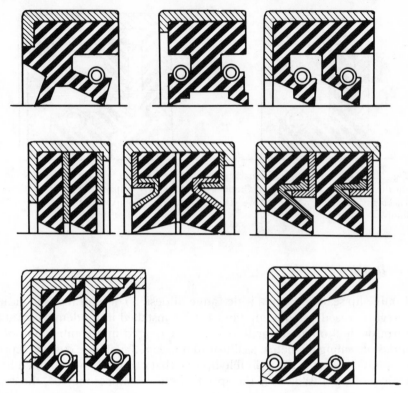

Fig. 3.5 Double-lip seal configurations (Courtesy of Crane Packing Company, Morton Grove, Illinois).

1. Design Principles

L. H. Weinand (20) describes the helixseal as a device in which the elastomeric member is provided with grooves to improve the sealing effectiveness. Figure 3.6 illustrates the basic concepts of the design.

The external back side of the lip ring is provided with ribs, spaced equally around the entire circumference of the ring. These ribs are designed so that they act as miniature hydrodynamic pump elements when they rotate and touch the shaft surface without interfering with the lubricant film when they ride along the shaft. The ribs have a twofold function. First, the lip has to prevent leakage of the oil across the lip surface. Second, the ribs act as a pump that forces the fluid back to the oil reservoir if it leaks through defective zones either in the seal lip or along the shaft contact surface.

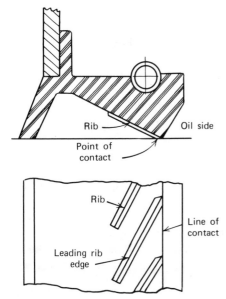

Fig. 3.6 Helixseal design [From (20), courtesy of *Machine Design*].

The ribs permit the seal to function at lower lip contact pressure than is possible for conventional lip seals with a sharp-edged lip. According to Weinand, for optimal film thickness the lower limit for lip pressure is on the order of 50 psi. Hydroseals can operate at lower lip pressure because they utilize dynamic action to provide sealing. The lip pressure has to supply only enough contact force to secure static sealing and follow shaft surface irregularities.

Helixseals function well at atmospheric pressure conditions or only slightly above atmospheric. The dynamic motion produced by the ribs has to supply only the pressure required to force the leakage flow to move back to the oil section. As far as seal life is concerned, the same principles are valid as indicated for conventional elastomeric lip seals. Normal seal practice shows that 0.003 to 0.010 in. of interference suffices to provide the necessary hydrodynamic pumping effect.

Weinand recommends that the seals be presoaked before assembly when they are made of materials that tend to swell. Presoaking counteracts dimensional changes that could be detrimental in service. The following important factors should be considered.

Diametral interference of the lip with the shaft should be 0.010 in. minimum, considering the dimensional corrections as a result of swelling and thermal expansion.

Radial load caused by interference of the lip should range from 0.3 to 0.7 lb/in. of circumferential length.

The helix angles for the ribs range from 20 to 30 degrees to provide optimal pumping effects. Larger angles are less effective. Smaller ones are inclined to plug when contaminating solids are floating in the environment.

Height of the ribs should range from 0.002 to 0.004 in. The outside lip surface forms an angle of 15 to 20 degrees with the shaft surface.

2. Advantages

The helixseal is capable of transferring fluid into a desired direction. In other words, leakage that may occur through the stationary seal lip is pumped back into the oil sump. This is not possible with regular lip seals. The stationary lip or the shaft surface can be slightly damaged, thus making some leakage possible. This leakage will be pumped back through the same openings. Consequently, the life expectancy of helixseals is good and the seal operates with high reliability and is less affected by variations in the film thickness.

The helixseal will tolerate a wide range of fluid viscosities and can withstand higher surface speeds than is possible for conventional lip seals. It requires very little space and compares favorably in cost with ordinary lip seals. Because of its relatively low lip contact pressure and thicker lubricant films, it is capable of tolerating larger amounts of shaft eccentricity, runout, and shaft deviation from roundness than is possible for mechanical end face seals.

3. Disadvantages

Helixseals function only when the shaft is rotating in one predetermined direction. The elastomer lip is subject to degradation when exposed to both high temperatures and chemical reaction of certain fluids. Dust, water, abrasives, and atmospheric contamination are other factors that deteriorate seal performance. It is, therefore, customary to use exclusion devices that prevent the contaminants from reaching the sealing lip.

4. Applications

Helixseals find their widest successful application in the automotive industry, where they are used to seal crankshaft openings of engines with the housing. Transmission pumps and differential pinions are other possible applications. Farm equipment, earth-moving trucks, and airplanes are other potential users of this special type of elastomeric lip seal.

B. Hydroseals

Unlike the helixseal, the grooves of the hydroseal are arranged in the shaft surface. The seal lip, however, has the same shape and function as in conventional lip seals.

1. Design Principles

Most applications of hydroseals involve static and dynamic sealing. Because of the static seal requirement the grooves must be shallow, generally less than 0.001 in. deep. The surface grooving can be accomplished by a number of methods, including machining, stamping, etching, and so on, either by providing a precise pattern for depth and width or by grinding or paper-polishing a random surface pattern. Because of the interference requirements, the sealing lip covers the grooves or whatever the pattern of surface roughness may be by elastic deformation when the shaft is stationary. The lip then bridges the surface irregularities when the shaft is rotating.

A typical hydroseal design is shown in Fig. 3.7, where an elastomeric lip ring seal is used with the lip contact enforced by a garter spring. The grooves in the shaft surface show a parallel pattern. They practically repeat the pattern of the helixseal with regard to depth, width, distance, and angles.

For equipment where the oil reservoir is not always filled, felt or rope packing seals can be used. The helical grooves in the shaft surface are longer than the contact area widths of the seals. In this case the seal function of the hydroseal has great similarity with the visco seal, which is discussed in Chapter 7.

2. Advantages

With the hydroseal oils of higher viscosity can be used, which reduces leakage. The grooves also improve backflow of the leakage fluid.

3. Disadvantages

The helical grooves favor a depletion of the lubricant film in the lip contact area. It is possible that the fluid film is completely eliminated in the lip contact area, so the lip may run dry. However, dry-run is det-

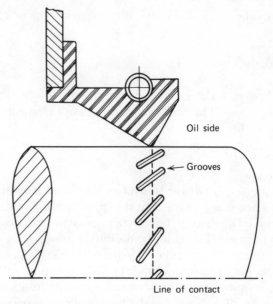

Fig. 3.7   Hydroseal design [From (20), courtesy of *Machine Design*].

rimental for the lip. Contact pressure load of the lip and groove geometry are critical factors that require careful evaluation. The entire seal design turns out to be a compromise for contact load in the lip, groove depth and width, and number of grooves.

4. Applications

Hydroseals are predominantly used in the automotive industry for sealing crankshafts. Knurled shafts are basically sealed by using rope packing devices.

C. Bidirectional Seals

Bidirectional seals are hydrodynamic seals whose direction of shaft rotation can be reversed any time without the seal losing its function (20).

1. Design Configuration

A bidirectional seal has a sharp-lip sealing element provided with helices on the atmospheric back side of the ring. Along half of the circumference of the ring helices are arranged with right-hand grooves; on the other

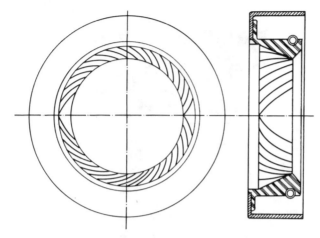

Fig. 3.8   Bidirectional seal [From (20), courtesy of *Machine Design*].

half there are left-hand grooves. When the shaft rotates, oil may leak out through the sealing lip on one half; it is immediately pumped back to the reservoir by the reversed other half. The principle of a bidirectional seal is illustrated in Fig. 3.8.

Experience shows that the back pumping action is more forceful than the leakage flow; this ensures a fluid balance in the oil reservoir.

2. Modification of Bidirectional Seals

A modification of the normal bidirectional seal is shown in Fig. 3.9. This design incorporates a triangular pattern of recessed helices on the atmospheric back side of the lip ring. The recessed helices in the seal ring start in close proximity to the seal edge, near the contact area. When the shaft rotates in the direction from the top to the bottom of the illustration, the triangular cavity section *a* fills with leak oil and a hydrodynamic pump action takes place. The leakage oil is forced back to the reservoir. Oil passes across the static lip and land *c*, separating grooves *a* from cavity *b*. Oil is pumped back to the reservoir because the resistance across the lip is smaller than the resistance across the land.

3. Advantages and Disadvantages

As a result of the close similarity to the helixseals of the conventional design, the advantages and disadvantages of the two seal types are ap-

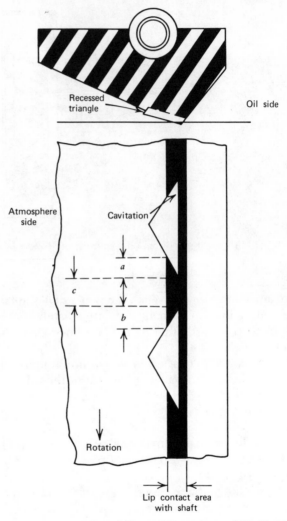

Fig. 3.9 Bidirectional lip seal [From (20), courtesy of *Machine Design*].

proximately the same. The seal capacity, however, is reduced because of the reversal of the flow directions.

D. Wave Seals

The wave seal is a recent development of the Chicago-Rawhide (CR) Company of Chicago, Illinois. R. V. Brink and L. A. Horve (4) report in detail on the development of this design and present actual performance data from the field. The various specific details are summarized here.

1. Design Principles

The wave seal is a hydrodynamic lip seal with a special elastomeric lip configuration that provides a static sealing edge that forms a sine wave as contact pattern along the surface of the shaft when rotating. The wave contact of the lip is designed into the seal to guide the lubricant and to increase the effective seal area along the shaft surface, which allows a better dissipation of the frictional heat generated by the lip. The lip is smooth and functions as a birotational seal. It is designed to pump lubricant back into the oil reservoir against the contact pressure, regardless of the rotational direction of the shaft.

The major contact pattern of the elastic sealing lip is illustrated in Fig. 3.10. The shaded area on the shaft surface represents an area range rather than a circular line concentric with the shaft surface. The area covers the running path of the static lip, reflecting a shallow sine wave on the shaft surface. Thus the swept width of the contact area is greater than the line produced by the line of the lip of a conventional lip seal, although the actual contact width is still rather narrow at any given point of contact. Consequently, the edge lip covers a larger shaft contact area, and the frictional effect is reduced, resulting in a lower lip contact temperature (4). This in turn means that a larger heat transfer area on the shaft surface is available to dissipate more of the frictional heat generated by the lip. With the reduction of the frictional heat, seal life is extended.

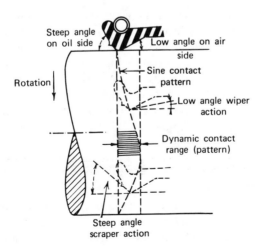

Fig 3.10 Wave seal [From (4), reprinted by permission of the American Society of Lubrication Engineers].

2. Hydrodynamic Lip Contact Mechanism

The design configuration of a wavy seal lip produces several interacting operational mechanisms that help to improve the performance of the seal. When the shaft rotates, the static lip permits an areal wetting pattern on the shaft surface.

The lip of the seal is provided with two different slopes, steep and low. The steep slope is on the oil side; the low slope is on the air side. The steep edge scrapes the oil back to the reservoir, whereas the low angle slope helps to wipe oil under the edge. The line of contact of the lip edge with the shaft surface is shown as a stretched sine wave.

The illustration of the contact range is drastically enlarged ₊to emphasize the principle. The contact region usually ranges from 0.005 to 0.010 in. The total width of the dynamic heat transfer region is on the order of 0.020 in., shown as a shaded area in the diagram.

3. Advantages of the Wave Seal

As reported by Horve, the pump rate was found to be moderate compared with birotational and hydrodynamic seals, delivering a somewhat constant rate over a wide range of shaft speeds. The seal has more lubrication than the conventional lip seal but noticeably less than the hydrodynamic seal devices. Dry-runs have not been observed.

Whereas hydrodynamic seals with ribs, vanes, or projections of various kinds tend to accumulate dust, the wave seal is less subject to dust ingestion.

A considerable advantage of the wave seal is its ability to absorb shaft eccentricity. Figure 3.11 illustrates the eccentricities, as functions of the shaft velocities, the wave seal can absorb before leakage occurs. Lower curve 1 designates the limit beyond which conventional lip seals will leak. The area between curve 1 and 2 designates the superiority of the wave seal over the normal lip seal. Eccentricities above limit curve 2 can no longer be taken by the wave seal. For comparison, hydrodynamic lip seals are less able to absorb mechanical tolerances than the conventional lip seals.

4. Conclusions

As a general rule wave seals deserve a higher rating than the conventional lip seals and the hydrodynamic sealing devices. Their service life is also better. Their pumping effect is three times greater than that for conventional lip seals.

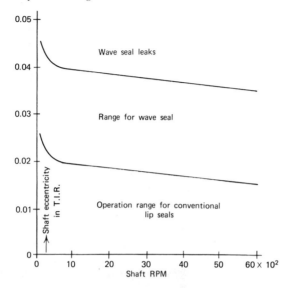

Fig. 3.11 Leakage behavior at various shaft eccentricities [From (4), reprinted by permission of the American Society of Lubrication Engineers].

Seal alignment is an important factor since cocking can have detrimental effects, increasing the contact force against the shaft surface, which in turn produces an increase of friction and a consequent rise in the lip contact temperature. Cocking does not provide a larger contact area with the shaft.

Frictional heat, and thus the temperature of the lip contact point, is a dominating factor, as reported by B. V. Brink (3).

The heat generated by friction at the static seal lip generally expressed as underlip temperature produces a chain reaction. Hot oil expands the seal, which then reduces the surface contact pressure with the shaft. The film thickness increases and friction diminishes. Heat also decreases the modulus of elasticity of the seal ring material, thus lowering the contact force. Consequently, the seal material must be evaluated with great care to obtain good service life, minimum drag, and tolerable leakage rates.

The influence of the shaft material in connection with the finish condition of the shaft surface is covered in a later section of this chapter.

All advantages, designed into conventional and hydrodynamic seals, are also found in the wave seal devices. For conventional seals these advantages are as follows: reasonable cost, minimal contamination by dust, absorption of relatively high shaft eccentricity, tolerance of a wide

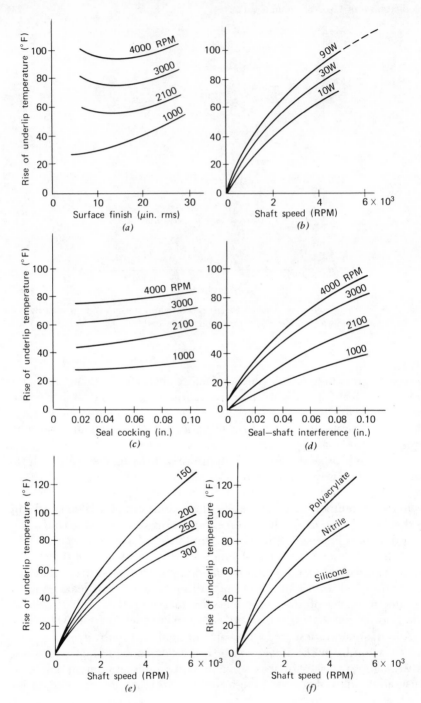

252

range of interference with the shaft surface, promotion of relatively good service life under normal operating conditions, and accommodation of favorable material selection.

For hydrodynamic seals the advantages basically are as follows: tolerance of light surface damage in shaft and seal lip, provision of good service life at high reliability, and ease of handling in the field.

The major factors that significantly influence seal function are summarized in Fig. 3.12: Fig. 3.12a shows the influence of surface finish of the shaft on the rise of underlip temperature of the seal for various shaft speeds; Fig. 3.12b gives the influence of fluid viscosity on the underlip temperature as a function of the shaft speed for a variety of different oils; Fig. 3.12c presents the influence of seal cocking on the underlip temperature for various shaft speeds; Fig. 3.12d gives the influence of lip interference with the shaft on the underlip temperature for a variety of shaft speeds; Fig. 3.12e gives the influence of sump temperature on underlip temperature as a function of shaft speed; and Fig. 3.12f gives the influence of the elastomer of the seal lip on the underlip temperature as a function of shaft speed.

The curves of Fig. 3.12f indicate the results of tests using polyacrylate, nitrile, and silicone with an oil of the composition of SAE-10-W-30 at a sump temperature of 250°F. Best results were obtained from silicone. Polyacrylate shows the highest underlip temperature. Nitrile is located between polyacrylate and silicone.

VI.   Basic Factors Affecting Seal Performance

Adequate seal performance is the result of a number of factors that must be thoroughly evaluated. They are discussed in detail to provide fundamental information for a full understanding of the complex subject matter. These factors include seal design concepts, shaft requirements, bore prerequisites, lip ring materials, environmental conditions, and installation rules.

A.   Design Concepts

As stated earlier in this chapter, design has a considerable influence on the effectiveness of the lip seal in general. Seal lip edge diameter,

Fig. 3.12   Variations of underlip temperature as function of major design factors. (a) Surface finish. (b) Oil viscosity. (c) Seal cocking. (d) Seal-shaft interference. (e) Sump temperature. (f) Elastomer composition [From (3), courtesy of J. Product Engineering].

geometrical interference, eccentricity, out-of-roundness, spring location, width of the contact area, flex-section thickness, and material modulus are the factors that determine the quality of the device.

Interference fit and flexibility produce the load the elastomeric lip must impart on the circumference of the shaft. A mechanical spring is added to compensate for lip contact wear and dynamic shaft irregularities. Eccentricity, and geometric irregularities of the shaft and the lip ring induce additional dynamic loading components of the lip. These factors can become critical and are responsible for most seal failures and the cause for ineffective operation. Most design factors are influential in forming the lubricant film whose proper supply and suitable thickness is a major condition for satisfactory operation of the lip.

The second major design parameter is the ability of the elastomeric seal lip to flex in accordance with the irregularities of the shaft movement without failing by fatigue.

When the lubricant film is supplied by wedging resulting from cocking of the seal during assembly or by eccentricity of the shaft, the seal will scrub in the axial direction and additional load will be put on the lip. This must be reduced to a minimum.

It should be noticed, however, that the seal would also fail because a lubricant film cannot be established if the shaft surface is absolutely perfect. Surface roughness of some degree, shaft runout and deflection, and some degree of eccentricity must be present to allow the development of an oil film. Under ideal conditions it is most likely that high-frequency stick-slip would occur as a result of temperature and torsional shear stresses.

B. Requirements for the Shaft

Adequate seal performance requires a shaft with a suitable geometry in accordance with the seal dimensions, adequate surface finish, surface hardness, axial straightness, and a proper material composition.

1. Surface Finish of the Shaft

A significant factor for satisfactory seal performance is the condition of surface finish of the shaft. A surface finish of better than 8 μin. (2 to 8 μin. rms) is not desirable. It has been found that such surfaces are too smooth and cannot establish satisfactory wettability for the lubricant film. A surface finish ranging from 10 to 20 μin. AA is a good prerequisite to develop a surface pattern on the shaft that will permit satisfactory wettability. Commonly accepted values for shaft surface finish range from 15 to 20 μin.

2. Measurement of Surface Finish

Surface roughness was discussed briefly in Chapter 2 in connection with interface preparations for mechanical end face seals. A short review of the subject may be in order.

Measurement of the degree of surface finish is concerned with determining the relative differences of the local surface topography. An instrument is used that incorporates a stylus that is moved across the surface to be measured. The stylus detects the surface asperities and converts the motion into an electrical potential that is then recorded.

At the present time the following instruments are in use to measure surface roughness:

1. Linear proficorder, which plots axial finish
2. Profilometer, which measures axial roughness
3. Talysurf, which measures axial roughness
4. Rotary proficorder, which records circumferential finish and indicates roughness.

For readout the three methods used are mean root square (rms), center-line average (CLA), and arithmetical average (AA). The rms method provides results that are approximately 11% higher than the other two methods. Center-line average (CLA) is the British designation for the American arithmetical average method (AA). This method is now preferred in U.S. industry.

ANSI-Standard B 46.1 1962 illustrates conditions for surface finish.

3. Shaft Surface Hardness

Good service life for lip seals can be expected when the shaft surface is hard. In addition to straightness and smoothness, a minimum surface hardness of R_c-30 is required. Field experience has shown that a Rockwell-C hardness of 45 provides excellent operating conditions for seal lip. In the presence of abrasives in the fluid to be sealed a shaft hardness of R_c-55 is recommended.

4. Shaft Material

Hardenable steels are by far the most common materials used for shafts sealed with lip seals. Stainless steels are preferred when the fluid is corrosive. Surface plating with chromium is also used. The relatively soft alloys, such as brass, bronze, aluminum, magnesium, zinc, and the like, are not recommended, except under unusual conditions.

Table 3.1 Shaft Tolerances

Nominal Shaft Diameter (in.)	Tolerance
Up to 4.000	± 0.003 in.
4.001 to 6.000	± 0.004 in.
6.001 to 10.000	± 0.005 in.
10.001 up	± 0.010 in.

Shaft sleeves are also in use that make the selection easy at reasonable cost.

5. Shaft Tolerances

In lip seal service the shaft tolerances shown in Table 3.1 are required, as recommended by the Rubber Manufacturers' Association:

Good seal performance can be expected if the shaft tolerances are considered carefully. An oversized shaft produces additional lip contact load, whereas an undersized shaft does not provide satisfactory seal pressure in the lip contact area. Overpressure results in early seal failure whereas insufficient contact pressure produces leakage.

6. Shaft Eccentricity

The performance of a lip seal is affected by two types of eccentricities:

1. SHAFT-TO-BORE MISALIGNMENT. This expresses the magnitude by which the shaft center deviates from the center of the bore of the housing. This is generally a result of inaccuracies imparted by machining and assembly. It can be measured by using a dial indicator that rotates slowly with the shaft during the measurement.

2. DYNAMIC RUNOUT. This represents the amount by which a point at the surface (lip contact point as an example) does not rotate around the true center. This occurs when the shaft is bent, misaligned, or unbalanced. It is measured by attaching the indicator to one point of the bore and then rotating the shaft. The total deviation from true round rotation is the dynamic runout.

Good seal performance can be expected if the flexibility of the elastomer lip can absorb these shaft irregularities. In seal practice the total eccentricities given in Table 3.2 (sum of 1 and 2) are generally tolerated.

Any amount of shaft eccentricity affects seal performance and gradu-

Table 3.2   Shaft Eccentricities

Maximum Total Eccentricity* (in.)	Maximum (ft/m)	Shaft Speed (RPM)
0.025	75	100
0.020	150	200
0.018	350	500
0.015	700	1000
0.013	1000	1500
0.010	1500	2000
0.009	1800	2500
0.008	2000	3000
0.007	3000	4000

* Shaft-to-bore misalignment plus dynamic runout.

ally increases the chance of fatigue failure caused by wear and increased hardening of the elastomer. Permanent deformation of the lip contour finally results in seal failure.

The various types of eccentricities usually encountered with rotating shafts are illustrated in Fig. 3.13, including eccentric housing bore, producing wear on one side of the seal. The figure in the center represents an eccentric shaft with shaft whip resulting in leakage especially at high rotational speeds. The worst case is an eccentric housing–shaft combination.

Because of the wide range of seal manufacturers, values of eccentricity that can or cannot be tolerated in lip seal service vary drastically. It seems feasible to make comparisons and then develop a personal judgment.

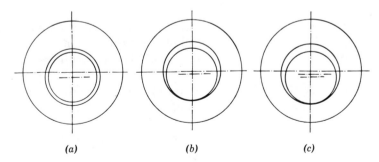

(a)   (b)   (c)

Fig. 3.13   Eccentricity of shaft-bore combinations.

7. Preparation of the Shaft for Installation of the Seal

The internal diameter of the elastomeric static sealing lip is smaller than
the OD of the shaft. The assembly of the seal, therefore, requires special
tools to install the seals safely without damaging the lip. The shaft is
provided with a chamfer or a radius. Minimum conditions for the shaft
face are generally provided by the seal manufacturer. More details are
given in Section VI.D for installation.

8. Wear Sleeves

Wear sleeves are used where hardened shafts are not available. Generally they are devices of mild steel that are pressed over the shaft of cast
iron, bronze, brass, aluminum, magnesium, or similar materials. They
permit easy replacement of the sealing surface and should be replaced
whenever the seal is replaced. For installation of the seal special tools are
used to prevent damage of the seal lip. Special mounting sleeves provide
protection from the sharp edges of the key grooves.

C. Requirements for the Bore

All lip seals, both bonded and cased, are embedded in metal cases that
are pressed into a bore in the housing. The seals are assembled with
press-fit; therefore, precise dimensions are required for the cavities in
the housing to prevent leakage. The housing must provide sufficient
strength and rigidity to accommodate a press-fit condition with the OD
of the lip seal casing.

1. Bore Tolerances

Press-fits require the assurance of certain tolerances that should provide
a prerequisite for proper seal installation. For assurance of the necessary
press-fit with the housing the bore should satisfy the tolerances indicated
in Table 3.3.

The bore edge should be chamfered to permit easy installation of the
seal without damaging the seal casing. The OD of the bore must be
smooth to prevent scoring of the seal casing and to permit seepage. In a
stepped-bore cavity the width of the recess section should exceed the
width of the seal itself by at least $\frac{1}{64}$ in.

Table 3.3   Bore Tolerances for Seal Installation

Bore Diameter (Steels)	Bore Tolerances	Nominal Press-fit	
		Seal w/ Metal OD	Seal w/ Rubber OD
Up to 1.000	±0.001	0.004	0.006
1.001 to 2.000	±0.001	0.004	0.007
2.001 to 3.000	±0.001	0.004	0.008
3.001 to 4.000	±0.0015	0.005	0.010
4.001 to 6.000	±0.0015	0.005	0.010
6.001 to 8.000	±0.002	0.006	0.010
8.001 to 9.000	±0.002	0.007	0.010
9.001 to 10.000	±0.002	0.008	0.010
10.001 to 20.000	+0.002 −0.004	0.008	—
20.001 to 40.000	±0.002 −0.006	0.008	—
40.001 to 60.000	+0.002 −0.010	0.008	—

Source: From Ref. (5a).

2. Bore Finish

To prevent oil leakage between the bore and the seal casing the internal surface of the bore should provide a surface finish of at least 63 μin. or better (but in no case less). Where this condition cannot be satisfied a coating should be used which can be applied either to the bore surface or to the seal casing. The coating can consist of synthetic rubber, which is well suited to fill the metallic surface asperities. Synthetic rubber is preferred, but sealants or cements are also useful.

3. Bore Configuration

The ID of the recess bore in the housing must be provided with a 15- to 30-degree chamfer to facilitate the installation of the seal casing without being damaged by burrs. The inside corner of the bore must be rounded using a minimum radius of $\frac{3}{64}$ in. Dimensional bore requirements are usually provided by the seal manufacturer.

4. Bore Material

The tolerances shown in Table 3.3 for the bore are valid for steel housing in general. Where other materials must be used the coefficient of thermal expansion must be taken into consideration, especially when larger seal sizes are involved.

D. Installation of Seals

Elastomeric lip seals, both bonded and cased, are mechanical devices of high precision. Great care in their assembly and installation is required to achieve the reliability for which they have been designed and chosen. The Rubber Manufacturers' Association, in association with SAE, has established a Guide to the Application and Use of Radial Lip-Type Oil Seals-SAE J-946.

The following installation rules are recommended:

1. Check dimensions of seal, shaft, and bore. Make sure they match the specifications.
2. Check seal for damage in the edge of the lip and the housing.
3. Check seal for dimensions, smoothness, bore surface, and chamfer.
4. Check shaft for dimensions, surface finish, burrs, or damage.
5. Check shaft face for chamfer and sharp edges and corners.
6. Check splines and key ways. Key ways require special sleeve.
7. Check direction of rotation for seal. As a rule the lip always faces the lubricant or fluid to be sealed.
8. Prelubricate seal and install immediately.
9. Check correct seal installation tool. Apply pressure for press-fit only to casing at OD.
10. Apply smooth driving force, preferably using an arbor press. Avoid shock to prevent popping off of spring from the seal.
11. Avoid cocking of seal. Use proper bottom tool.

E. Environmental Conditions

Elastomeric lip seals have limitations for application with respect to temperature, shaft speed, and pressure. Most major seal manufacturers state the following limit conditions:

For temperature	400°F max, −60°F min
For pressure	100 psi max
For shaft speed at contact lip	6000 ft/min

These are guidelines only and may vary somewhat, depending on material used. As an example, the average lip seal is used for a pressure range of up to 30 psi. With special design, certain seal configurations are used for pressures in excess of 100 psi.

Temperatures of up to 400°F require the use of silicones, polyimid, elastomeric Kel-F, or PTFE materials. It should be noticed that these conditional operation limits must be considered in combination with fluid compatibility of the lubricant.

F. Corrosion Resistance

In addition to the other factors—such as load, speed, pressure, and temperature—corrosion can drastically limit the application of lip seals. Two types of corrosion affect seal life, namely, chemical and galvanic corrosion. Since this involves an intricate and rather complicated mechanism, consultation with the seal manufacturer is recommended.

G. Selection of Lip Seal Materials

The selection of lip seal materials is governed by mechanical, dynamic, and environmental conditions. Once these have been evaluated materials can be selected, based on compatibility and life characteristics.

The most commonly used seal materials are nitrile, polyacrylate, silicones, and fluoroelastomers. For specialty applications other materials, such as urethane and Butyl, are also in use. Each of these has limitations. The advantages and disadvantages may be outlined briefly.

Today a number of base materials are available from which a large number of finish materials can be compounded to provide a desired seal life. A typical listing of elastomeric base compounds is shown in Table 3.4 as reported by R. L. Dega (5a) of General Motors Corporation.

1. Temperature Characteristics

In selecting the proper materials for lip seals, both the low- and high-temperature limitations have to be taken into consideration. At low temperatures two characteristics are important, the brittle point of the material and its stiffening point, where loss of flexibility becomes critical. The brittle point indicates the temperature at which cracking or fracture occurs.

Materials with approximately the same brittle point can have quite different stiffening characteristics. Important is the value of the mod-

Table 3.4   Elastomeric Lip Seal Materials (5a)

Chemical Designation	General Useful Temperature Range	
	Minimum °F	Maximum °F
Polychloroprene	−60	212
Styrene butadiene	−30	225
Polyurethane	−60	225
	−60	225
Butadiene-acrylonitrile	−40	225
Epichlorohydrin	−20	225
	−40	225
Polyacrylate	−20	300
Polydimethyl siloxane	−60	300
	−60	300
Fluoroelastomers	−40	300
	−20	300
	−60	300

ulus of elasticity, which may vary significantly with the temperature for similar materials, as reported by Ostmo (13).

As the temperature rises the modulus decreases, which is accompanied by a change of the load of the lip against the shaft surface. With rising temperature the material undergoes thermal expansion, reducing load and interference conditions. Load and lip interference also change when the fluid causes a swelling of the lip material. When hardening or softening of the lip material takes place as a result of temperature changes, load and interference deteriorate and the geometry of the lip contact area changes.

If the modulus is high, excessive wear occurs because the oil film thickness decreases. On the other hand, if the modulus is too high, the spring can no longer function to maintain the flexing load of the lip to compensate for eventual higher eccentricities. If the fluid viscosity is too low the oil film becomes too thin and leakage results.

At high temperatures the effects on the material of the seal are noticeably more complex than what is described for the low-temperature region. With rising temperature the rubber modulus decreases, reducing the radial load of the lip against the shaft surface. As the heat influence continues the rubber may undergo a vulcanization process, making it harder and thus producing an increase in the rubber modulus. The heat may also cause the sump fluid to decompose and form by-products that may destroy the lip of the seal.

Finally, elastomers deteriorate as a result of tension and compression, which produces a permanent set and in turn affects the radial load of the lip interference with the shaft.

2. Fluid Resistance

A major factor in the selection of elastomers for sealing materials is their compatibility with the fluids to be sealed. This information, however, is readily available in numerous publications, papers, books, and manufacturer brochures and catalogs; therefore it is not discussed here.

3. Specific Lip Seal Materials

The materials used in present-day seal manufacture for elastomeric lip seals and their various configurations are selected from the following components (13).

a. Nitrile. Perhaps the most extensively used material for elastomeric lip seals is Nitrile. It has good resistance to oils, provides good wear, and can be fabricated at low cost as a copolymer of butadiene and acrylonitrile. Its resistance to oil and low temperature is imparted by the percentage of acrylonitrile in the compound.

Nitrile is used wherever possible, particularly because of its low fabrication cost. It provides excellent service when hydrocarbon fluids must be sealed at reasonably moderate temperatures. It is suited for low-molecular-weight fluids, such as gasoline or highly aromatic fluids.

Nitriles cannot be used in the presence of aromatics with low molecular weight, such as benzene or polar aliphatics.

The general application range for nitrile temperatures is from -40 to $225°F$.

b. Polyacrylates. Polyacrylic polymers, which have a somewhat higher cost than nitriles, have an inherent advantage because of their resistance to higher temperatures. They can be used for temperatures up to $300°F$ when handling engine or transmission oils. They replace nitriles in all cases where hardening and cracking is a problem.

Polyacrylates, however, are very susceptible to water, acids, bases, and all polar solvents, such as ketones, esters, and the like.

Polyacrylates have two major disadvantages—low wear resistance and poor low-temperature properties.

c. Silicones. Of all lip seal materials discussed thus far the silicones offer the widest temperature range, from -100 to $400°F$. Silicones are recommended for petroleum-based lubricants, such as engine oils and transmission oil, and for nonpetroleum-based brake fluids, diesters, water, inorganic acids, and bases. Silicones should not be used for chlorinated aromatic solvents, gasolines, and the like. Oxidized lubricating oils also will affect the rubber. Their dry-running characteristics are poor. The rings should be soaked in oil prior to installation. This prevents dry-running during the start-up or break-in operation. Special care is required for assembly and installation because of poor tear characteristics.

d. Fluoroelastomers. In the family of fluoroelastomers two types of materials are distinguished, including fluorocarbons and fluorosilicones. The fluorocarbons cover at present a wide range of applications; the fluorosilicones have just begun to enter the market.

The fluorocarbons have excellent resistance to a wide range of chemicals and to elevated temperatures. They are well suited for heavy-duty service because of their excellent toughness. However, high stiffness at low temperatures is a major shortcoming. The garter spring becomes ineffective at temperatures below $-10°F$, although the brittle point is at around $-40°F$.

The fluorosilicones combine the chemical and high-temperature resistance of the fluorocarbons with the excellent resistance to low temperatures of the silicones. Tear strength is low, however, and the wear characteristics are below those of the fluorocarbons.

Table 3.5 shows a summary of the temperature ranges and limitation characteristics of these materials.

Table 3.6 presents important details for comparison of mechanical and environmental requirements. It should be noticed that leather is used for packings only, as in reciprocating plunger service, where shaft runout is not a problem. Agitator shafts generally operate with more shaft runout than leather rings could tolerate.

Table 3.7 offers a somewhat more detailed list of important information on rubber-based materials for the general seal industry.

e. Viton Fluoroelastomers. Viton fluoroelastomer is a product of the DuPont Company that is similar to Fluorel, which is made by the 3-M Company. Although fairly expensive, a seal made of this material is generally worth its expense because of its reliability and performance.

Viton is suitable for up to 400°F service where some hardness and resiliency are still retained at this level. Viton has excellent high-temperature characteristics and provides a much better service life than silicones, nitrile, polyacrylate, and so on. At a temperature of 400°F the

Table 3.5 Temperature Ranges and Limitations for Synthetic Seal Materials

Material	Relative Cost	Temperature Range (°F)	Advantages	Limitations
Nitrile	100	−60 to 225	Good wear and oil resistance at moderate temperature	Poor resistance to EP additives
Polyacrylates	115	−40 to 300	Good oil resistance, generally resistant to EP additives	Only moderate wear properties
Silicone	130	−80 to 325	Very broad temperature range	Poor resistance to oxidized oil and some EP additives
Fluorocarbon	200	−40 to 350	Excellent oil and chemical resistance, good wear properties	Poor followability at low operating temperatures

Source: Boyce [2]. Courtesy of *Plant Engineering.*

Table 3.6 Comparison of Seals

Condition	Homogeneous Rubber Seals	Fabric-Reinforced Rubber Seals	Leather* Seals
Working Temperature			
High	450°F	500°F	180°F
Low	−130°F	−20°F	−65°F
Finish required	16 μin.	32 μin.	40 μin.
Extrusion resistance	Fair	Good	Excellent
Clearances required	Very close	Close	Liberal
Wear resistance	Good	Good	Excellent
Coefficient of friction	High	High	Low
Rotary motion	Poor	Fair	Good
Reciprocating motion	Good	Good	Good

* For packings only.

Source: Boyce [2]. Courtesy of *Plant Engineering.*

Table 3.7 Relative Properties of Various Base Polymers for Seals

Property	Type of Rubber Base						
	Nitrile (Buna N)	Neoprene	Styrene (Buna S)	Viton*	Butyl	Thiokol*	Silicone
Tear resistance	Fiar	Good	Poor	Good	Good	Fair	Poor
Abrasion resistance	Good	Excellent	Good	Good	Fair	Poor	Poor
Aging							
Sunlight	Poor	Excellent	Poor	Excellent	Excellent	Good	Good
Oxidation	Fair	Good	Fair	Excellent	Good	Good	Good
Heat (maximum temperature)	300°F	300°F	250°F	500°F	300°F	160°F	450°F
Static (shelf)	Good	Good	Good	Good	Good	Fair	Good
Flex cracking resistance	Good	Excellent	Good	Good	Excellent	Fair	Fair
Compression set resistance	Good	Excellent	Good	Excellent	Fair	Poor	Good
Lubricant resistance							
Low aniline mineral oil	Excellent	Fair	Poor	Excellent	Poor	Excellent	Poor
High aniline mineral oil	Excellent	Good	Poor	Excellent	Poor	Excellent	Good
Silicones	Fair	Fair	Poor	Excellent	Poor	Good	Poor
Diesters	Fair	Poor	Poor	Good	Poor	Poor	Fair
Phosphate esters	Poor	Poor	Poor	Good	Excellent	Fair	Poor
Silicate esters	Fair	Poor	Poor	Good	Poor	Fair	Fair
Solvent resistance							
Aliphatic hydrocarbon	Good	Fair	Poor	Excellent	Poor	Excellent	Poor
Aromatic hydrocarbon	Fair	Poor	Poor	Excellent	Poor	Good	Poor

	1	2	3	4	5	6	7
Halogenated solvent	Poor	Poor	Poor	Good	Poor	Poor	Poor
Ketones	Poor	Poor	Poor	Poor	Poor	Poor	Poor
Gasoline resistance							
Aromatic	Good	Poor	Poor	Excellent	Poor	Good	Poor
Nonaromatic	Excellent	Good	Poor	Excellent	Poor	Excellent	Good
Acid resistance							
Dilute (under 10%)	Good	Fair	Good	Good	Good	Poor	Fair
Concentrated	Poor	Poor	Poor	Good†	Fair*	Poor	Poor
Alkali resistance							
Dilute (under 10%)	Good	Good	Good	Good	Good	Poor	Fair
Concentrated	Fair	Poor	Fair	Poor	Good	Poor	Poor
Low temperature flexibility (maximum)	−65°F	−65°F	−70°F	−65°F	−65°F	−65°F	−120°F
Resistance to gas permeation	Fair	Good	Fair	Good	Good	Good	Fair
Water resistance	Good	Fair	Good	Good	Good	Fair	Fair
Resilience	Fair	Good	Fair	Good	Poor	Poor	Good

* Poor in nitric and sulfuric.
† Poor in nitric and hydrochloric.
Source: Boyce [2]. Courtesy of Plant Engineering.

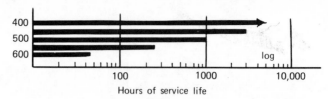

Fig. 3.14 Service life of fluoroelastomers (Viton) [From (3), courtesy *J. Product Engineering*].

diagram of Fig. 3.14 shows a running life far in excess of 1000 hr. Service is considered possible up to 500°F, although this statement may be too optimistic.

Viton also retains a high degree of flexibility at high temperature. At 300°F, for instance, the retained sealing force of Viton fluoroelastomer far exceeds the capability of nitrile, polyacrylate, and fluorosilicone. A comparison of wear life of Viton-fluoroelastomer with nitrile, polyacrylate, and silicones proves the superiority of Viton (23).

VII. Working Life of Elastomeric Lip Seals

The life of a lip seal, including the wave seal, is directly proportional to the rate of which the seal can absorb the effects of friction and is inversely proportional to the rate of friction it develops in the lip contact area. Seal life thus is a direct function of the rotational speed of the shaft. Conventional lip seals can provide several thousand hours of service life if the shaft speed is kept below 2000 RPM. Figure 3.15 illustrates service life of seals for two shafts, 3 in. Φ and $\frac{1}{2}$ in. Φ, as a function of the shaft rotational speed. It is obvious that an increase in shaft speed in excess of 2000 RPM will lead to rapid deterioration of seal life and subsequent rapid failure.

The rise of temperature at the seal lip interface with the shaft is the result of frictional heat. The magnitude of work developed is also a function of viscosity. It should be kept in mind that viscosity decreases with rising temperature and that reduced viscosity diminishes the strength of the lubricant film. When the film at the lip interface becomes too thin, temperature rises, thus decreasing seal life. The film thickness is a function of the radial load. On the other hand, if the film grows too thick, leakage occurs. Too thin a film reduces seal life.

The chemical composition of the various lip materials has a strong influence on seal life. Although nitrile is a very frequently used material for lip seals, its life expectancy for normal service conditions is lower

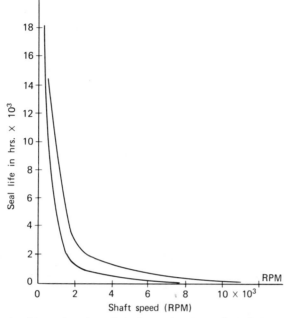

Fig. 3.15 Service life as function of shaft rotational speed. Ref (2) courtesy *J. Plant Engineering.*

than that of polyacrylic and fluoroelastomers. For life expectancy, silicone is far superior to fluoroelastomers.

Sump temperature is another significant factor affecting seal service life. Below a temperature of approximately 200°F seal life is practically indefinite. With rising sump temperature service life deteriorates drastically, reaching a life of hardly more than 100 hr at 300°F. From there the life expectancy curve approaches the abscissa asymptotically toward zero. A temperature of 400°F for conventional seals is not acceptable. For this range fluoroelastomers conditionally or silicones unconditionally or Teflon, Kel-F, or polyimids should be chosen.

VIII.   Special Seal Configurations with Teflon as Seal Material

In lip seal design the elastomer characteristics constitute the limiting factor. By replacing the elastomer with Teflon (PTFE) or a combination of elastomer and Teflon and substituting the elastic memory effect by a metallic spring an entirely new family of rotary seal configurations has become available. Material compatibility, friction losses, and wear resistance are greatly improved; however, the interference requirements and subsequently the automatic radial load conditions have been changing. This in turn means that dynamic runout and eccentricity considerations have forced the designer to ask for closer tolerance limits.

A. High-Pressure Oil Seal by Johns-Manville

The Johns-Manville Company recently marketed a new type of lip seal that can be designated as a J-M lip seal. The manufacturer claims that this design is capable of handling pressures up to 300 psi and temperatures ranging from −40 to 300°F. These operating conditions are supposed to be met at shaft speeds as high as 4000 fpm.

1. Design Concept

The design principle of the J-M high-pressure oil lip seal is shown in Fig. 3.16. The lip ring consists of a fluoroelastomer bonded to a metal washer. The lip against the shaft surface is backed by a Teflon cone wedge ring that supports the lip. The seal is held in an outer metal case that is inserted with a press-fit into the seal cavity of the housing. A metal snap ring ensures the steady position of the seal unit in the bore.

The J-M seal is available in sizes from ½ through 4 in. Operating conditions are as follows:

Shaft speed	4000 fpm maximum
PV value	300,000 maximum
Bore tolerance	for ΦΦ up to 3.000 in.: 0.001 in. maximum
	for ΦΦ 3.001 to 6.000 in.: 0.0015 in. maximum
Bore finish	15 to 60 μin.
Shaft-to-bore misalignment	0.003 in. T.I.R. maximum
Shaft runout	0.003 in. T.I.R. maximum
Temperature range	−40 to 300°F
Pressure range	atmospheric to 300 psi

2. Advantages

The J-M seal has excellent resistance to a very wide range of chemicals, oils, solvents, and fuels at relatively high temperatures. The seal can be effective at both low and high pressures at low torque.

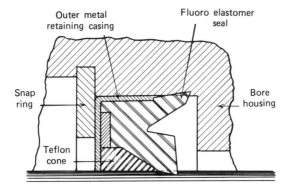

Fig. 3.16 Johns-Manville lip seal with fluoroelastomer.

3. Applications

The J-M seal is suitable for pumps, motors, propeller shafts, road wheels, fuel injectors, and the like. It has great potential for widening its range of application.

B. The Teflon Lip Seal in General

Single- and double-lip seals are now available with PTFE materials for the sealing lip. The lips come in all kinds of configurations, including reinforced single lips with and without garter springs. With the use of PTFE an entirely new sealing concept has developed. Teflon (DuPont), Halon (Allied), and similar products from other manufacturers have two key properties that deviate from all other sealing materials. They are self-lubricating and permit operation in the cryogenic temperature range and up to 500°F.

A prime disadvantage is creep, generally referred to as cold flow. Another shortcoming is the difficulty of limited techniques for forming parts to any particular desirable geometry. The cold flow can be encountered by designing the seal either to trap the permanent flow, thus containing the material at the seal area, or to add springs of metal or elastomers to maintain a continuous sealing contact force.

By adding suitable reinforcing compounds a combination can be achieved that strengthens the material in various ways. Additives are graphite, glass fiber, metal oxide, bronze, and molybdenum disulfide, giving the Teflon base a wide range in strength increase and other desirable operating characteristics.

PTFE lip seals are designed to satisfy the following service conditions:

Compatibility.   Resistance to most chemicals and oil additives.
Friction.   Very low, can withstand dry run.
Wear.   Low wear rate for lip ring and shaft surface.
Temperature range.   From -120 to $500°F$.
Pressure.   From atmospheric to in excess of 350 psi.
Rotational speed.   Can take surface speeds in excess of 6000 fpm.
Life expectancy.   Excellent, practically unlimited; very good in dynamic and static sealing.
Leak prevention.   Excellent.

C. BAL-Seal as Special Design

In the discussion of the BAL-Seal the manufacturer's designation must be used, since no other identification is available. The BAL-Seal is a U-shaped ring with the opening directed toward the pressure of the system with a special spring embedded inside the U-cavity. It is applied like an O-ring and can be used equally for static, rotating, or reciprocating services with a few design modifications, depending on the service required. The seals are fabricated by the BAL-Seal Engineering Company of Tustin, California.

The compulsory spring provides the contact pressure for the automatic seal interfaces at all times. By providing toothed contact face profiles multiple line contacts are produced with low friction values. The teeth subdivide the seal spaces into differential pressure chambers, thus facilitating lubrication and the sealing mechanism and eliminating low-pressure seepage.

The spring has canted coils so they can readily deflect when subjected to load, resulting in lower frictional forces and longer service life of the seal. The basic design concepts of the BAL-Seal are illustrated in Fig. 3.17.

The seal body consists of either PTFE or a combination of PTFE with glass, graphite, or molybdenum disulfide. The graphite- or moly-filled seal will outlast the PTFE seal by a factor of 10 to 20, depending on the severity of the operating conditions in connection with rotational shaft speed.

The manufacturer claims that these seals are capable of withstanding dynamic runouts of up to 0.025 in. T.I.R. because of the resiliency built into the design of the U-cavity section and the efficiency of the metal spring. In addition to the spring action the BAL-Seal also utilizes the self-energizing sealing principle.

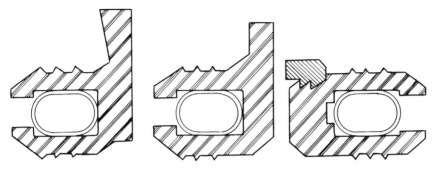

Fig. 3.17 BAL-Seal configurations for rotary sealing.

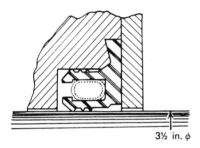

3½ in. φ

Fig. 3.18 Application of a BAL-Seal for agitator shaft. Courtesy *BAL-Seal.*

Under otherwise equal operating conditions the BAL-Seal has a life expectancy of a multiple of that experienced by conventional elastomeric lip seals.

Figure 3.18 illustrates the author's application of a BAL-Seal to an agitator shaft of 3.500 in. diameter, for moderate agitator rotation and pressure. By assembling the seal the spring is compressed and then provides continuous automatic seal contact.

D. The FCS-Seal by Fluorocarbon

The FCS-Seal is manufactured by the Fluorocarbon Company, Los Alamitos, California. The seal is designed for static, rotating, and reciprocating service. A helical spring provides the contact force. The spring either is inserted in a C-shaped Teflon casing or is fully enveloped by Teflon. The spring is immune to the fluid of the system when enveloped, whereas the selection of the material for an open spring depends on the corrosiveness of the fluid involved.

1. Design Concept

The design of the FCS-Seal is based on the nature of the spring employed, which is fabricated from a flat strip of metal and wound into circular spirals. The spring is inserted into a Teflon housing that has a C shape, thus establishing two points of contact with the areas to be sealed. The depth of the cavity is smaller than the height of the seal device. Consequently, when the seal is installed it must be precompressed to fit into the seal cavity. This amount of compression provides the necessary automatic contact force. The resiliency of the spring maintains automatic contact of the seal with the interface area during operation. The seal must be inserted so that the C always opens against the system pressure, thus providing self-energizing seal service. Basic standard design configurations of FCS-Seals are illustrated in Fig. 3.19.

If for any reason a metallic spring is not desirable, an elastomeric O-ring may be used instead. However, over a period of time the O-ring material may develop a permanent set, thus reducing some of the resiliency and seal efficiency.

2. Materials of Construction

The FCS-Seal is currently available in two basic types of Teflon combinations, forming alloys that are designated as Fluoroloy S and Fluoroloy SL.

a. Fluoroloy S.   Fluoroloy S is the material predominantly used for standard FCS-Seals and consists of a Teflon-base plastic material. The manufacturer claims that Fluoroloy S has a wear resistance 10 to 100 times better than that of virgin TFE.

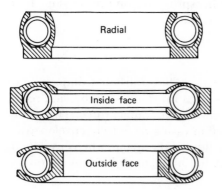

Fig. 3.19 Basic FCS-standard seals.

b. Fluoroloy SL.   Fluoroloy SL is also a Teflon-base plastic mixed with carbon and graphite. For this material the manufacturer claims a wear resistance of 1000 times that of nonalloyed Teflon. It is recommended for applications with water or fluids where high PV values are required. Its cost factor is slightly above that of Fluoroloy S.

3. Service Life

As far as seal service life is concerned, it can be stated that the FCS-Seals provide a life expectancy essentially above that of silicone seals. When they are combined with molybdenum disulfide their life span is even greater.

Like the BAL-Seal, the temperature range for applications spans from −100° to 500°F. Pressure may go as high as 10,000 psi for moderate shaft speeds. The higher the pressure, the lower the shaft speed.

4. Applications

FCS-Seals are found in static, rotating, and reciprocating applications for a wide range of service conditions, both in the cryogenic region and for high temperatures.

E.   Omniseal by Aeroquip

Omniseal was formerly a division of Aeroquip Company. The manufacturing facilities for lip seals have recently been sold to Fluorocarbon Company of Anaheim, California. The old designation of Omniseal may be continued here because it is well known among users.

The Omniseal design provides the same type of sealing mechanism and operating principles as described for both BAL-Seals and FCS-Seals.

1.   Design Concept

The Omniseal was also a C-shaped device that seals at two point contacts at reinforced corners of the C-shaped geometry, with the opening directed against the pressure.

The success of the Omniseal is rated equal to the BAL- and the FCS-Seals for both static and dynamic applications for rotating and reciprocating service conditions. During rotation small particles of the Teflon casing actually deposit along the contact path of the seal on the shaft surface, gradually forming a microscopic film deposit along which the C-shaped seal can ride with a minimum of friction and wear, providing a Teflon versus Teflon condition. This is the basic explanation for the low wear in all PTFE-base seals. The seal acts on the self-energizing sealing principle with a certain degree of self-lubrication, which is why there can be intermittent dry-run conditions without immediate seal failure. This principle applies to BAL- and FCS-Seals alike.

2. Applications

Pressure, temperature, and corrosion ranges are the same as those described for the BAL- and FCS-Seals. The same statement can be made with regard to the wide range of feasible construction materials.

F. Wedge Seal

The wedge seal is a special design frequently used for moderate operating conditions when rotating seals are involved. For reciprocating shafts the conditions can be somewhat more rigorous.

1. Design Configuration

The design shown in Fig. 3.20 represents a wedge seal with two Teflon wedge rings, activated by a spring. As a result of the spring action the circular inner wedge ring is compressed against the shaft surface, thus simultaneously pushing the outer wedge ring outward, forcing it to move against the housing for sealing. The inner ring is forced against the shaft; the outer ring is forced against the housing.

2. Applications

Although the wedge seal is preferably used for sealing reciprocating shaft operation, it is also used for sealing rotating shafts. In equipment with rotating shafts the operating conditions are relatively mild. For reciprocating shaft operations the seal has been successfully used for pressures on the order of 4000 psi at temperatures of up to 500°F.

IX. Summary Conclusions for Lip Seal Configurations

Of all dynamic seals for rotating shaft equipment the lip seal covers the largest range of industrial applications. The elastomeric lip seal is well suited for numerous service conditions for relatively moderate requirements. Its basic limitations are pressure, temperature, and corrosion.

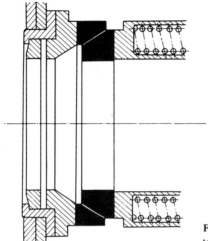

Fig. 3.20 Wedge seal for high-pressure and temperature (Courtesy of *Machine Design*).

Considerable improvements are imparted when Teflon is used as a base material. When it is further alloyed with carbon, graphite, or molybdenum disulfide, the temperature limitation can be considerably improved and the seal life may increase by a factor of 5 to 10 or more.

With the use of PTFE-base material the dynamic runout tolerances must be noticeably narrowed. The dynamic shaft irregularities can be reduced by placing ball bearings in close proximity to the seal, as indicated in Chapter 2 for mechanical end face seals.

Among the elastomeric lip seal configurations the wave seal represents a noticeable step forward, increasing normal seal life by at least 25%. Fluoroelastomers permit an even greater step forward.

With the application of Teflon and its corresponding combinations with carbon, graphite, and molybdenum disulfide, resistance to corrosion, temperature, and wear has been markedly improved. Wear especially has been reduced.

The standard materials listed in Table 3.8 are customarily used.

PART 2.   Exclusion Devices

I.   Introduction

Exclusion devices represent a category of seals exclusively designed to prevent entry of foreign particles through openings in equipment with

Table 3.8 Materials for Lip Seals

Material	Purpose of Application
Elastomers	
Leather	Very general use. Has inferior heat resistance but excellent and low friction values.
Nitrile Rubber	Very generally used in wide range. All kinds of practical design configurations.
Polyacrylic rubber	Applicable to same range as nitrile rubber, but to a higher and extended range of temperature.
Silicone rubber	Excellent low-temperature stability. Good resistance to hot oil, covering certain lubricant types.
Fluoroelastomer	Best of all elastomers, high temperatures, low friction values, excellent resistance to lubricants.
Plastics	
PTFE	All-purpose use, considering pressure, temperature, and
PTFE plus	wide-range chemical corrosion with narrow tolerances.
carbon,	Best life expectancy of all lip-type rotating seal
graphite,	configurations.
molybdenum	
disulfide	

rotating and/or reciprocating shafts. This type of seals is categorized as wipers, scrapers, axial seals, and boots, often in combination with bellows.

II. Wipers

Wipers are exclusion devices that strongly resemble radial lip seals. The wiping design components are made of leather, elastomers, or plastics, commonly encased by metal. Wipers are applied to both reciprocating and rotating shafts, with primary emphasis on reciprocating shafts.

A. Basic Configurations

Some basic wiper designs are presented in Fig. 3.21. Figure 3.21a illustrates a single-piece wiper with a strong wiper lip used for light- and medium-duty service, typical for reciprocating shafts. This design is also applied to rotating shafts, in which case the device must be kept from rotating.

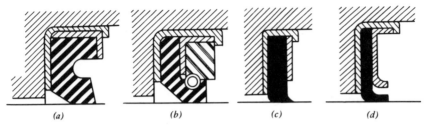

(a)   *(b)*   *(c)*   *(d)*

Fig. 3.21   Wiper design configurations (Courtesy of *Machine Design*).

Figure 3.21*b* represents a single-piece wiper with garter spring for heavy-duty operations for both reciprocating and rotating shaft. The garter spring is backed up by a soft rubber ring protecting the spring from penetration of foreign particles and from contaminating atmospheres.

Figure 3.21*c* shows a wiper with a self-forming elastic lip suited for light- to medium-duty service. This design allows some side play of the shaft.

Figure 3.21*d* also represents a wiper with a self-forming lip suited for dry or moist dirt application in slow movement. Because of the abrasive nature of the dirt the wiper is often provided with a multiple-lip arrangement.

For wiper action leather is a good material; it has excellent resistance to abrasion and a low friction factor. Leather is capable of absorbing and releasing lubricant, depending on need. Temperature limitation is rated at 200°F maximum, whereas elastomers go to higher limits but are accompanied by lower abrasion resistance.

Leather is suitable for most oils and a wide range of chlorinated solvents. Leather is not recommended for acids and alkalies.

With a hardness durometer of 70 to 90 Shore A hardness elastomers provide good toughness and abrasion resistance as wipers. They can be recommended for temperatures ranging from −70 to 250°F.

Temperature application range drops to −140°F in the cryogenic region and rises to 350°F in the higher range for silicone rubber. When used in a dry atmosphere the temperature may be raised to 500°F. Abrasion resistance is lower than that for other synthetic materials.

In recent years polyurethane has also been used in larger quantities for wipers because of its low friction values and good abrasion resistance against metals. The application range of polyurethanes for temperature is −90 to +250°F.

PTFE wipers are gaining increasing importance and have a low coefficient of friction, high temperature range, and excellent resistance to corrosion. The newest member of the plastics family for wiper application is polyimid, which allows temperatures in excess of 500°F.

B. Mechanical Requirements

Proper wiper action can be expected if the mechanical conditions meet similar requirements as those indicated for lip seals, such as shaft material, shaft condition, shaft speed, temperature, pressure, lubrication, type of motion, materials encountered, and the like.

Shaft Condition. Straightness, roundness, surface finish, and surface hardness are of great significance to seal life. Unlike conventional lip seals, wipers call for a shaft hardness of $R_c \sim 50$ as a minimum and a surface finish of 10 to 20 μin. rms. The higher the surface hardness of the shaft, the longer the service life of the wiper.

Operating Pressure. Wipers are not designed to take pressures. They function well under atmospheric pressure conditions and they are capable of holding a vacuum. If the vacuum is constant it is possible to design the wiper with an extra special lip.

Shaft Speed. Wipers for rotating shafts are capable of withstanding normal speeds similar to those for lip seals. Excessive speeds develop frictional heat and result in premature lip failure. For a reciprocating shaft the speed is never a problem.

Lubrication. The wiper lip must be lubricated, which is often a problem in rotating shafts. In reciprocating shafts lubrication is mandatory, and control of the lubricant is essential.

Temperature. Temperature generally is not a problem for wiper seals.

Motion. Reciprocating and rotating motions are not critical to wiper seal lips; however, the situation becomes critical if rotating motion is reversed or restarted after long intervals of shutdown. In reciprocating motion, runout, deflection, and eccentricity can finally become detrimental if the phases of irregularities are prolonged.

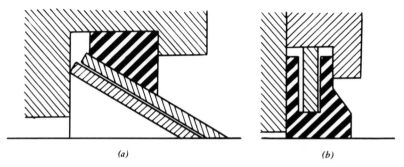

(a)   *(b)*

Fig. 3.22 Scraper designs (Courtesy of *Machine Design*).

III. Scrapers

Scrapers are actually wipers, preferably made of metal. The metal lip is in direct positive contact with the shaft. Scrapers are used exclusively for reciprocating shafts. Scrapers are often combined with wipers installed behind the scraper to prevent penetration of fine particles or traces of fluid that may pass the scraper edge.

Two scraper configurations are illustrated in Fig. 3.22. Figure 3.22*a* shows two conical metal scrapers. Each of the scrapers may be provided with three or more slots to increase flexibility of the scraper lip, providing a spring action for maintaining continuous surface contact with the shaft. The two scrapers are installed in staggered positions in order to prevent particles from passing all the way through the slots. An external rubber ring acts as a seal and simultaneously corrects scraper position in shaft side plays, providing automatic contact force also.

Figure 3.22*b* represents a grooved scraper ring for insertion of a wave spring for augmentation of the contact pressure. The design provides shielding from particle penetration. This design is also often referred to as a scraper ring.

Conical scrapers usually have sharp edges to remove foreign matter from the shaft surface. The scraper ring has a flat top, ranging from 0.030 to 0.062 in. in width. This extra width provides additional strength.

FCS-Fluorocarbon Company has marketed a new scraper, shown in Fig. 3.23, especially designed for equipment handling abrasive or viscous fluids. The shear action of the highly loaded lip prevents fluid buildup between the seal and the mating hardware.

For the selection of scraper materials, toughness, wear resistance, and resistance to abrasion play an important role. Conical scrapers are selected with regard to the position of optimal spring action.

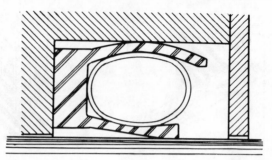

Fig. 3.23 PTFE scraper (Courtesy of *Machine Design*).

IV. Bellows and Boots Devices

Bellows and boots are exclusion devices that operate without direct contact with the movable hardware. In these devices the motion is absorbed by the flexing of the boot itself.

Bellows and boots are used to protect universal joints, ball joints, hinges, and shift levers. They are usually made of leather, elastomers, or plastics.

V. Axial Exclusion Devices

There are cases where the conventional lip seal configuration of the radial-type design is not applicable. For this purpose the axial-type seals have been developed to fill the gap. These seals are similar to the axial mechanical end face seals in that they use two flat faces for an interface contact in rotating motion. A typical representative of this type is shown in Fig. 3.24. The illustration is self-explanatory.

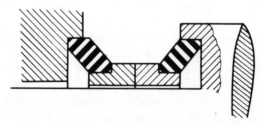

Fig. 3.24 Axial exclusion device (Courtesy of *Machine Design*).

VI. Summary

Exclusion devices are classified into four basic groups: wipers, scrapers, axial devices, and boots. They provide protection of systems by preventing penetration of foreign particles into the systems with rotating or reciprocating shafts. They are not designed to provide a direct sealing function.

References

1. Bingham, T. G. and Callaghan, G. P.
   "Fluid Resistant Elastomers"
   Lubr. Eng. (August 1971).

2. Boyce, H. L.
   "Selection Factors for Seals and Packings"
   Plant Eng. (November 1972) basic materials; (January 1973) types and shapes.

3. Brink, R. V.
   (a) "Reducing the Friction in Seals Is Active Goal of Engineers"
       Prod. Eng. (January 1975).
   (b) "The Heat Load of an Oil Seal"
       Company Report, Chicago Rawhide Company, Chicago, Illinois.

4. Brink, R. V. and Horve, L. A.
   "Wave Seals a Solution to the Hydrodynamic Compromise"
   Paper presented at the 27th Annual Meeting of the ASLE, Houston, May 1972.

5. Buchter, H. H.
   "Apparate und Armaturen der Chemischen Hochdrucktechnik"
   Springer-Verlag, New York, 1967.

5a. Dega, L. R.
   "Fluid Film Sealing"
   ASLE Education Course

6. Field, G. J.
   "Seals That Survive Heat"
   Mach. Des. (May 1975).

7. Gillespie, L. H., Saxton, D. O. and Chapman, F. M.
   "New Design Data for Teflon"
   Mach. Des. (1960).

8. Horve, L. A.
   "Fluid Film Sealing—Elastomeric Lip Seals"
   ASLE Education Course, Chicago, IL.

9. Hyde, G. F. and Fuchsluger, J. H.
   "Materials for High-Temperature Seals"
   Mach. Des. (December 1961).

10. Jackson, D. B.
   "Rotary Shaft Seals"
   Mach. Des. (June 1969).

11. King, W. H.
    "A Fresh Look at Elastomers"
    Mach. Des. (January 1973).

12. Malcolmson, R. W.
    "Elastomers for Sealing"
    Mach. Des. (October 1964).

13. Ostmo, O.
    "How to Select Shaft Seal Materials"
    Lubr. Eng., **29** (1973).

14. Payne, D. C.
    "Choosing Lip Type Shaft Seal Materials"
    Chem. Eng. (February 1972).

15. "Dealing with Lip Type Shaft Seal Problems"
    Chem. Eng. (May 1972).

16. Schnürle, F. and Upper, G.
    "Influence of Hydrodynamics on the Performance of Radial Lip Seals"
    Paper presented at 28th Annual Meeting of the ASLE, Chicago, May 1973.

17. Symons, J. D.
    "Seal Design Parameters and Their Effect on Seal Operating Temperatures"
    ASME Paper 66-MD-10, 1966.

18. Stephens, C. A.
    "Matching Seals and Lubricants"
    Mach. Des. (January 1965).

19. Taylor, E. D.
    "Birotational Seal Designs"
    Lubr. Eng. (October 1973).

20. Weinand, L. H.
    "Radial Hydrodynamic Seals"
    Mach. Des. (June 1969).

21. Design Guide
    "Elastomers"
    Mach. Des. (August 1965).

22. Reference Issue
    Mach. Des. (August 1965).

23. "Custom-Designed Fluid Seals Are Unique, and Can Be Cheap"
    Prod. Eng. (June 1974).

Numerous manufacturers' publications, papers, catalogs, and brochures were also consulted, as were various seal reference issues of *Machine Design* (1973, 1974, 1975).

CHAPTER 4

Circumferential Split Ring Seals

I. Introduction

Circumferential split ring seals represent a modified version of the commonly used piston ring seal. Although the conventional split ring seal resembles the piston ring design, the sealing mechanism comes closer to that of mechanical end face seals. The design is characterized by the fact that the seal ring contacts always the same section of the shaft. Consequently, lubrication, wear, and temperature distribution are similar to those found in mechanical end face seals. Piston ring design in particular is discussed in connection with reciprocating rod seals in Chapter 6, Part 4.

II. Design Concepts

Circumferential split ring seals are available in a variety of design configurations classified into two basic groups. One group comprises a category paralleling piston ring seal design with one-ring, two-ring and three-ring configurations. The second category represents the segmental ring design.

A. Piston Ring-Type Seal

Unlike piston rings for sealing reciprocating shafts circumferential split ring seals perform their dynamic sealing function at the ID of the seal rings. Consequently, they require a preload device in a radial direction to force positive contact on the rotating shaft surface in the absence of fluid pressure. The contact force is generally provided by the spring tension of the piston rings in addition to external springs that supply that automatic contact force.

Piston rings in service for sealing rotating shafts have one-piece or multipiece designs. The material used for piston rings as well as the application requirements may vary considerably for each specific type of design. Multipiece rings may have their individual parts made of different materials, depending exclusively on application requirements. Further details on piston ring seals can be studied in the *Engineer's Handbook of Piston Rings and Seal Rings*, published by Koppers Company, Inc., Metal Products Division; Baltimore, Maryland.

The most common and most frequently applied piston ring is the standard compression ring with a plain rectangular cross section. There are many modifications of the rectangular cross section serving specific purposes that cannot be discussed in detail here. For rotating shafts, the one-piece contracting ring is used, which is designed to contract to form a seal on the ID of the ring; the standard compression ring, on the other hand, expands, in a self-energizing fashion.

A one-piece seal ring has only one gap through which leakage may occur. For this reason the location of the rotation lock is a problem. Frictional torque and torque restraint produced by the presence of the rotation lock may result in "winding" or "unwinding" of the seal ring. When the ring winds, the seating force on the shaft surface increases. On the opposite when the ring unwinds the seating force will diminish and the joint may finally fail by excessive leakage.

The shape of the joint is, therefore, a significant factor for a satisfactory function of the seal ring. The step can either be flat or stepped as indicated in Fig. 4.1, where only three configurations of many possible designs are illustrated.

For multiring designs combinations of two or three rings are used in all kinds of industrial applications. A two-piece seal is shown in Fig. 4.2 and a three-piece device is illustrated in Fig. 4.3. The two-piece self-energized contracting ring (Fig. 4.2b) consists of a step joint inner ring

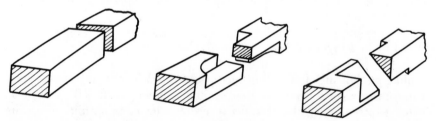

Fig. 4.1 Joint steps of single-piston seal rings (Courtesy of Koppers Company, Inc., Baltimore, Maryland).

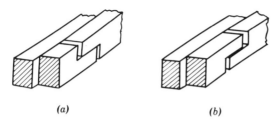

(a)   (b)

Fig. 4.2 Steps of two-piece rings (Courtesy of Koppers Company, Inc., Baltimore, Maryland).

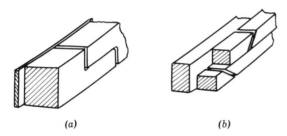

(a)   (b)

Fig. 4.3 Steps of three-piece rings (Courtesy of Koppers Company, Inc., Baltimore, Maryland).

and a straight-cut outer ring. The inner ring fits on the shaft, sealing its circumference except for a minute clearance at the joint of the ring. The clearance is sealed by the outer ring, which also provides inward tension to exert contact force against the inner ring and on to the shaft.

The three-piece design (Fig. 4.3) consists of a straight-cut joint inner ring and two outer rings with joints of opposite angles. The inner ring supplies the required tension and seals the joints of the outer two rings. Joints with opposite angles do not line up in service; consequently locking pins are not required.

The overall wall of a three-piece ring set is thicker than the wall of a one-piece ring of the same shaft diameter. As a result the corresponding depth of the piston ring groove must be increased.

It is obvious that a multipiece arrangement provides more joint strength than a single-piece arrangement.

B.   Segmental Ring Design

Segmental rings consist of various interchangeable segments that are regularly furnished with well-filleted step joints. They are available in a variety of design configurations. Step seal joints are straight-cut joints

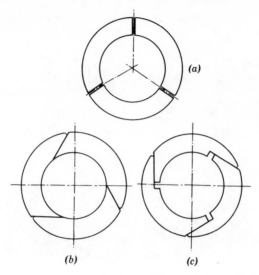

Fig. 4.4 Special rod packing rings.

and can be furnished when desired. The segment application permits the use of a much heavier wall than is possible with a one-piece ring, which must be sprung over a shaft. Consequently, service life can be improved proportionally. An elliptical outer ring provides tension for the segmental ring. The segmental ring shown in Fig. 4.4a is designed for external sealing, as is the case for reciprocating motion of a rod. Further details on segmental rings are discussed in Chapter 5. As a basic rule, piston rings are used or designed for both internal and external sealing purposes. When they are used for the rotating shaft or reciprocating rod, the type of sealing is internal and the rings are made as contracting rings. When they are used in a piston that reciprocates in a cylinder, the rings must seal externally. For external sealing the rings must be able to expand outwardly to exert the required sealing contact force.

A different kind of segmental seal rings is shown in Fig. 4.5, which is also often referred to as the rod packing ring assembly. This design usually consists of a tangential segmental ring, a radial segmental ring, and two garter springs. Each ring is assembled as a unit using three segments provided with an external groove to insert the garter spring, which holds the segments on the shaft surface with a uniform radial contact force. As the illustration indicates, the joints of the tangential ring are designed so that as wear takes place in the contact area with the shaft on the ID of the ring segments the two mating surfaces of the joints

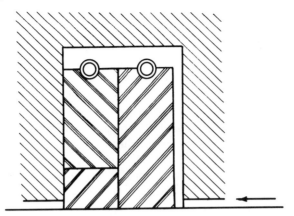

Fig. 4.5 Assembly of set of rod packing rings.

remain in contact, thus establishing a constant seal at all times. The external leg of each joint must be undercut to prevent the formation of a shoulder that could interfere with the free inward movement of the segment with progressing wear. Further details of the rod packing ring assembly and specific applications are discussed in Chapter 5.

Advantages of the segmental ring set are obvious. A one-piece seal ring must be assembled on the shaft by slipping the ring over the end face of the shaft. Segmented seal rings can be assembled directly on the shaft without dismantling any parts of the system to be sealed.

Figure 4.4 illustrates three basic types of segmented rings, including a radial cut ring (*a*) and tangentially cut rings (*b*) and (*c*).

A set of rod packing rings is shown in Fig. 4.6 in assembled condition. Further details on this subject are presented in Chapter 6.

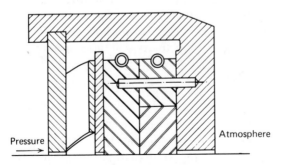

Pressure
Atmosphere

Fig. 4.6 Basic segmental shaft seal with wave spring [From (2), courtesy of *Machine Design*].

III. Sealing Mechanism

Unlike conventional piston rings for reciprocating sealing action in pistons, circumferential split ring seal rings perform their dynamic sealing action at the ID, along the surface of the rotating shaft. To maintain steady contact with the interface in the absence of fluid pressure a garter spring is used to provide the automatic contact load.

The rings must also seal against the housing, which is often referred to as secondary sealing action, accomplished either with the help of multiple-coil springs or by the use of a wave spring. The sealing mechanism becomes obvious in the sketch of Fig. 4.6. In sealing, the segmental seal is unidirectional. For installation the wave or coil springs must always be on the pressure side.

When the seal is in operation, fluid pressure acts and maintains the necessary contact loading for primary and secondary sealing action. The seal is free to move radially thus permitting the segments to follow irregularities in the shaft movement without producing excessive leakage. Simultaneously the straight cylindrical geometry of the primary sealing interface has the freedom to move axially. In addition to the fact that segmental seals can be assembled with a minimum of machine disassembly, they offer a distinct advantage over the axial mechanical end face seals.

The rings are in steady contact with the same surface area of the shaft. Because of the segmental design characteristic the circumferential split ring seals are capable of developing hydrodynamic lift at the seal interface. The fluid then produces a pressure underneath the contact surface. As a result the seal segments are lifted up slightly, thus raising the spring force until this increased spring force counteracts the lift and equilibrium is established again. Once the seal segments separate from the shaft surface, leakage increases. The tendency of lifting increases with an increase of the fluid viscosity. For this reason circumferential split ring seals are not suited for sealing liquids. However, they are effectively applied in sealing gases and vapors. They are subject to high-velocity rubbing at the primary seal faces and, therefore, should be cooled.

IV. Design Features

The design is characterized by rings that are designed as single one-piece devices or as segmental rings. The one-piece seal ring requires a strong joint. This is achieved by groove and tongue design. Two-piece seals use step joint rings around a straight-cut segmented ring with a tangentially cut segmented ring. The garter spring, the multiple axial springs, and the rotation stops in the form of pins are the additional components to complete the circumferential split ring seal.

Interface loading is proportional to the fluid pressure acting around the seal rings. By increasing the fluid pressure wear will also increase. With the application of hydrostatic pressure balancing very similar to the balancing of mechanical end face seals, shaft speed and pressure range can be increased. An example for hydrostatic pressure balancing is illustrated in Fig. 4.7.

Here the contact surfaces are vented so that full fluid pressure is acting only against a proportionally reduced area, whereas the entire surface of the seal is exposed to the fluid pressure. The degree of balance can be designed to any desirable contact load, thus reducing wear to a minimum and increasing seal life. The amount of balance is limited because balance is achieved at the expense of the seal interface area, which can be seen readily from Fig. 4.7. The limit of interface reduction is given by the strength requirements for the seal contact area.

Since the seal contact always occurs at the same location of the shaft surface, damage of the shaft in this area must be expected. It is, therefore, advisable to use a shaft sleeve to protect the shaft. A shaft sleeve simplifies the selection of proper shaft material and the seal material.

The sleeve area can also be used to remove frictional heat generated by the ring contacts either by introducing a cooling medium or by

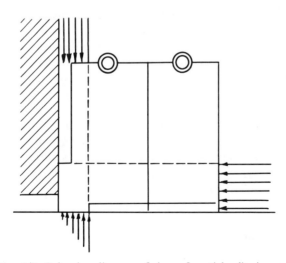

Fig. 4.7   Balancing diagram of circumferential split ring seal.

establishing metal-to-metal contact between the sleeve and the shaft. Cooling can also be accomplished by using a jet of oil directed toward the sleeve or underneath it. Great care should be taken to provide uniform cooling of the entire seal unit to minimize distortion, keeping in mind that centrifugal force tends to move the oil away from the cooling area.

V. Modes of Leakage

The rings of circumferential seals must provide a sealing function axially along the shaft surface and radially against the contact face of the housing. They have a high degree of built-in flexibility in both the radial and axial directions and easily accommodate moderate irregularities of the shaft motion, considering eccentricity, runout, or other inaccuracies.

VI. Materials for Circumferential Split Ring Seals

The nature of the design of the seal permits the use of low-strength materials, particularly since the garter springs hold the segments toward the shaft and provide the necessary contact force to establish con-formability with the rotating shaft surface. When used as a gas seal the contact force is on the order of 0.25 to 0.75 lb/in. of circumference.

The selection of materials for the seal rings depends on the service conditions. Carbon graphite is almost entirely used for seal rings. Carbon is further modified with PTFE, molybdenum disulfide and/or a combination of carbon-PTFE and glass fiber. In recent years polyimid resins have entered the field because of their extreme resistance to high temperatures and excellent low-friction characteristics.

In many cases the carbon-graphite matrix is impregnated with various resins, metallic salts, or such metals as silver, lead, Babbit, or copper to combat temperature and corrosion.

Hardenable steels are well suited for shaft or sleeve design, especially hardenable stainless of the 400 series. Hardness, surface finish, and perfect surface geometry are all important factors. Surface hardness of $R_c \sim 35$ to 45 provides excellent service for mating with carbon-graphite. A surface finish of 10 to 20 μin. is satisfactory.

VII. Practical Seal Applications

The most widely used design is the unit shown in Fig. 4.6. The seal employs three sealing-ring elements, each ring consisting of several segments. In conventional design it is customary to use three segments per ring for seals up to 4 in. shaft diameter; more segments are required for seals matching shafts with diameters larger than 4 in. The gap against the housing is sealed by two concentric rings staggered against each other to cover the gap of the joints in the radial direction. The secondary seal ring is of the one-piece design, arranged to cover the gaps of the joints of the primary rings. The only possibility of leakage occurs in the passage along the shaft surface across the contact area with the seal rings and with the housing.

A garter spring provides the automatic contact force for both rings against the shaft. The axial force for the contact with the housing is provided by a wave spring. The fluid pressure enhances the seal contact force by a larger amount than is possible by the garter springs alone.

In the modified version shown in Fig. 4.8 the secondary seal ring is eliminated, making the axial length of the seal unit somewhat smaller. The contact surfaces within the tongue-and-groove-type joint of the inner ring segments provide a satisfactory length for sealing the primary ring. The primary outer ring, lapped in the contact areas, prevents radial leakage along the housing gap. This design is somewhat complex, difficult to machine, and, therefore, rather costly. Axial and radial fits must match with a high degree of accuracy (2).

Primary outer ring, segmented

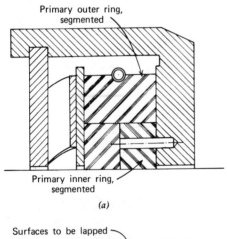

Primary inner ring, segmented

(a)

Surfaces to be lapped

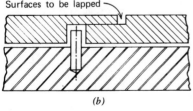

(b)

Fig. 4.8 Modified standard split ring seal design [From (2), courtesy of *Machine Design*].

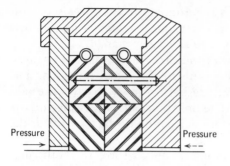

Fig. 4.9 Bidirectional seal [From (2), courtesy of *Machine Design*].

L. J. Cawley and R. M. Duncan (2) describe a seal in which the system pressure can act from either direction. This bidirectional design is shown in Fig. 4.9. This seal is characterized by the use of unsprung rings in the axial direction. The rings are assembled so as to permit a move in the axial direction by an amount of up to 0.005 in. between the limiting radial walls of the housing. Once the system pressure is applied, the rings are pressed against the opposite radial wall of lower pressure, where the faces form a tight seal. On reversal of the pressure the ring combination is moved in the opposite direction and seals against the other limiting wall.

A tandem seal arrangement is illustrated in Fig. 4.10 (2). This configuration is used frequently for systems where gases that cannot tolerate contamination by lubricant oil must be sealed.

The tandem arrangement can also be used where an inert gas purge of nitrogen or helium must be applied to keep the system gases completely separated.

Vent

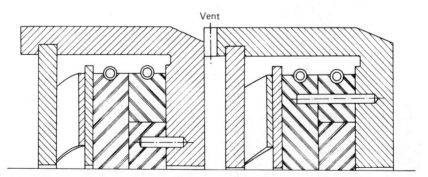

Fig. 4.10 Tandem seal arrangement [From (2), courtesy of *Machine Design*].

VIII. Summary Remarks

Circumferential split ring seals are used for systems handling gases, such as air, carbon dioxide, hydrogen, helium, nitrogen, or even oxygen. They can also be used to retain fluids. The hydraulic balance limitation may create a problem, but one that is not serious as long as the seal is operated at low pressure differentials. When operation at high speed and high temperatures is involved, the pressure must be reduced correspondingly. Speeds of 400 ft/sec at temperatures of 850°F and a pressure differential of 35 psi can be handled without difficulty, providing the shaft is cooled. By reducing temperature and speed, the pressure can be raised correspondingly and cooling may not be required.

Shaft or sleeve waviness may become critical at high rotational shaft speeds because the annular clearance gap with the shaft is increased, which increases the likelihood of leakage. Deflections caused by fasteners and nonuniform temperature distribution can create problems.

The criteria for the selection of suitable seal ring materials are quite similar to those valid for mechanical end face seals. Good wear and antifrictional properties are essential. High-temperature resistance and chemical compatibility for the operating environment are mandatory for ring and shaft-sleeve materials to ensure long seal life. A low modulus of elasticity, unlike that required for mechanical end face seals, is desirable to provide flexibility, making it possible for the seals to conform closely to the shaft surface.

Carbon-graphite is by far the best substance to satisfy most of the requirements for a good ring material. It can be used for temperatures on the order of 900°F, although grades are available that allow temperatures up to 1200°F and continuous service.

Additional information on segmental seal ring components is presented in Chapter 5.

References

1. Bauer, P., Glickman, M. Iwatsuki, F.
   "Analytical Techniques for the Design of Seals for Use in Rocket Propulsion Systems" Technical Report AFRPL TR-65-61, Air Force Rocket Propulsion Laboratory, Edwards, CA, 1965.
2. Cawley, L. J. and Duncan, R. M.
   "High Performance Seals"
   Mach. Des. (April 1, 1971).

3. Ruthenberg, M. L.
   "Mating Materials and Environment Combinations for Specific Contact and Clearance-Type Seals"
   Paper presented at 27th Annual Meeting of the ASLE, Houston, TX May 1972.
4. *Mach. Des.*
   (a) Seals reference issue, 1973–1974.
   (b) Reference issue, 1975: "Mechanical Drives and Seals."
5. *Engineers Handbook of Piston Rings, Seal Rings, Mechanical Shaft Seals,* 9th ed.
   Koppers Company, Inc. Metals Products Division, Baltimore, Maryland, 1975.
Numerous brochures, catalogs, papers, and other manufacturers' publications, including extensive NASA literature, were also consulted.

CHAPTER 5

Packings as Shaft Sealing Components

Packings are fundamental components widely used in all branches of industry. They are used to seal both rotating and reciprocating shaft equipment. With valve stems the sealing must be achieved against both rotating and reciprocating motion.

The subject of sealing by packing devices is somewhat intricate and demands a good deal of practical experience. It was, therefore, considered important to subdivide the chapter into four parts, with Part 1 dealing with fundamentals, Part 2 with the rotating shaft, and Part 3 with reciprocating shaft equipment. Special packings of metals for extreme high-pressure service are presented in Part 4.

PART 1. FUNDAMENTALS OF MECHANICAL PACKINGS AS SHAFT SEALING DEVICES

I. Introduction

Packings are without doubt the oldest devices used to seal static and dynamic components. For dynamic sealing packing is generally inserted in a stuffing box cavity and then tightened by a gland ring plate attached to a casing through which the shaft is passed.

It is an unfortunate fact that although packings are the most widespread sealing devices, they are the least understood. It is common to tighten a packing gland, once leakage is observed, regardless of condition of the packing. In reality the packing may be "jammed," thus becoming subject to increased wear and perhaps accelerated failure. Very few plant people responsible for services and maintenance have ever learned the basic function and the mechanism of a mechanical packing. Their actions usually create more damage than many plants can tolerate.

This chapter, therefore, attempts to shed some light on this subject. This chapter provides a wealth of concentrated information on packing basics not readily found elsewhere.

II. Types of Motion to Be Sealed

The shaft in a pump or in a piece of chemical operating equipment can undergo four basic types of motion. The shaft may reciprocate, rotate, rotate and move simultaneously in axial direction, as is the case in a valve stem, and finally swing in an angle 360 degrees around the shaft axis while rotating. In each of these four principal motions the seal of the shaft is subject to an entirely different sealing mechanism. These problems must be fully understood before a proper solution can be achieved.

First of all a true shaft cannot be sealed, as indicated earlier. A certain amount of runout, eccentricity, surface roughness, and surface hardness must exist to provide traction for the lubricant film to adhere to the shaft surface. A perfect shaft develops stick-slip and the seal fails because of lack of proper lubrication.

Rotating lip seals and packing operate differently. It is generally stated that a packing must stop flow. This is not completely true. A packing of the common design, not including the TFE-resin rings, merely throttles the flow of fluid and does not stop or completely block it. This is the reason for the old saying that a "packing seals when it leaks."

III. Packing Classification

Packings may be classified in two major groups, the jam type and the automatic type. The jam type may assume any packing shape that can be inserted or jammed into a stuffing box chamber and compressed by tightening the bolts of the gland ring plate. Jam-type packings may be circular, square, wedge, or cone shaped and are either braided, twisted, woven, or laminated of rubber, leather, Viton, fiber, fluoroelastomers, TFE resins, or a combination of them.

The automatic-type packings are designed primarily as lip-type components where the lips have the task of sealing after establishing contact with the corresponding seal areas in the housing or with the shaft surface. After insertion in the stuffing box, once the system is under pressure the automatic-type packings need no tightening force. They are installed in the stuffing box with the lips directed toward the pressure of the system fluid to utilize the self-energizing sealing principle. The major members of this large family of packings are designated as follows: U-cup rings, V-rings, C-rings, M-rings, and flange rings.

IV. Jam-Type Packings

In the early years, before TFE resins and O-rings were available, packings were made by the jam-type design. They have undergone a significant development since and are still a strong competitor on the sealing market in spite of Teflon, Viton, and so on, or perhaps because of them.

A. Sealing Mechanism

All packing design configurations for industrial sealing purposes have been developed empirically. Despite the high state of the art, there is no reliable analytical design theory that can predict jam sealing behavior.

The average jam-type packing of the braided configuration has approximately 40 to 45% interstructural voids when made from fiber. These voids are filled with lubricant. When it is installed in the stuffing box chamber and tightened by hand, the packing has a certain length within the stuffing box chamber. When the bolts of the gland are tightened the packing rings are compressed, the lubricant is squeezed out, the voids are less, and the length of the packing is reduced, making this compression possible. The deformation of the ring volume has two major effects. First, it decreases the internal diameters of the packing rings; second, it simultaneously increases their external diameter. This produces an increase in contact force against the shaft surface and against the stuffing box wall. In other words, the static and the dynamic sealing function of the packing are enhanced. When saturant in the packing is completely forced out, the pressure in the packing fades away and the gland must be retightened. This cycle is repeated until no more saturant is left and the packing becomes dry and finally fails. Here practically all the 40 to 45% of the voids are used up, the saturant is drained out, and the packing begins to damage the shaft surface by scoring it.

It should be noticed that it is the lubricant film that seals the fluid from leaking out to the atmosphere and not the solid packing. Once the lubricant is drained out after reaching the compression limit of the packing, leakage takes place and the shaft may be damaged.

B. Influence of the Shape of Jam-Type Packings

The number of rings in the stuffing box, their shape, and the material of which they are made greatly influence the efficiency of the packing. At first it might seem appropriate to insert into the stuffing box as many rings as possible. Practice shows, however, that this is not true, for the force imparted by tightening the packing decreases parabolically with increasing number of rings as a function of the length of the packing. The last ring at the stuffing box throat may not obtain any contact force for tightening the seal or may not squeeze out any saturant for lubricating the shaft, whereas the first ring is subject to the full tightening force.

It appears obvious that the highest contact pressure exerted by the gland is acting at a point where it is required the least. As is shown later in this chapter, the number of rings reaches an optimal limit of five. Additional rings are of little value.

A comparison of jam-type packings with square and conical rings reveals that the square rings reflect a steeper drop rate at the gland than conical rings do. The system fluid pressure is a maximum at the stuffing box throat and decreases from there to zero at the gland ring face. The pressure curves are parabolas for both square and cone rings. The ideal case would be a straight line across the entire length of the packing, designated generally as contact length.

The curve for the cone-shaped packing ring is located closer to the ideal distribution curve. The contact pressure effect produced by the gland on a square and a conical ring exerts force components in all three principal directions, such as shaft, housing, and stuffing box bottom, giving the cone ring an ideal chance for proper function.

The cross section of the cone ring is arranged so that it forms a slope angle with the shaft surface and the contact area of the stuffing box. The system pressure acts along the shaft surface against the ID's of the packing rings. The gland contact force can be analytically resolved into thrust components acting as seal contact forces. The static seal against the stuffing box wall is accomplished by additional force components.

Best sealing results and lowest frictional losses are achieved by using packing ring cone angles of 45 degrees.

Certain designs suggest the arrangement of beveled ends in the stuffing box throat to enhance sealing efficiency. Experience has shown that beveling stuffing box bottoms has no advantage. The packing is actually designed for direct contact in a direction perpendicular to shaft and stuffing box wall. Beveled ends are justified for cone rings only and for special sealing and process requirements.

Clearance between gland ring and stuffing box throat with the shaft is critical. Excessive clearance facilitates extrusion of the packing. Clearance values are a function of shaft speed, fluid, and machine or equipment characteristics.

C. Influence of Gyration Effects

Shaft gyration is common where there is a rotating shaft. Runout, deflection, misalignment, and certain amounts of eccentricities cause shaft vibration effects, generally termed *shaft gyration*, that produce larger openings in the shaft packings. To demonstrate the effects of shaft gyration on the seal efficiency, it may be assumed that a shaft of a centrifugal pump operating at 3600 RPM shaft speed may gyrate by 0.003 in. radially. At each revolution the shaft increases the packing diameter by 0.006 in. The shaft whips the packing 3600 times every minute. It is readily apparent that there is hardly any packing material capable of maintaining steady contact with the shaft surface under such severe conditions.

Figure 5.1 depicts vibration measurements as a function of shaft speed for a variety of system pressures, ranging from 50 to 1000 psi.

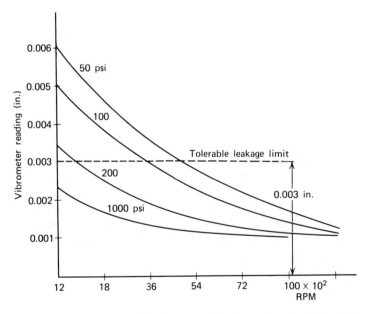

Fig. 5.1 Influence of gyration of shaft on rate of leakage through packing [From (12), reprinted with permission from *Power Magazine*, copyright McGraw-Hill, Inc.].

In most cases a total gyration reading of 0.003 in. can be tolerated without excessive leakage. Any vibration greater than 0.003 in. T.I.R. must be corrected by shaft or bearing adjustment, pump balancing, or some other means. Replacing the packing will not improve the situation.

D. Types of Jam Packing Rings

Jam-type packing rings are available in a wide range of design configurations satisfying the diversity of sealing requirements. They can be categorized into two major groups, braided and twisted rings.

Perhaps the simplest packing is the twisted design, made of twisted fibers strongly resembling steel cables. This packing type is very flexible and has the advantage because a given size can be used to match several sizes of stuffing box cross sections. New strings can always be added or taken off at the plant site without affecting the efficiency of the packing as a whole.

The braided packing is made in three basic groups, square, braid over braid, and interlocking. The various shapes of jam-type packings are readily available in extensive fabricator literature. Braided-type packing comes in rayon, nylon, cotton, asbestos, flax, jute, Teflon, Kel-F, glass, and such metals as copper, aluminum, lead, Monel, and others.

Metal braids require good lubrication. They are susceptible to high gyration effects. Best service life and corrosion resistance is provided by Teflon braids and its many combinations. At higher temperatures the coefficient of thermal expansion must be considered. Teflon, impregnated with carbon-graphite or molybdenum disulfide, offers up to 10 times the service life of all other materials.

Metals are widely used in crimped, spiral-wound, and braided designs. They have better resistance to higher temperatures than the ordinary fiber and they operate with lower coefficients of friction. They are fabricated in an extensive range of combinations, such as all-metal packing and metal-foil with resilient core and lubricated fibers, wound with foil cover. Metal foils have excellent bearing properties.

Where temperature levels are low and the system fluid tolerates it, leather is very widely used because of its toughness and excellent resistance to abrasion. The triple U-coil and stitched wedge configuration are the shapes preferred.

E. Loose-Fill Packing

Loose-fill packings are made from loose particles of asbestos fiber, separate pellets, shreds, or particles of lead or other suitable materials.

Loose-fill packings are suitable for higher temperatures and can easily be shaped to match the geometry of any stuffing box and the shaft surface. They are available in bulk, coils, spirals, or stiff bonded rings.

Excellent results are offered by rings made from shredded Teflon resins; they resist high temperatures, corrosion, and wear. Teflon is ideal for strong concentrated acid, such as sulfuric, chromic, nitric, and hydrofluoric acids, and for alkaline solutions. If it is used without a binder, Teflon resists fluorine, chlorine, bromide, sulfuric chloride, hydrogen fluoride, oxygen, and concentrated hydrogen peroxide, if excessive shaft speeds are not involved.

When made up as ring sections it is feasible to insert a solid Teflon ring as spacer between each filled ring. The Teflon ring then acts as a backup ring preventing the filled rings from welding together. A thickness of $\frac{1}{16}$ in. for the Teflon ring is sufficient to provide the desired stiffness.

F. Grafoil Packing

Early centrifugal pumps were sealed by filling the stuffing box chamber with fibrous packing rings, which consist primarily of hemp or flax. Over the years these original packing materials have been replaced by more effective and more durable materials. In present-day industry numerous types of packing materials are available that provide excellent sealing service and that resist high pressures, temperatures, and a wide range of corrosive materials.

New packing materials have recently been introduced into the market that can resist temperatures on the order of 2700°F. They are fibrous on the basis alumina and silica refractory components and offer a high degree of service reliability.

With the development of Grafoil, a Union Carbide product, another valuable packing material has joined the large family of successful sealing components. Grafoil comes in a fine sheet about 0.015 in. thick that can be used to fabricate packing rings in a variety of thicknesses and sizes for insertion in a stuffing box of any shape.

The rings of smaller size are made from Grafoil tape 0.015 in. thick, which is usually available in widths of 0.500, 0.750, and 1.000 in. These tape sizes are well suited for producing packing rings of various categories of small shafts up to 1 in. diameter. For larger shaft sizes the sheets can simply be cut to strips of any desirable width. From these tapes or strips packing rings can be made using a die, which is shown in Fig. 5.2.

The die has a bore with the dimension of the stuffing box diameter and a smaller bore corresponding to the diameter of the shaft. The tape

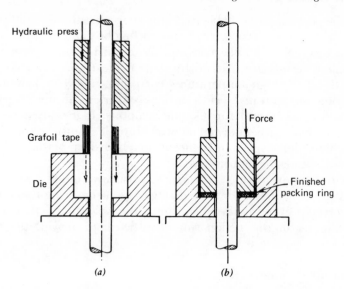

Fig. 5.2 Device to make packing rings from Grafoil tape. (*a*) Preparation of tape. (*b*) Fabrication of packing ring.

is wrapped around the shaft four or five times and then is pushed into the bore of the die with the piston and compressed to a thickness of about 0.100 to 0.125 in. Depending on compression force, any desirable ring thickness can be made. When a ring thickness of 0.125 is produced, the cross section has enough internal voids for a bore of ½ in. that a desirable flexibility is still available. Compression force, ring thickness, and cross-sectional ring density have to be developed empirically for each case in a matter of minutes.

During the compression a considerable amount of air is trapped in the cross section, which gives the ring a desirable amount of flexibility. It is preferable for the ring bore to have a diameter smaller than the diameter of the shaft to be sealed; this makes possible an interference fit when the packing is assembled.

Since Grafoil consists predominantly of graphite, it provides lubricity against the shaft without scorching despite the interference fit. When inserted in the stuffing box with a slight interference fit with the stuffing box bore, the ring segments provide a heat sink, since graphite has favorable thermal conductivity characteristics.

The author has operated a centrifugal pump with a Grafoil packing seal, using a coil spring for automatic loading. The pump was in operation for more than 1500 hr without leakage. The Grafoil packing can also be used for sealing reciprocating rod motion.

G. Basic Applications

Jam-type packings are almost universally effective as sealing devices for rotating and reciprocating shaft motion. They are available in an almost endless variety of design configurations, materials, and combinations. They constitute strong competition for mechanical end face seals.

Jam-type packings are applicable in equipment with large shaft diameters, where a replacement may necessitate complete disassembly of the machine. Jam packing rings are split and can therefore be assembled with a minimum of downtime and cost.

With the advent of fluorocarbons, TFE resins, polyimids, and combinations of these materials with carbon-graphite or molybdenum disulfide there are relatively few problems left that cannot be solved satisfactorily.

Packings will always leak to a certain degree. It is important to decide how much leakage can be tolerated without jeopardizing the process, profitability, environment, and operating personnel.

V. Automatic-Type Packings

Automatic-type packings are composed of two major groups, the lip-type devices and the squeeze-type packings, as indicated earlier.

A. Lip-Type Devices

The lip-type sealing devices have lips that press against the corresponding seal contact areas through the system pressure for both static and dynamic sealing. Each ring is designed for a specific purpose. The lips are provided by rings with V-, U-, C-, or M- cross sections. They are installed in the stuffing box like the jam-type rings. Other configurations are the flange ring and the U-cup ring. These designs have been available ever since the Teflon resins conquered the market. Newer material combinations with carbon-graphite and molybdenum disulfide have substantially widened the range of applications.

1. Sealing Mechanism

If one looks at the shape of the ring cross sections, it becomes obvious how the sealing mechanism functions. The rings have a built-in interference with the seal cavity and the shaft. This interference fit provides initial sealing at zero system pressure. Once hydraulic pressure is applied, the lips are spread outward and establish contact with the corresponding sealing areas. With rising system pressure the sealing effect is increased, because of the self-energizing sealing principle. This function is automatic, making springs or other mechanical devices obsolete.

With the advent of TFE resins and the corresponding low-friction combinations, which offer excellent durability, these devices have become much more widely applicable. Special attention must only be given to purely mechanical factors, such as closer tolerances, surface finish, surface hardness, shaft straightness, roundness, and minimal gyration effects.

2. Basic Applications

Automatic lip-type packing devices must satisfy the same basic sealing requirements as those discussed for jam-type packings. During many years of industrial field experience in the chemical and petrochemical industries the author has found very little difference between the basic behavior of jam- and lip-type sealing devices. There are two basic factors that must be considered before satisfactory sealing can be achieved. The first factor is the mechanical conditions to be met, including minimum gyration effects, good alignment of shaft and bearings with the housing, and shaft surface conditions that do not deviate from the tolerance requirements, including the shaft bearings. When these requirements are strictly met, the seal problem is usually 80% solved. The second factor is that the seal function must be carefully analyzed and fully understood in order to draw the appropriate conclusions about materials, shape, lubricant, installation, and maintenance. When these two factors are adequately coordinated, the seal problem no longer exists in many cases.

In general, automatic lip-type packings require closer gyration control than is necessary for jam-type packings. Speed of the rotating shaft also can be a critical factor. With TFE resins and their combinations with graphite and molybdemun disulfide as impregnations lubrication becomes less critical, although cooling always remains a significant requirement. Where the lubricant cannot dissipate the heat generated, the stuffing box must be cooled through the jacket. Because of graphite's excellent thermal conductivity, the graphite-impregnated rings contribute to heat removal.

Specific applications are discussed in detail later, in the sections dealing with rotating and reciprocating shaft sealing.

B. Squeeze-Type Automatic Packings

The category of automatic squeeze-type packings comprises O-, T-, X-, square, and delta rings. These devices, with the exception of the O-ring, have been developed for special purposes and each ring has its particular advantages and disadvantages.

1. Sealing Mechanism

Squeeze-type packing rings are fabricated from elastomers and highly resilient materials with a significant degree of memory. All rings have a cross-sectional height that is greater than the depth of the cavity into which they are inserted. For installation they must be stretched when they have to seal a shaft, or they must be compressed when they seal statically. After compression they tend to regain their initial shape. They, therefore, exert a rebounding force that is utilized as the initial sealing contact force. Once pressure is applied the rings close the leakage gap even further, thus enhancing the seal efficiency. On release of the system pressure the rings reassume their initial shape.

2. Basic Applications

The first member of the squeeze-type packing family was the O-ring, which has had a considerable history of success in both static and dynamic sealing applications. O-rings were primarily used in dynamic sealing for reciprocating shafts. They can also be applied to oscillating and rotating shafts; however, the shaft speeds must be reduced drastically. The pressures cannot be very high when conventional O-rings are used for rotating shafts.

Since squeeze-type packing configurations are preferably used for reciprocating rods, they are discussed in detail in the section on that subject at the end of the chapter.

PART 2. PACKINGS AS SEALING DEVICES FOR ROTATING SHAFT

I. Introduction

The most conventional method of sealing rotating shafts has long been and still is the application of packings in a standard stuffing box. For moderate pressures, four or five rings of jam-type packing suffice to establish a seal, providing the operating temperature is not a problem.

As indicated earlier in this chapter, sealing rotating shafts is more difficult than sealing reciprocating shafts. Speeds of rotating shafts, as in centrifugal pumps, compressors, blowers, centrifuges, and the like, are considerably higher than the speeds of reciprocating rods. Packings that seal rotating shafts are in continuous contact with the same section of the shaft surface, making the dissipation of heat more difficult. In addition, the rotating shaft is always subject to gyration disturbances of some amplitude, producing an annular gap in the packing through which leakage may occur. In order to seal rotating shafts, special designs must be provided, such as external cooling, the use of lantern rings, and so on.

II. Modification of the Packing Designs

Improved sealing efficiency of rotating shaft motions has come from the use of cooling and adequate lubrication of the packing during operation. This is possible by fully utilizing the entire stuffing box chamber and designing a lantern ring for insertion inside the stuffing box cavity.

A. Packing with Lantern Ring

It is a widespread but very erroneous opinion that a lubricant that is imparted into the stuffing box through the gap at the throat will work its way through the entire packing automatically along the surface of the shaft until it reaches the face of the gland ring. As a result of the parabolic distribution of the gland face force within the packing, the packing is essentially improved by dividing it into two halves and installing a lantern ring in the center between the two packing sections. The lantern ring not only splits the packing but also allows the introduction of fluid to lubricate the two packing halves. By recycling the lubricant the fluid simultaneously serves as a coolant. A design of this nature is shown in Fig. 5.3. When the lubricant is introduced with the same pressure level

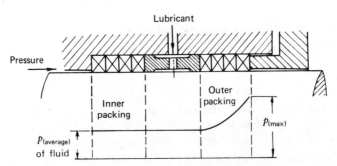

Fig. 5.3 Conventional packing arrangement [From (9), excerpted by special permission from *Chemical Engineering*, November 1967, copyright © 1967 by McGraw-Hill, Inc., New York, N.Y. 10020].

as exists in the system fluid of the pump, the pressure in the lubricant film along the shaft surface remains constant between the stuffing box throat and the face of the packing gland. System fluid and lubricant establish an equilibrium and leakage does not occur. The maximum pressure in the stuffing box is observed at the first ring of the packing facing the gland flange.

B.   Modified Lantern Design

A practical rule in packing design states that lubrication should be applied at the point of highest pressure, where it is most needed. The design of Fig. 5.4 follows this rule and directs the lubricant pressure to the first packing ring in contact with the lantern ring.

This design uses only one half of the packing of Fig. 5.3 for sealing the system fluid, whereas the outer half actually retains the lubricant inside the stuffing box. In the modification of Fig. 5.4 the outer set of the packing for the stuffing box is replaced by a filler sleeve, and the lubricant is sealed against the outer atmosphere by a small additional auxiliary packing.

The filler sleeve is sealed at the stuffing box wall by a soft metal wedge

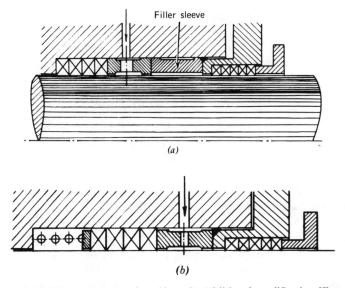

Fig. 5.4 (a) Modified conventional packing. (b) Additional modification [From (9), excerpted by special permission from *Chemical Engineering*, November 1967, copyright © 1967 by McGraw-Hill, Inc., New York, N.Y. 10020].

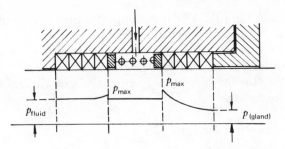

Fig. 5.5 Optimal packing arrangement [From (9), excerpted by special permission from *Chemical Engineering*, November 1967, copyright © 1967 by McGraw-Hill, Inc., New York, N.Y. 10020].

ring. When temperature is not high an elastomeric O-ring can also be used to replace the metal wedge ring.

C. Replacement of Lantern Ring

In both designs shown in Figs. 5.3 and 5.4 the tightening of the gland is done by hand. When the lantern ring is replaced by a helical spring, equally pressurizing both packing halves, a design is obtained that exerts automatic contact pressure against the packing rings, thus maintaining constant contact pressure against the packing to substitute for packing wear. The diagram of Fig. 5.5 illustrates that a practically constant pressure is acting across the entire length of the stuffing box.

The highest pressures are developed at the points of contact with the spring washers, where lubrication is simultaneously introduced. The coil spring is an ideal component to compensate automatically for variations in the packing geometry.

Packings of this design have been operated by the author for many years, applied to autoclaves at pressures in excess of 20,000 psi. This design is also successfully used in reciprocating shaft operations.

III. Packings for Extreme Internal Pressures

When the sealing principle illustrated in the design of Fig. 5.5 with two spring-loaded packing sections is applied to autoclave operation, the arrangement may look like the sketch shown is Fig. 5.6.

A coil spring replaces the lantern ring. At this location the lubricant is introduced with a pressure equivalent to the system pressure inside the autoclave. The two packing rings below the coil spring have the purpose

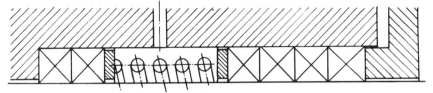

Fig. 5.6 Automatic packing for autoclave shaft [From (7), excerpted by special permission from *Chemical Engineering*, September 1965, copyright © 1965 by McGraw-Hill, Inc., New York, N.Y. 10020].

of protecting the autoclave from contamination by the lubricant oil. The four packing rings above the spring must seal the system against leakage to the atmosphere. Depending on pressure, three rings may suffice. Whatever the pressure level, more than five rings are not required.

The packing of Fig. 5.7 is applied to autoclaves for extreme internal pressures of 20,000 psi and over. This design has been reported by W.

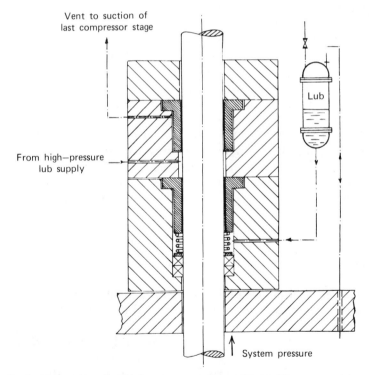

Fig. 5.7 Special packing for high-pressure autoclave [From (7), excerpted by special permission from *Chemical Engineering*, September 1965, copyright © 1965 by McGraw-Hill, Inc., New York, N.Y. 10020].

Coopey (7), who developed it for special laboratory services. Simultaneously the author worked with the same design prior to Coopey's report with pressures in excess of 10,000 psi.

The design utilizes the principle of the "free-piston gage," first recognized by P. W. Bridgman. In this system sealing of extreme pressures is achieved by means of a lubricant film, making use of the resistance to displacement of a viscous lubricant in a narrow annular ring space between a piston and a cylinder wall. This principle is applied to achieve a seal for a rotating shaft against very high pressures with high differentials. The design scheme shown in Fig. 5.7 was tested by Coopey. It is an ideal model to demonstrate that it is the oil film that seals the shaft and not the type or the design of the particular packing ring.

Two special sealing sleeves are fabricated from metal and machined to very close clearance. Extreme care is applied to align the shaft, the sleeves, and the stuffing box components with the greatest possible accuracy. Shaft finish should be no less than 10 μin. at the best possible straightness.

Lubricant is supplied between the two sealing sleeves at a pressure of 4 to 5 ft above the autoclave pressure to ensure lubrication supply. For this purpose the lubricant reservoir is located 4 to 5 ft above the bore supplying lubricant to the autoclave packing. The reservoir has a connection with the interior of the autoclave gas phase. The altitude difference provides the driving force for the lubricant.

A spring-loaded packing of two rings prevents leakage of lubricant into the autoclave. A very important feature of this design is the application of the principle of balanced pressure for both sealing sleeves and the staging of the pressure drop, which facilitates reduction of the pressure level. It is obvious that the viscous lubricant film in the sleeve contact areas is responsible for achieving the leakage-free seal performance.

IV. Packing Versus Mechanical End Face Seal

Packings and mechanical end face seals have been strong competitors ever since mechanical end face seals reached their high level of performance efficiency. Better materials and more sophisticated design have resulted in longer seal life on both sides, and industrial competition between the two continues with even greater intensity.

Numerous published reports have been studied for this work and have been combined with many personal plant studies to find a reliable answer to the question of whether plants are using packings or whether they prefer mechanical end face seals and why. A conclusive scientific answer has not been found. One wonders how industry can be so efficient despite the lack of scientific "reasoning."

Nevertheless, packings have proved effective and highly efficient, which is also true of mechanical end face seals. It is amusing that the proponents of both sides use the same strong arguments to defend their objectives and to justify their opposition to the other side.

Packings can be made to perform efficiently as mechanical end face seals. The selection of either one is always, or should always be, made after a thorough evaluation of the operating conditions.

The author was fortunate enough to be responsible for chemical equipment including more than 3000 centrifugal pumps where mechanical packings were applied long before the mechanical end face seal had reached its present competitive position. His experience was that packings can operate satisfactorily without leakage for long periods of time when selected, handled, maintained, and treated properly.

With the appearance of TFE resins as fibers, impregnated with carbon-graphite or molybdenum disulfide, packing design has gained a new significance. Many new shapes, configurations, and applications have now become available. When it is woven as fiber and braided to its final cross section, a packing can be compressed to one third of its initial volume. It thus offers an excellent flexibility with sufficient voids. When it approaches its final phase of compression, it even acts as a heat transfer medium to dissipate frictional heat. Because it acts as a lubricant it does not scorch the shaft. Because TFE-graphite is a self-lubricating material, its sealing function continues even when all voids have disappeared unlike asbestos ring packing.

V. Mechanical Factors That Influence Packing Performance

Earlier it was stated that a packing may fail even though there is nothing wrong with it. In such an unusual case there is nothing that can be applied to improve performance. Such cases exist when a variety of mechanical factors of the system are wrong, and they must be corrected before the packing can function properly. Without such a correction of mechanical factors even an ideal packing cannot work effectively. Such factors typically include shaft misalignment, eccentricities, and other gyrational disturbances in the form of deflections, runout, and the like. A very important factor that can cause packing failure is operation of the pump beyond the point of maximum efficiency or at a critical or subcritical speed.

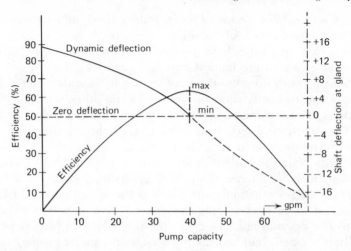

Fig. 5.8 Influence of pump efficiency on shaft deflection [From (18), excerpted by special permission from *Chemical Engineering*, September 1956, copyright © 1956 by McGraw-Hill, Inc., New York, N.Y. 10020].

A. Influence of Shaft Deflection

The farther apart the point of operation is from the point of optimal efficiency, the more likely a pump packing is to leak. The reason is related to the deflection of the shaft. In other words, the pump shaft moves away from the position of least gyrational disturbance. In pump design it is known that at a given rotation the deflection of the shaft using a pump of the volute type varies inversely with the fourth power of the shaft diameter. Shaft deflection is theoretically zero at the point of maximum efficiency, as can be seen in Fig. 5.8.

The deflection curve reaches a maximum near the shut-off point, after which it steadily decreases to the point of maximum efficiency, where it reaches zero. Shaft deflection distorts packing, which leads to eventual failure of the packing.

B. Operation of Pump at Point of Maximum Efficiency

Many process designers specify a centrifugal pump with excess capacity so as to have it available in case of an unexpected need, although this need may never arise. This means that such a pump will never be operated at maximum efficiency. An example reported by R. D. Norton (18) may demonstrate the error of this reasoning and its detrimental results on the packing.

A pump may be required with a capacity of 10 gpm at a head of 100 ft. If a pump is ordered with a capacity of 15 gpm and a head of 125 ft, it is overcharged in capacity by 50% and in head by 25%. In other words, the pump must be throttled down to 10 gpm. At this point of operation the packing definitely does not function properly, as the pump H-Q-curve shows, in Fig. 5.9.

Point A represents the actual design point of the pump capacity on a curve for an impeller with a 5-in. diameter delivering 10 gpm at 100 ft of head. By specifying point B instead with a capacity of 15 gpm at a head of 125 ft, located on a curve for an impeller with $5\frac{5}{8}$ in. diameter, the pump actually operates at point C for the curve with the larger impeller. This point represents a capacity of 10 gpm at 125 ft of head after being throttled. As a result, deflection rises from 0.009 in. for point A to 0.013 in. for point C. By operating at point D the deflection would be 0.001 in; however, the capacity would rise to 40 gpm. Excessive waste of electric power would result.

Figure 5.9 shows how detrimental mechanical factors can adversely affect the performance of a packing. It further demonstrates that the

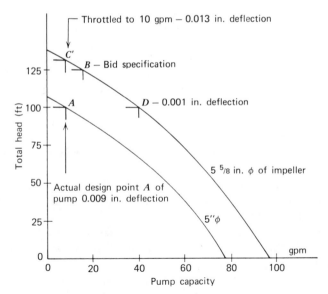

Fig. 5.9 Influence of size of pump impeller on shaft deflection [From (18), excerpted by special permission from *Chemical Engineering*, September 1956, copyright © 1956 by McGraw-Hill, Inc., New York, N.Y. 10020].

situation cannot be improved by modifying the packing. Even the best packing design cannot withstand detrimental mechanical factors. The basic mechanical prerequisites must be satisfied before the pump is started.

VI. Packing Handling

Packings for sealing shafts with rotary motion can be successfully used for long-term leakage-free operation when the design incorporates the following major requirements:

1. The design configuration must recognize the fundamentals of dynamic sealing.
2. The packing components must satisfy minimum standards of materials and mechanical assembly conditions.
3. The mechanical requirements of the machine to be sealed must be satisfied in all possible respects before a seal is installed. On this basis only can optimal seal operation be expected.

The major fundamental sealing principles are as follows:

1. In a properly functioning packing it is the oil film that establishes the seal.
2. The lubricant must be introduced at the points of highest pressure and friction.
3. Packing functions satisfactorily with an optimal number of rings. More than five rings per set of packing does not improve sealing efficiency because the rings farthest away from lubricating supply are not reached.
4. Lubrication must be applied at a pressure slightly above the system pressure.
5. The lubricant must be compatible with the system fluid.
6. Packing should be chosen for automatic operation, replacing the lantern ring by a coil spring.
7. Packing material should be sufficiently flexible to adjust automatically to the lubricant pressure and establish contact with the corresponding contact areas for the seal.
8. Packing material should be compatible with lubricant and system fluid.
9. Packing must be able to absorb a minimum amount of shaft gyration without increasing the leakage rate.

10. Major mechanical conditions of the system must be satisfied before the packing can function properly.
11. Under no circumstances should a packing be used as a bearing for the shaft or for shaft alignment.

The mechanical conditions that must be satisfied before packing assembly is started are as follows:

1. *Alignment.* The alignment of shaft and bearings for optimal true rotation within minimum tolerance values is one of the most significant requirements for satisfactory seal service, before the packing is installed.
2. *Material quality.* The shaft or sleeve material must be chemically compatible with lubricant and process fluid.
3. *Surface conditions.* The shaft must be round, have a surface finish ranging from 20 to 10 μin., a surface hardness of $R_c \sim 30$ min., and a straightness within 0.001 T.I.R.
4. *Lubrication.* Packing performance is a direct function of the lubricant supply.
5. *Operating point.* Pump should be operated at maximum efficiency to ensure minimum shaft deflection.

Selecting a top mechanic for handling packing and other seals in the plant is one of the most important decisions the plant engineer can make. If a single pump shutdown stops production, simple arithmetic will demonstrate how many mechanics could have been employed for the cost of one hour of lost production.

VII. Packing Selection

Compression packing functions when it is compressed in the stuffing box between the gland ring and the stuffing box throat. Compression forces the rings to expand toward the sealing contact areas, establishing a seal with the shaft and the seal cavity wall. Compression packing requires adjustment of the gland contact force to compensate for loss of voids and lubricant.

Semiautomatic packing is a combination of compression jam-type packing and a mechanical spring, which is substituted for the lantern ring. The spring maintains constant contact of the two packing sets. The spring replaces manual adjustment.

Automatic packing uses lubricant or system fluid pressure to establish a seal. Automation eliminates manual gland adjustment. Interference fit of the lips or the elasticity of the ring materials provides initial sealing at zero pressure.

Compression packing can be chosen from an endless variety of greatly diversified materials, including rubber, leather, asbestos, fluorocarbons, fluoroelastomers, graphite, nylon, Viton, Teflon, and Kel-F. Sheets, fibers, metals as braided packings, metallic and plastic packings, and many kinds of fabric packings are available. Consequently, the choice of suitable packing materials can be quite complex.

Generally valid rules for choosing packing materials cannot be given. The operating conditions are much too diversified and could fill books. Only guidelines for selecting packing materials can be given, especially since the packing that matches all process conditions on a universal basis does not exist. Each case must be evaluated individually and then analyzed to reach acceptable solutions.

As far as temperature limitations are concerned, Table 5.1 offers some helpful information.

For the selection of compression packings, manufacturers offer the guidelines shown in Table 5.2.

There is a practical relationship between packing size and shaft diameter, reflecting the requirements of the stuffing box. If the packing rings are too small, the clamping force induced by the gland ring becomes too high and excessive compression results. The width sizes in Table 5.3 have been accepted by the Hydraulic Institute for given shaft diameters.

In the selection of suitable packing materials there is no difference

Table 5.1 Temperature Limitations for Packing Materials

Material	Temperature (°F)	Material	Temperature (°F)
Cotton	220	Graphite filament	650
Flax	220	Asbestos plastic	500
White asbestos	500	Lead	450
Blue asbestos	500	Aluminum	1000
TFE—impregnated with white asbestos	500	Asbestos with Inconel wire	1200
TFE—impregnated with blue asbestos	500	Alumina fibers depending on composition	2500
TFE filament	500		

between rotating or reciprocating shaft applications. Therefore, materials are covered in more detail in Part 3.

PART 3. PACKING AS SEAL FOR RECIPROCATING ROD

I. Introduction

In Part 2 it was repeatedly pointed out that packings work equally well for rotating and reciprocating shafts. In packings for rotating shafts it is likely that lubricant fluid will pass along the shaft and enter the pump housing through the clearance gap in the stuffing box throat if the lubricant is compatible with the system fluid. In packings for reciprocating rods it is not desirable for the lubricant to penetrate the compression chamber in measurable quantities to mix with the system gas.

Because materials for packings have not been covered in detail previously, this is done in this section. Only guidelines for packing selection are given, because full coverage would fill books.

In rotating shaft equipment, mechanical end face seals provide a sealing alternative, even though the face seal will never completely replace packing for all applications. For reciprocating rods no alternative exists; hence, a solution must be found to satisfy seal requirements with packings.

II. Classification

Packings may be classified in many ways. One possibility was discussed in Part 1 of this chapter, considering jam-type and automatic-type packing configurations. Figure 5.10 illustrates a similar classification with more detailed information.

Piston rings, designated as floating-type packing seals, are covered in Chapter 6.

III. Materials for Packing Seals

Packings are available in an endless variety of materials. To provide order for this wealth of information the categories may be discussed in detail using the classification in Fig. 5.10.

Table 5.2 Compression Packings for Various Service Conditions

		Service Condition		
Fluid Medium	Reciprocating Shafts	Rotating Shafts	Pistons or Cylinders	Valve Stems
Acids and caustics	Asbestos (blue) Metallic Plastic (pliable) Semimetallic TFE fluorocarbon resins and yarns	Asbestos (blue) Plastic (pliable) Semimetallic TFE fluorocarbon resins and yarns Graphite yarn	TFE fluorocarbon resins	Asbestos (blue) Plastic (pliable) Semimetallic TFE fluorocarbon resins and yarns Graphite yarn
Air	Asbestos Metallic Plastic (pliable) Semimetallic	Asbestos Plastic (pliable) Semimetallic	Leather Metallic	Asbestos Plastic (pliable) Semimetallic
Ammonia	Duck and rubber Metallic Semimetallic	Asbestos Semimetallic	Duck and rubber	Asbestos Duck and rubber Semimetallic
Gas	Asbestos Metallic Semimetallic	Asbestos Semimetallic	Leather Metallic	Asbestos Semimetallic
Cold gasoline and oils	Asbestos Plastic (pliable) Semimetallic	Asbestos Plastic (pliable) Semimetallic	Leather	Asbestos Plastic (pliable) Semimetallic

Service				
Hot gasoline and oils	Asbestos Plastic (pliable) Semimetallic	Asbestos Plastic (pliable) Semimetallic Graphite yarn		Asbestos Plastic (pliable) Semimetallic Graphite yarn
Low-pressure steam	Asbestos Duck and rubber Metallic Plastic (pliable) Semimetallic	Asbestos Metallic Plastic (pliable) Semimetallic	Duck and rubber Metallic	Asbestos Duck and rubber Plastic (pliable) Semimetallic
High-pressure steam	Asbestos Metallic Plastic (pliable) Semimetallic	Asbestos Metallic Plastic (pliable) Semimetallic	Metallic	Asbestos Metallic Plastic (pliable) Semimetallic
Cold water	Duck and rubber Flax, jute, or ramie Leather Plastic (pliable) Semimetallic	Asbestos Cotton or rayon Flax, jute, or ramie Plastic (pliable) Semimetallic	Duck and rubber	Asbestos Duck and rubber Flax or cotton Plastic (pliable) Semimetallic
Hot water	Duck and rubber Leather Plastic (pliable) Semimetallic	Asbestos Plastic (pliable) Semimetallic	Duck and rubber	Asbestos Duck and rubber Plastic (pliable) Semimetallic

Table 5.3 Widths of Packing Rings

Shaft Diameter		Width of Packing	
(in.)	(mm)	(in.)	(mm)
$\frac{1}{2}-\frac{5}{8}$	12–15	$\frac{5}{16}$	8
11/16–1$\frac{1}{2}$	15–40	$\frac{3}{8}$	10
1$\frac{9}{16}$–2	40–50	$\frac{7}{16}$	11
2$\frac{1}{16}$–2$\frac{1}{2}$	50–65	$\frac{1}{2}$	12
2$\frac{9}{16}$–3	65–75	$\frac{9}{16}$	14
3$\frac{1}{16}$–4	75–100	$\frac{5}{8}$	15

A. Materials for Compression Packings

Compression packings are available in fabrics, plastics, and metals.

1. Fabric Materials

Fabric materials can be categorized as yarn and cloth. The various members of this large group are shown in Fig. 5.11.

On the basis of the discussion of fundamentals in Part 1, Fig. 5.11 is self-explanatory. Flax, jute, cotton, and asbestos are considered as base materials; they also come in a variety of modifications with TFE resins.

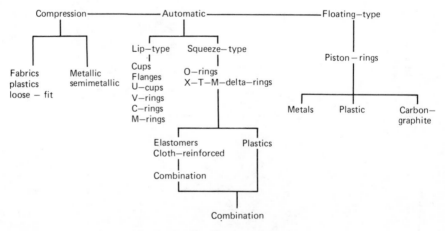

Fig. 5.10 Classification of packings.

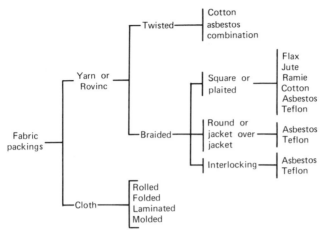

Fig. 5.11 Jam-type fabric packings.

2. Plastic Packings

Plastic packings are made from materials shown in Fig. 5.12. They are fabricated and supplied as lubricated or as dry-bonded packing components.

3. Metal Packings

Metal packings are fabricated as foil and as wire. Metal packings classify as shown in Fig. 5.13. The chart illustrates category, geometry, and

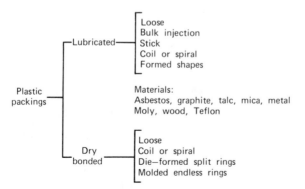

Fig. 5.12 Jam-type plastic packings.

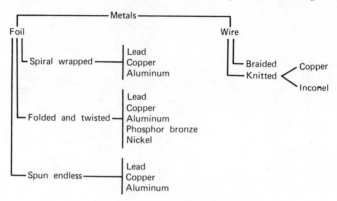

Fig. 5.13   Jam-type metal packings.

materials. Specific high-pressure metal packings for extreme pressure
levels are discussed as a separate subject in Part 4 of this chapter.

4.   Semimetallic Packings

Semimetallic packings are generally those packings that are fabricated as
combinations of metals and any other suitable packing material, selected
from any group presented in this section.

B.   Materials for Automatic Packings

Automatic-type packings are specifically, but not exclusively, designed
for sealing reciprocating shafts. They are partially used for installation
in the piston and move with it. They are equally installed in the stationary housing to seal the moving rod or plunger. They utilize the self-
energizing sealing principle by incorporating the system pressure to
enhance the seal effect.

1.   Lip-Type Automatic Packings

Lip-type automatic packings are available in a wide range of materials.
They comprise piston-cups, flange rings, U-cups, and V-rings, recruiting
from all kinds of packing materials, as indicated in Fig. 5.14.
    All lip components are designed to be endless in shape (i.e., solid
complete rings, not stepped like piston rings); they have dimensions that
create interference fits when the rings are assembled inside the corresponding seal cavity.

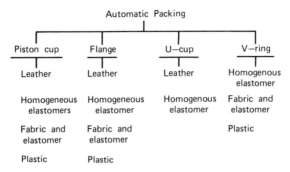

Fig. 5.14 Automatic lip-type packings.

2. Squeeze-Type Automatic Packings

Squeeze-type automatic packings are molded devices with a high degree of elasticity and flexibility. Their sealing effect results from a simple compression during installation. Their major representative is the O-ring, which is further modified to form a square, X-, T-, or delta cross section. They are fabricated from all kinds of homogeneous elastomers, plastics, and composite materials.

Homogeneous elastomers are described in Chapter 1 in connection with O-rings as static sealing devices. Information is also presented in Chapter 2 in the section on secondary sealing components.

IV. The Design of Automatic Packings

Automatic packings consist of two groups, the lip-type and the squeeze-type packings.

A. Lip-Type Packings

All lip-type packing components are designed as endless rings without a cut of any kind. With the advent of TFE resins an ideal material for automatic lip-type packing seal rings became available. Lowest coefficient of friction of any solid material, excellent durability, and near elimination of lubrication have made the TFE resins ideal for reciprocating rod seals. It should be noticed, however, that the individual fabricators of TFE resins have their own variations of designs. Thus consultation with the manufacturer is generally recommended.

1. Piston Cup Rings

Piston cup rings are used on hydraulic and pneumatic pistons. As reported by L. H. Gillespie (11), Teflon piston cups are capable of static and dynamic sealing against pressures up to 80,000 psi.

A typical piston cup design is shown in Fig. 5.15. In Fig. 5.15a the cup is operating as single-action device, whereas in Fig. 5.15b a seal for double action is illustrated, using the cups in either direction. The lips are installed with an interference fit, so the wall of the cylinder is in steady contact with the lips. Since the cup is shaped conically the heels have little support, forcing the cup actually to provide its own strength to withstand the system pressure without being extruded. The heel must be designed with the necessary rigidity to bridge the gap against the fluid pressure.

The lip cannot lose contact with the cylinder wall; otherwise leakage occurs. The lips must be installed in such a way that they are not damaged during assembly. Defective lips cannot prevent leakage. For centering the piston cup the bottom of the piston should be recessed with a shoulder that will prevent the cup from working loose. According to Gillespie, the shoulder height should be such that a compression of the cup thickness beyond 10% does not occur. If the compression ex-

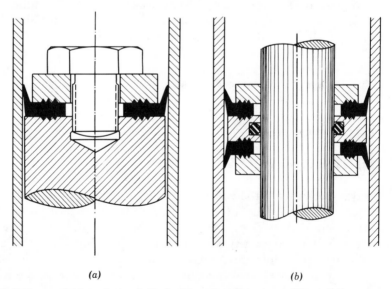

(a)   (b)

Fig. 5.15 (a) Single-acting and (b) double-acting piston cups [From (11), courtesy of *Machine Design*].

Table 5.4 Clearance Gap Values for Piston Cups

Cylinder Diameter (Φ) (in.)	Clearance (in.)	
	Up to 600 psi	Greater than 600 psi
1 to 3	0.006	0.004
3⅛ to 8	0.008	0.006
8½ to 10	0.010	0.008
10½ to 12	0.012	0.010

ceeds the 10% limit, the cup may bulge in the heel and force it excessively against the wall of the cylinder. Serrations on both sides of the cup bottom limit compression to the desired value and force the cup to maintain the desired location, simultaneously preventing leakage. Preferable serration dimensions are 0.007 to 0.010 in. in depth and form an angle of 90 degrees (11).

In the design of Fig. 5.15b the O-ring in the center ring spacer on the surface of the shaft stops leakage along the shaft. The spacer must have adequate clearance against the cylinder wall. Excessive clearance allows extrusion of the cup heel. Adequate clearance gap values are given in Table 5.4. Clearance values decrease with rising pressure levels.

The external diameter of the inside follower is critical with regard to the ID of the piston cup. Fabricators usually use a clearance value of 25% of the thickness of the lip. In other words, the OD of the inside follower is smaller than the ID of the piston cup to prevent pinching of the lip against the cylinder wall.

The inside spacer must be provided with a radius at the corner contacting the cup bottom. A sharp corner will cut the cup at a very critical location. Surface finish at the cylinder contact area should be no less than 16 μin.

For satisfactory operation it is very important to know the coefficient of thermal expansion of the TFE cup material. If the temperature differential exceeds a value of 130°F, the lip may move away from the cylinder wall. A spring is thus required to maintain the contact in case of undesirable heat buildup. The spring prevents the lip from expanding outwardly. The piston cup, therefore, moves inwardly toward the heel. When the suction stroke cools the cup, the TFE resin shrinks and tends to move the lip away from the wall. The spring expander prevents this and maintains steady lip contact with the cylinder wall.

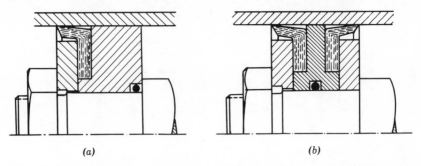

(a)   *(b)*

Fig. 5.16 (*a*) Single-acting and (*b*) double-acting leather piston cups [From (11), courtesy of *Machine Design*].

By using piston cups formed from flat sheets, stresses are built in that become effective as temperature rises, making the use of an expander spring obsolete. With this method temperatures up to 450°F can be handled without losing lip contact.

For operation in low-temperature regions the coefficient of thermal expansion is lowered and the TFE resin contracts. This can be partially eliminated by adding glass fiber. It is reported that adding 15% glass fiber reduces the coefficient of thermal expansion by one-half.

Piston cup designs, using rings made from leather, are shown in Fig. 5.16 for a single- and a double-acting piston cup seal configuration. Only the lips contact the cylinder wall. With increasing pressure more and more of the lip portions are forced against the cylinder. The final critical limit is reached when the heel is forced against the wall, where the strength of the cup material must prevent an extrusion into the clearance gap with the spacer ring.

2. *Flange Ring Packing*

A practical reversal of the piston cup is the flange ring design, often referred to as hat packing. It is generally used as a stationary seal installed in the housing to seal reciprocating rods. This is usually accomplished with one ring, which seals so that the lip of the flange ring is in contact with the rod at the lip ID, whereas the piston cup lip seals at its OD.

An example of a flange ring seal is illustrated in Fig. 5.17. The bottom of the flange ring is inserted in a recessed cavity of the housing and is supported by the gland ring flange. The lip always faces the pressure, in order to utilize the self-energizing sealing principle. It rides freely on the

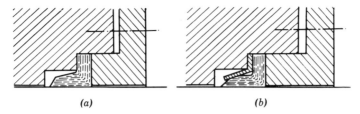

(a) (b)

Fig. 5.17 (a) Flange ring packing assembly. (b) Flange ring packing with contractor ring.

shaft surface. At low and moderate pressures a contractor device is used, which acts like a spring, forcing the lip against the shaft at all times, as shown in Fig. 5.17b.

The flange ring packing is unbalanced and seals only on the ID. It is made mostly of leather and is used for medium-pressure ranges. Considerable improvement is reached by using rings of PTFE resins, compounded with carbon-graphite or molybdenum disulfide.

Flange cup rings must be well supported. This is generally accomplished by backup washers with close clearance gap between rod and gland to prevent ring extrusion. The cavity surrounding the lip should be larger than 25% of the flange thickness to provide sufficient space for penetration of system fluid to energize the lip for automatic sealing contact.

The flange ring packing rod seals are generally used for small-diameter shafts where U-cup or V-rings either are impractical or do not have enough space to permit other ring configurations. The bottom area acts as a static seal once the follower is tightened.

3.   U-Cup Rings

U-cup seals represent a combination of the piston seal and the flange ring, which can either be installed in the moving piston or be arranged so that it is stationary in the housing. Figure 5.18a illustrates a U-cup seal moving with the piston. A modified U-cup design using an O-ring as pusher force is shown in Fig. 5.18b. Both lips seal automatically when the system pressure is acting. A gland adjustment is not required.

U-cup packings are balanced seals, which means that they seal on both the internal and external diameters. With mechanical expanders the seal can take advantage of the wide range of temperature applications when TFE resins are used.

Support rings are applied in a variety of design modifications to protect the seal against collapse or distortion. The U-cup must be in-

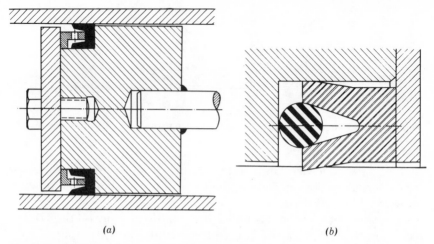

(a)   (b)

Fig. 5.18 (a) U-cup ring moving with piston. (b) U-cup ring with elastomeric O-ring modified [From (11), courtesy of *Machine Design*].

serted in the seal cavity in such a way that the lips never touch the pedestal ring. Clearance should be $\frac{1}{16}$ or $\frac{1}{8}$ in. A clearance of at least 25% of the lip thickness is required between pedestal and lips to allow the system fluid sufficient access to apply the self-energizing sealing principle. Horizontal holes in the vertical web of the pedestal ring will secure equalization of the pressure inside the U-opening.

In Fig. 5.18b the initial seal effect is enhanced by the insertion of an O-ring, increasing the static and the dynamic contact forces. The special Johns-Manville lip design cited in Chapter 3 may also be designated as a modified U-cup seal, which in this case is used for sealing a reciprocating rod.

U-cup rings are generally used for pressures of up to 1500 psi, when they have an elastomer with a durometer of 70. For pressures exceeding the 1500 psi the durometer should be increased to 80 or 90. Backup rings are recommended to prevent extrusion. For higher pressures the clearance gap must be decreased proportionally.

Like piston cup rings the lips of the U-cups are provided with a flare to produce an interference fit during installation and to supply initial sealing under static conditions. Any squeeze of the heel by the pedestal is detrimental to the seal ring. Wherever the heel of U-cup rings undergoes noticeable deformation, the lips tend to move away from the seal contact areas.

The use of springs enhances automatic sealing action. Metal adapters

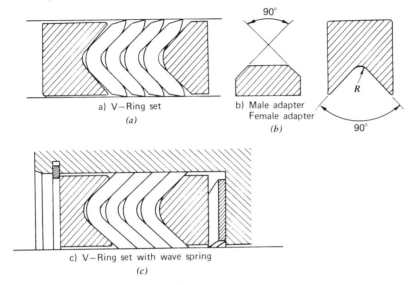

a) V—Ring set

(a)

b) Male adapter
   Female adapter

(b)

c) V—Ring set with wave spring

(c)

Fig. 5.19   V-ring packing sets and details. (*a*) V-ring set. (*b*) Male adapter and female adapter. (*c*) V-ring set with wave spring.

facilitate the establishment of uniform forces for dynamic and static sealing contact. Radial clearances are approximately 0.003 in. at the ID.

4. V-Rings

Perhaps the most popular ring of the automatic-type packing configurations is the V-ring. Basic details of this design are shown in Fig. 5.19. Because of the geometry of the ring, male and female adapters are required. Springs can be used to improve the automatic contact force. The V-ring is available in a variety of design configurations.

Unlike all other automatic lip-type packing designs, the V-ring is applied in multiple numbers in one particular set of packings, thus increasing seal efficiency and service life.

The lips have a curved inside contour line; the outside contour is a straight line, which provides a point contact between two consecutive rings, thereby establishing a high degree of flexibility. Lip curves restrict contact of nested rings to a single line, which forces each ring to flex with minimal friction against the 45-degree angle leg of the adjacent ring.

A female adapter is required and prevents deformation of rings when the pressure rises. The male adapter facilitates automatic lip action and provides a positive seal under pressure.

Eccentricity, runout, and other gyration disturbances of the shaft exert a major influence on the design of the width of the packing. With increasing runout or eccentricity the width of the seal in the radial direction must be increased to maintain optimal flexibility. A rule of thumb for small shaft diameters is that a good packing should function well with a width of no more than ⅜ in. Surface finish of the shaft should be about 16 μin. At pressures below 500 psi a packing set with three rings is satisfactory.

The application of mechanical spring devices has become increasingly popular because they eliminate the human factor in retightening the packing gland. For the determination of the adequate spring size it is customary to assume 5 lb/linear inch of the mean ring circumference for homogeneous and leather V-rings, 10 lb for fabric reinforced-type rings, and 10 to 15 lb for TFE resins and their corresponding combinations with glass, graphite, or molybdenum disulfide.

V-ring packings are installed in sets of multiple rings, each set representing a number of rings with a male and a female adapter. The number of V-rings applied to each set is governed by the pressure level in the system fluid and the construction material. Table 5.5 contains practical information derived from field experience.

V-rings made from leather give excellent performance against hot and cold esters and especially against oil. In recent years leather has been impregnated with synthetic materials to increase the range of application.

Homogeneous V-rings are available with close tolerances and have a smooth, uniform surface that enables the ring to serve as a wiper ring running against a clean and dry shaft.

Homogeneous V-rings are suitable for pressures of at least 3000 psi.

Table 5.5 Number of V-Rings Used per Set of Packing

Pressure to Be Sealed		Number of Rings Per Set in a Packing		
(atm)	(psi)	Leather	Homogeneous	Fabricated
Up to 35	Up to 500	3	3	3
35–100	500–1500	4	4	4
100–200	1500–3000	4	5	4
200–350	3000–5000	4	5	5
350–700	5000–10,000	5	—	6
>700	>10,000	6	—	—

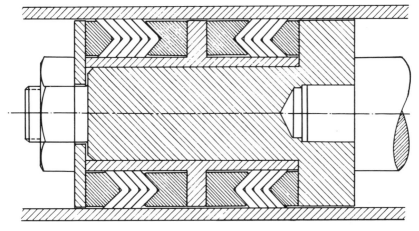

Fig. 5.20   V-ring packing riding with piston.

For higher pressure levels combinations are made with TFE resins, reinforced fabrics, or leather.

Much consideration must be given to installing the packing sets without damaging the lips. Sharp edges of the cavities must be avoided. It has become customary to provide chamfers to facilitate damage-free installation of the packing rings.

For cavities with an internal thread the bore for the packing should be smaller than the ID of the thread pitches. Any damage of the sensitive seal lips defeats the purpose of the ring.

It is also customary to install the V-ring sets in the moving piston. In this design it is important how the rings are installed. The rings are to be inserted back to back with a fixed female adapter to prevent a transfer of loads between the two sets, giving both sets a chance to operate independently, as shown in Fig. 5.20.

By assembling the two sets of packing rings on a separate sleeve their independence is readily ensured. Although V-ring packing sets function on an automatic basis it is sometimes necessary or preferable to preload the packing with single-spring or multiple-spring arrangements.

The springs eliminate the necessity of manually retightening the gland ring, and the metal springs further compensate for the overall tolerance. A conical spring design should be used with a single spring, as suggested by most spring fabricators.

For automatic packing arrangements the same surface finish conditions must be applied as required for all other shaft sealing devices. Straightness and the best possible shaft alignment are mandatory. Surface hardness should be no less than $R_c{\sim}45$ or better. As a rule the service life of a packing increases proportionally with surface smoothness of the shaft.

A straight run of the plunger must be assured with suitable alignment procedures of the machine. Packings cannot serve as bearings. Deflections, buckling, runout, or bending are detrimental and diminish the effectiveness of the packing within a short time. Plunger alignment within close tolerance limits is an absolute must.

B. Squeeze-Type Packings

The predominant squeeze-type packing is the O-ring. It seals very effectively within a wide range of temperatures and pressures with the ideal property of sealing in both directions and requiring a minimum of space. Modifications of the conventional O-ring are X-, T-, D-, square-, and delta-rings, which are discussed in some detail.

1. The O-Ring as Seal for Dynamic Shaft Motion

O-rings were designed primarily for use of static and dynamic sealing motion and particularly for sealing reciprocating shafts. The ring is installed with an interference fit, which means the ring must be stretched over the shaft for installation and then inserted into a standardized groove, where it functions without adjustment and with a minimum of gland contact force.

Prerequisites for proper functioning of the O-ring for sealing reciprocating shafts follow:

1. The metal contact area serving as sealing surface for the O-ring must have a surface finish of no less than 16μin. The groove can have a minimum of 32 μin. rms; however, it must be free from nicks, burrs, or scratches.
2. The surface hardness of the metal contact areas should be at least $R_c{\sim}35$ or better. Metals like aluminum, brass, or bronze cannot be used and should be avoided.
3. The annular clearance gap between the shaft and the housing section with the O-ring groove should be held to the maximum tolerance recommmended by manufacturers. Breathing of cylinders is also a factor and must be considered.
4. Concentricity between plunger and bore must be held within recommended practice. O-rings should not be used as bearings.
5. O-rings are very susceptible to damage from dirt and abrasive particles. Cleanliness is, therefore, a top requirement.

6. For good sealing performance the minimum diametral squeeze should be applied. For low pressures and low friction the squeeze may be decreased somewhat. But then leakage is likely.

7. O-rings used in pneumatic systems must be well lubricated. If they run dry, they abrade and twist.

8. Groove design is a very significant factor for successful sealing. Grooves with backup rings differ from grooves without. Two types of O-ring grooves are used. One is the recommended practice for military aircraft as indicated in MIL-R-5514; the other is the recommended practice for industry applications.

O-ring stretch, applied to the ID of the ring, is from 0 to 2% for military designs and from 1 to 5% for industrial applications.

In some special cases grooves show a V shape. Such a groove does not allow the use of backup rings of any kind to prevent ring extrusion; it also results in higher friction.

Figure 5.21 shows how O-rings behave under pressure, which ranges from atmospheric to 3000 psi, used to seal a reciprocating shaft. Behavior of the O-ring is illustrated when the ring is installed in a standard groove, a V-groove, and a groove with backup rings.

As a general rule, dynamic O-rings are used without backup rings for pressures up to 1500 psi. Above this pressure limit antiextrusion rings are recommended.

It has been established that a ring material with a durometer of 80 to 90 withstands extrusion better than a ring with durometer of 70; however, friction increases. It is better to use a ring of durometer 70 with antiextrusion rings than a harder ring without.

A single O-ring in a groove can be used for both single-acting and double-acting cylinders. More than one O-ring in one groove is not recommended. With two rings in one groove pressure can be locked in between the two rings.

One of the oldest materials used for antiextrusion rings is leather, which has high resistance to extrusion and the ability to compress under pressure and deform radially to bridge the metal clearance gap. Leather is nonabrasive and durable. It is susceptible to heat, strong acids, and caustic solutions. Temperature limitation is 200°F.

Surface finish, pressure, and clearance being equal, O-rings have a shorter service life than V-rings. This is because O-rings seal as long as they maintain positive contact with interference. If the compression effect is lost the seal loses its effectiveness and starts to leak.

Experience shows that in dynamic service pressures up to 4000 psi can be handled by conventional rubber O-rings, supported by backup rings of TFE resins. With higher pressures harder O-rings are preferred.

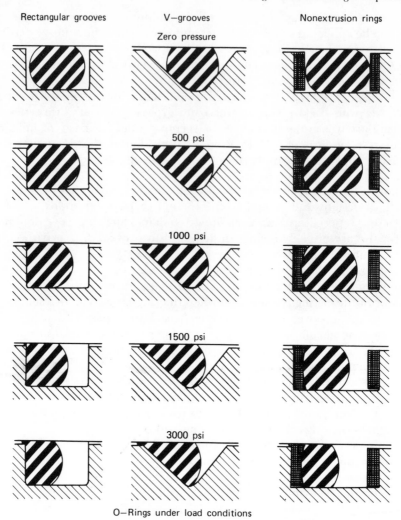

O—Rings under load conditions

Fig. 5.21 O-rings under load conditions.

However, the clearance values with the shaft have to be reduced correspondingly.

Enveloped O-rings, made from TFE resins with an elastic rubber core, are also available. The rubber supplies the elasticity and the flexibility; the TFE offers the wear characteristics, chemical compatibility, and antifrictional capacity.

2. *Other Squeeze-Type Seals*

Besides the conventional O-ring there are other elastomeric ring configurations using the same rectangular groove. Their cross sections are illustrated in Fig. 5.22. They are known as X-, T-, D-, square-, and delta-rings.

The T-shaped ring is used with or without two backup rings on both sides. The behavior under pressure is shown in Fig. 5.23, indicating a clearance between the flange section of the resilient T-ring and the backup rings, thus holding friction at a minimum. The T-ring deforms with rising pressure and produces a swelling in the radial direction of the backup rings on the down-pressure side. This dimensional increase establishes contact with the cylinder wall, narrowing the piston ʻclearance. Once full pressure is reached the backup rings contact the cylinder wall firmly. The resilient T-ring is squeezed against the backup rings, preventing extrusion of the elastomer into the annular clearance gap. Therefore, it can be concluded that the backup rings function automatically in their task as antiextrusion devices.

Fabricators report that this design assembly is suitable for pressures up to 20,000 psi without extrusion of the T-ring material. As is the case in all other ring designs, the clearances must become smaller as the pressure rises. With extra-heavy backup rings, packing can function with a diametral clearance of about 0.050 in. for both stationary rod seals and moving piston seals in hydraulic and pneumatic service at low pressure levels.

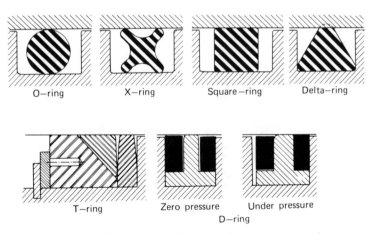

Fig. 5.22 Modifications of elastomeric squeeze-type seals.

Fig. 5.23 Pressure behavior of T-rings.

X-shaped rings offer four contact areas for sealing instead of the two of the conventional O-ring. The X-ring uses rectangular grooves of regular round O-rings. As pressure varies they show less movement from side to side, as can be seen in Fig. 5.24. The shaft surface finish for a stationary sealing function can be 32 μin. For dynamic sealing the finish should be about 15 μin. X-rings are not subject to spiral twist.

The rings seal better against higher pressures; however, lubrication and friction may create problems.

The triangle-shaped delta-ring is the newest member of the family of squeeze-type packing rings. Friction of this ring is greater; service life, therefore, is correspondingly shorter. The ring has little tendency to roll. The application of this ring is rather limited.

C. Special Design Configurations with Teflon

In Chapter 3 a series of elastic lip-type sealing devices was described that were applied to rotating shaft equipment, designated as elastic C-seals, fabricated by the BAL, Aeroquip, and FCS companies. These seals are equally suited for application to reciprocating shafts.

1. BAL-Seal

The BAL-Seal offers excellent service for static, rotary, and reciprocating operating conditions. Its function is based on point or line contact, using one or more lines for simultaneous sealing action. In addition to

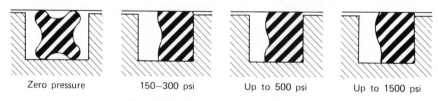

Fig. 5.24 Pressure behavior of X-rings.

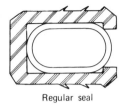

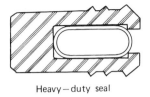

Regular seal   Heavy—duty seal

Fig. 5.25   BAL-Seals for reciprocating shaft.

the BAL-Seal configurations shown in Chapter 3 for rotating shaft conditions, the designs of Fig. 5.25 are preferably used for reciprocating load conditions.

The multiplicity of sealing lips in one design is based on the principle of breaking the total pressure down into several differential increments. With TFE resins and combinations as a base material—especially with graphite and/or molybdenum disulfide—excellent service can be expected.

The sealing principle and shaft conditions are practically the same as those discussed in Chapter 3 for rotating shaft sealing. Pressure enhances the seal efficiency. Initial compression of the spring component provides the static seal at zero system pressure. It is a significant fact that sealing reciprocating shafts is simpler than sealing rotating shafts.

2. FCS-Seal

FCS is the abbreviation of fluorocarbon, and is used as the company name of the fabricator who manufactures these seals. The principles of this seal have also been discussed in Chapter 3. The designs for reciprocating shafts incorporate a variety of mechanical springs of circular cross-section and U-shape configurations, either open or enveloped by Teflon or TFE combinations with graphite or molybdenum disulfide. Rubber and other appropriate materials are also available.

Some of the basic FCS-Seal configurations are illustrated in Fig. 5.26.

Teflon   Teflon   Rubber   Grafoil

Aluminum
Bronze

Fig. 5.26   FCS-Seals for reciprocating motion.

The designs indicate the versatility of the FCS-Seals for solving a wide range of sealing problems. The devices must be installed with compression of the spring to provide initial sealing at zero pressure. With increasing pressure the seal efficiency is enhanced because of the self-energizing sealing principle.

3. Omniseal

The Omniseal, also discussed in Chapter 3, is primarily used, with excellent success, for sealing reciprocating shafts. It was widely employed in hydraulic and pneumatic applications. Since the Seal Division of the Aeroquip Company has been sold to the FCS-Fluorocarbon Company, a further discussion of the Omniseal design is obsolete.

4. Poly-Pak Seal

Poly-Pak is a patented seal, precision-molded, self-lubricating device that combines the design of a V-ring with an elastomeric O-ring. The V-ring is molded from Teflon resin and is an automatic lip-type packing seal. The O-ring, inserted in the opening of the V, acts as a spring to provide initial preloading of the lips for static and dynamic sealing. The O-ring also prevents lip distortion under vacuum or extreme pressure. The general design principles of the Poly-Pak seal are shown in Fig. 5.27. The Poly-Pak seal can be used as a single-ring device or with a metallic male adapter, as shown in B of Fig. 5.27b, but without the elastomeric O-ring. The addition of a wave or helical spring improves seal effectiveness and makes the seal function independently on manual adjustment.

Poly-Pak seals are frequently also used in multiple arrangements as stacked seals in sets of two or three rings, as shown in Fig. 5.27c. When three rings are applied the set may replace a set of conventional V-ring packings of four V-rings, illustrated for comparison in Fig. 5.27d with male and female adapters. It will be noticed that the O-rings make the application of male and female adapters obsolete. The O-ring as a spring hydrostatically converts the system pressure on the first seal ring to lip sealing pressure on the following ring, this process continues from one ring to the next. Gland force is not required as long as all rings are endless and not split. Other mechanical spring devices are not needed.

5. U-Cup Ring Seal by Halogen Seal Corporation

The Halogen Insulator and Seal Corporation of Chicago fabricates a U-cup seal that is used for a wide range of pressures and is provided with

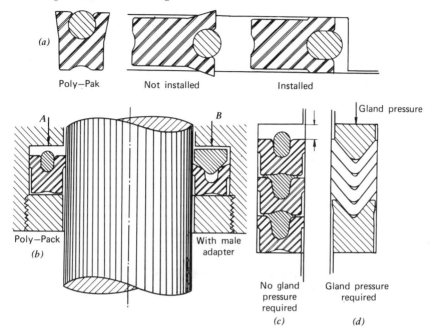

Fig. 5.27   Poly-Pak seal basics and applications.

springs for preloading at zero pressure. Made of TFE resin the ring has all the qualities of Teflon with regard to friction coefficient (0.04) and wear; in addition, it is suitable for commercial and military hydraulic fluids as well as for oxygen, water, and steam. It resists temperatures up to 450°F. The U-cup rings were originally developed for missile nozzles, where a short-time intermittent exposure to 850°F is constantly required.

The Halogen U-cup ring is illustrated in several sketches of Fig. 5.28 showing U-cup ring applications, using coil springs and/or wave springs for static sealing. The lip edges are carefully trimmed to 360 degree sharpness for true sealability. The rigidity of the heel prevents roll-over, which usually occurs with rubber cups. The seals are intended as primary seals but can be used also as dirt wipers or as suction seals to prevent air from penetrating into a system or for vacuum systems.

6.   Rev-O-Slide Seal

Rev-O-Slide seals were developed by C. E. Connover & Company, Fairfield, New Jersey, using the Revonoc O-ring as a basis. Revonoc is

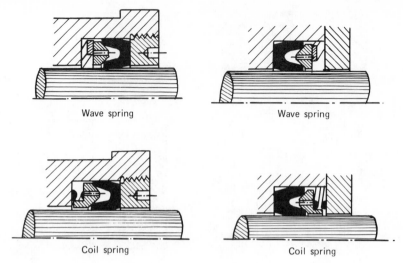

Wave spring   Wave spring

Coil spring   Coil spring

Fig. 5.28 U-cup seals by Halogen Insulator Seal Company (Courtesy of Halogen Insulator Seal Company).

used for applications where temperature extremes, friction, wear, corrosion, compatibility, and excessive pressure make the use of conventional elastomeric O-rings impractical. Revonoc, a TFE-base material, has all the TFE properties that make Teflon so attractive. It also has the negative characteristics, such as cold flow, resistance to stretch, and hardness of about 125 on the durometer. High forces are required for compressing the O-ring. However, the force of the compressed part (seal) against the seal contact surface constitutes a large portion of the sealing ability of the compound. To compensate for this condition the groove is designed to eliminate as much vacant space as practical.

Revonoc is impervious to most commercial chemicals, with a few exceptions. With a TFE base it does not require lubrication, and it is classed as self-lubricating. Its friction factor is approximately two-thirds of the running friction of the elastomeric O-ring. It is also repeatable and predictable, with a high degree of resistance to twist.

Aging, compression set, swell, explosive decompression, diffusion, leaching, and vacuum outgassing are not problems when using this material.

The basic behavior of Revonoc is explained in Fig. 5.29. Figure 5.29*a* indicates that the square groove is not filled because of the material's resistance to stretch. If the groove were filled completely, the friction force would be detrimental to the dynamic operation. In Fig. 5.29*b* the

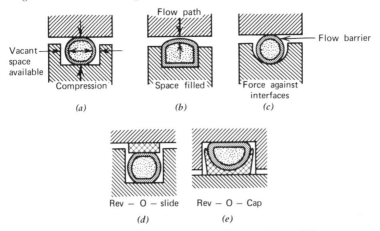

Fig. 5.29   Renovoc seals for static and dynamic conditions.

groove space is filled, showing a situation usually encountered in static sealing. A better approach is shown in Fig. 5.29c, where the groove matches the O-ring geometry. This is costly and makes the groove useless for conventional O-ring application. An ideal compensation for Revonoc's lack of stretch because of high hardness can be found in the elastomeric O-ring. A thin web of Revonoc is placed between the O-ring and the sliding surface of the shaft. By keeping the web thickness small enough the resiliency of the elastomer can be transferred to the Revonoc seal surface. The thickness of the web controls the leakage rate at various pressure levels, as shown in Fig. 29d, the Rev-O-Slide seal.

A Rev-O-Slide seal design for reciprocating shaft applications is illustrated in Fig. 5.30, showing rod and piston seal in the same machine.

Revonoc claims that backup rings should be used for pressures below 4000 psi. When exceeding this limit the Rev-O-Slide seal should be applied. Revonoc reports that pressures up to 50,000 psi have been sealed with Rev-O-Cap seals, as shown in Fig. 5.29e. In this case the clearance gap is 0.001 in.

7. *MiniGroove Seal*

A device that utilizes the same sealing principle as the Revonoc seal is called the MiniGroove seal, fabricated by Royal Industries, El Segundo, California. This seal is available in a wide range of configurations, as illustrated in Fig. 5.31. The web, consisting of TFE resins, comes in a variety of shapes, accommodating square-, T-, and O-rings and other squeeze-type packing components.

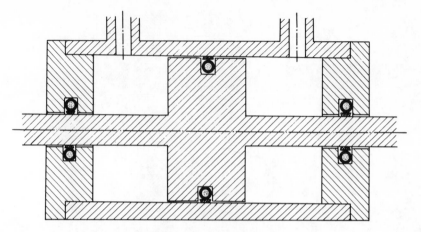

Fig. 5.30 Rev-O-Slide seal by Conover & Company.

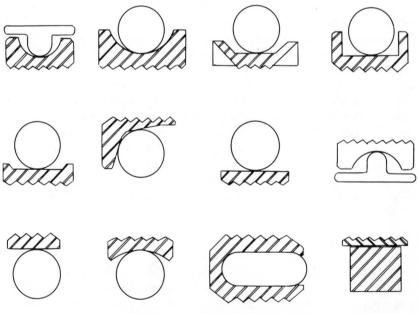

Fig. 5.31 Configurations of MiniGroove seal.

Another deviation from the Rev-O-Slide design is the incorporation of serrations to subdivide contact areas. The miniature circumferential grooves form a multitude of single chambers, similar to a labyrinth, allowing a pressure drop across each groove, thus breaking the total pressure into increments of a series of differentials. The tiny grooves act as reservoirs that retain small volumes of the lubricating fluid. This effect enables the seal device to ride constantly on a fluid film, minimizing friction and wear.

V. Selection Factors

The selection of the proper sealing system is related directly to the selection of adequate mating materials. Once the suitable materials are chosen types and shapes can be determined and in turn the packing style can be selected. Consequently, the selection of seals is the result of a series of factors that may be summarized as follows.

A. Basic Materials

The major problem of selecting a seal or packing material from a broad range of currently available materials is to choose one or a system of materials that offers the greatest performance and economy. This requires simultaneous evaluation of many interacting factors affecting the seal behavior.

Compression packings use fabrics, plastics, metals, and semimetallic materials.

Automatic packings primarily use leather, homogeneous elastomers, fabrics, plastics, and numerous combinations of these groups for lip-type packing systems.

Squeeze-type packings recruit primarily from homogeneous elastomers, plastics, and composite materials of all kinds.

Each of these large categories of materials has its place in the wide range of applications.

B. Shapes and Types

Packings are available in three basic categories, jam- or compression-type, automatic lip-type, and squeeze-type packings.

The compression-type packing, often referred to as jam-type packing seals, exerts a mechanical force when the gland ring is tightened.

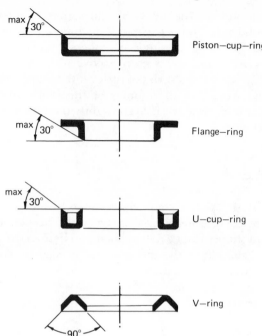

Fig. 5.32 Cross-sectional view of basic packing types [From (3), courtesy of *Journal of Plant Engineering*].

The automatic lip-type packing is custom molded and seals by exposure to the system pressure. In some cases springs are used to establish a seal at zero system pressure without excessive gland tightening force.

The squeeze-type automatic packing is purely pressure actuated. Its initial sealing effect at zero pressure is produced by inherent elastomeric properties that embody a high degree of resilience and flexibility. Seal efficiency is enhanced with increasing system pressure.

As stated earlier in this chapter the basic shapes of the automatic lip-type packing configurations are piston cup, flange ring, U-cup, and V-ring. Geometric details of these four types of seal rings are illustrated in Fig. 5.32.

C.   Packing Style

When sealing reciprocating rods the chosen seal can be installed as either inside packed or outside packed. Some of the packings can be used either way; some can be applied only one way.

An inside-packed seal is characterized by the fact that it is assembled on the piston and moves with it. Typical inside-packed devices are the piston cup, U-cup, V-ring, and O-rings.

The outside-packed device is located stationary in the housing, while the rod moves through the packing. Typical members of this group are the flange ring, U-cup, V-ring, and O-ring. It will be noticed that the U-cup, V-ring, and O-ring can be used in both types of installation.

Typical inside configurations are illustrated in Fig. 5.15 for single- and double-acting piston cup seals, in Fig. 5.16 and Fig. 5.17 for piston cup packings of leather, in Fig. 5.18 for U-cup packings, and in Fig. 5.20 for double-acting V-ring packings.

In the following section some basic concepts are presented concerning the four major types of automatic lip-type packing components.

The piston cup ring can be attached to many types of piston design configurations. When a boss is used, boss height should be equal to the thickness of the bottom of the cup plus 0.005 in. to restrict excessive compression. The bevel angle of the lips ranges between 15 and 30 degrees, forming a sharp external edge where the contact with the cylinder wall ends. For high pressures a harder cup is more resistant to extrusion. The best cup material is TFE resin or its combinations with graphite and molybdenum disulfide.

The U-cup ring is a balanced design, generally used in single units only. When fabricated in leather, the walls are straight, not beveled. For leather cups the U-opening is filled with flax, hemp, or rubber. In most cases a metal pedestal is inserted into the recess.

Fabric-reinforced U-cup rings are fabricated by molding with flared lips, so a preload is established when the rings are installed. The walls are stable enough that a filler is not required to support the walls.

Homogeneous U-cup rings are primarily used for pressures up to 1500 psi. For higher pressures antiextrusion rings are required, made from leather. When using pedestal rings care must be given that the pedestal does not exert excessive pressure on the bottom of the packing ring. This would force the heel to bind on ID and OD. Support of U-cup rings by pedestals is generally applied to stationary outside-packed seals, seldom to inside-packed piston cup packings.

The U-cup ring is widely used because of its simplicity in design, fabrication, and installation and its low friction and wear characteristics. In general, the U-cup ring is preferred over piston cup rings because of lower cost and ease of handling.

Among all automatic lip-type packings the V-ring is the most popular and most frequently used seal configuration. Symbol V can be considered an abbreviation for versatility. The V-ring covers a wide range of pressures in inside- and outside-packed arrangements and single- or double-acting cylinders. Its popularity is based on outstanding performance characteristics. It outperforms all other lip-ring components of the automatic lip-type seal family. The V-ring is applied to pressures up to 25,000 psi and over.

Unlike other lip-type packings the V-ring is used in sets of multiple rings with male and female adapters. Sometimes springs are used to make contact adjustment independent of human error.

It is common practice to use a number of rings per set of packing, as indicated in Table 5.6. Table 5.7 offers information on the overall height of each set.

Table 5.6 Number of Leather Rings Per Packing Set

Pressure (psi)	Leather	Synthetic Rubber	
		Homogeneous	Fabric Reinforced
Zero to 500	3	3	3
500 → 1,500	4	4	4
1,500 → 3,000	4	5	4
3,000 → 5,000	4	5	5
5,000 → 10,000	5	not applied	6
10,000 → up	6	not applied	not applied
on the basis of solid (endless) rings			

Table 5.7 Overall Height per Set of Packing

Cross Section	Nominal Overall Height per Set in Inches Number of Rings per Set of Packing			
	3	4	5	6
$\frac{1}{4}$	$\frac{5}{8}$	$\frac{23}{32}$	$\frac{51}{64}$	$\frac{7}{8}$
$\frac{5}{16}$	$\frac{55}{64}$	1	$1\frac{9}{64}$	$1\frac{9}{32}$
$\frac{3}{8}$	$\frac{31}{32}$	$1\frac{1}{8}$	$1\frac{9}{32}$	$1\frac{7}{16}$
$\frac{7}{16}$	$1\frac{5}{32}$	$1\frac{23}{64}$	$1\frac{35}{64}$	$1\frac{3}{4}$
$\frac{1}{2}$	$1\frac{7}{32}$	$1\frac{13}{32}$	$1\frac{39}{64}$	$1\frac{13}{16}$

Source: Reported by (3).

Flange ring packing is used the least among the automatic lip-type seal devices. Although available in fabric-reinforced and in homogeneous rubber, the flange ring packing is essentially made of leather and is exclusively installed as an outside-packed seal, designed for moderate pressure levels. It is applied in all cases where U-cup or V-rings are impractical. Flange packings are suited for both rotating and reciprocating sealing service.

The O-ring is the most popular device of the automatic squeeze-type packing seals. As a circular ring with a circular cross section the O-ring is capable of maintaining its excellent resilience for a long period of time, because it has a memory and stores large quantities of deformation energy. As an incompressible device it incorporates a combination of properties that produce an initial seal without gland force.

The O-ring provides long-term reliable service; however, the service life is shorter than that of lip-type packing rings. Tolerances and surface requirements are closer than those for lip seals. They are sensitive to damage by heat, abrasives, and other solid particles.

VI. Final Remarks Concerning Spring-Loaded TFE Seals

Spring-loaded TFE seals, using line contact, such as BAL-Seal, FCS-Seal, and Omniseal, offer an advantage in applications where long service life is required. These seals are distinguished by chemical inertness, wide temperature range, and very favorable friction and wear characteristics even at high pressure.

The seals under consideration have serrations, either one, two, or more per ring. Under system pressure the serration peaks flatten slightly and the contact lines flatten to a small area, deviating from the line, because of cold flow.

The design principle of the serrated seals is shown for the BAL-Seal, as an example with three serrations in Fig. 5.33. The arrows in (a) indicate the location where the initial line contacts take place. Because of cold flow and deformation under pressure in connection with wear the serration peaks flatten out and the deformation pattern reaches the shape illustrated in Fig. 5.33b.

The seals do not roll or twist during reciprocating sealing action, thus permitting smaller cross sections than exist for the O-ring.

There is low friction and wear in Teflon seal ring devices because under pressure a microscopic film of Teflon is deposited on the running surface, permitting the seal to ride on this film. This situation explains

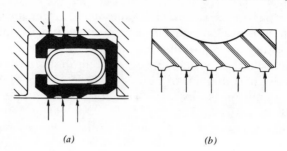

(a)   (b)

Fig. 5.33 Design concept of BAL-Seal with load and deformation pattern. (*a*) Scheme. (*b*) Load and deformation pattern.

why the TFE seals have a breakout friction force only one-third that of the O-ring of homogeneous elastomer materials.

Figure 5.34 represents estimated breakout and running friction curves for TFE resins, as reported by the BAL-Seal Company (1). The seal provides high unit loading at specific points, forcing the sealing areas together and achieving positive seal effects. By adding additional serrations, multiple sealing is provided that insures greater reliability. Trapping the lubricant fluid in the serration grooves causes a mode of fluid friction to take place, responsible for the unusual service life.

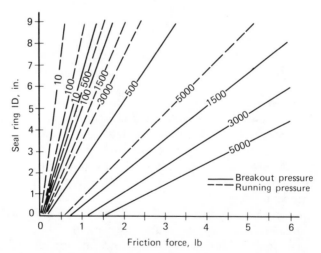

Fig. 5.34 Estimated breakout and running friction for BAL-Seals of PTFE, Series 301 A.

PART 4. SPECIAL METAL PACKINGS FOR HIGH-PRESSURE SERVICE

I. Introduction

Most high-pressure operations are conducted within pressure levels that can be satisfactorily sealed by devices extensively covered in Parts 1 to 3 of this chapter. This is particularly true when components of TFE and its combinations with graphite and molybdenum disulfide are utilized.

In service with higher pressures of extreme levels it is generally customary to use metal packings. Two basic types of metal packings are applied, soft metal packings and those predominantly using hardened steel. Combinations of metals with TFE resins, leather, rubber, and others are also found, regardless of the pressure level.

II. General Design Considerations

High pressures in excess of 350 atm (~5000 psi) are produced by reciprocating pumps, compressors and intensifiers. For higher levels, on the order of 1000, 2000, or 4000 atm (15,000, 30,000, or 60,000 psi), such as encountered in the low-density high-pressure polyethylene process, the sealing of plungers in commercially operating production units becomes really an art of a very high degree. Commercial production units of this process are expected to be in operation flawlessly for periods of 5000 to 6000 hr without interruption. In such installations the seal of the plungers is the most important factor in maintaining continuity of the process for longer periods of time. The author was extensively involved in the early development of this process as a plant engineer and can offer a firsthand view.

Design details of packing configurations for pressures in the 4000 atmosphere range are still considered proprietory and cannot be freely discussed here in order to protect owner rights. The discussion to follow offers only basic generalities.

A. Soft-Metal Packings

In compressor and intensifier design where extreme pressures must be handled, it is customary to use plungers instead of pistons. As indicated earlier, plungers are defined as straight steel rods usually sealed by outside packings. They have neither piston rings nor inside packings of the type described earlier in this chapter. The packings for high-pressure plungers are arranged in groups of rings in the housing.

In early phases of development of high-pressure low-density polyethylene the author used a two-stage compressor with a suction pressure of 350 atm and a discharge pressure of 1500 atm. The second stage then compressed the gas in one step up to 4000 atm. Both stages applied the same packing configuration to seal the plungers, using a combination of soft metal and steel, as shown in Fig. 5.35.

The basic seal ring had the shape of a U-cup ring, made from an alloy with 50% lead and 50% tin. The total packing had three sets with nine rings in each set, six of them using a U-opening angle of 90 degrees for the male adapter and three cup rings with an opening angle of 120 degrees. The male adapter rings were split to be flexible, permitting the gas to enter the U-opening of the cup ring and pressurize the cup

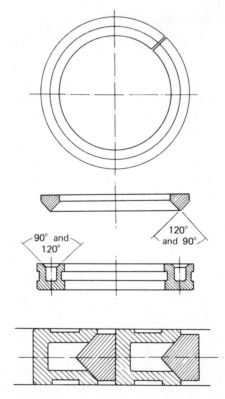

Fig. 5.35 Details of soft-metal packing for two-stage HP-gas compressor.

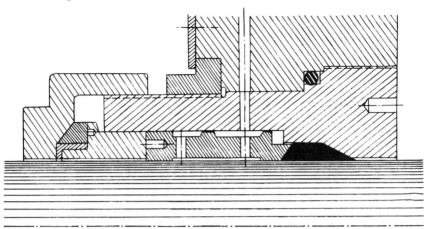

Fig. 5.36 Gas recycle pump sealed by soft-metal cone ring.

internally, preventing it from collapsing and thus enhancing the self-energizing sealing principle for lips. Solid male adapter rings could crush the cup rings or at least produce excessive deformation.

Figure 5.36 shows the seal of a rod for a gas recycle pump, using a soft metal cone ring of Pb-Sn alloy. A gland nut is tightened, forcing a guide sleeve against a lantern ring in contact with the seal ring. The pump operates at a pressure of 300 atm. Because of the low wear of the soft-metal cone ring average operating periods of 3000 hr are customary. The lubricant is supplied to the lantern ring at moderate pressure. Again, alignment and mechanical tolerances require high standards for satisfactory service. It is preferable to tighten the seal nut at startup and to release the nut when the pump is shut down. The diameter of the rod of the pump is 60 mm (~2.36 in.).

For compressors of larger capacities and essentially higher pressures the packings must be designed differently. These machines usually operate at higher piston velocities and require certain allowances for radial minute displacements of the rods during operation. For these types of compressors multiring arrangements are used. This requires that the stuffing box chamber be subdivided into a series of separate chambers, occupied by special sets of packing ring arrangements. The number of chambers is a function of the final pressure level, rod diameter, and rod velocity. It is a rule of practice to arrange the number of individual chambers as shown in Table 5.8.

**Table 5.8   Number of
Chambers per Individual
Set of Packing**

Pressure Atmospheres	Number of Chambers
Up to 100	3 to 4
100 to 400	4 to 5
400 to 800	5 to 6
800 to 1200	6 to 7
1200 up	8 to . . .

B.   All-Metal (Steel) Packings

For service with extreme pressures, in excess of 700 atm, rings of special design are used that represent a combination of radial and tangential rings, as shown in Fig. 5.37. The pressure level is an arbitrary figure and must be considered as a guideline only.

As discussed briefly in Chapter 4 the ring is split and cut in three segments in a radial and/or tangential direction. The tangential cuts are made in such a way that they fall along the three sides of an equilateral triangle with the sides of the triangle parallel to but displaced outward from a tangent to the inner circle. As wear takes place on the circumference of the inner circle, the three segments move simultaneously at equal distance toward the center, thus compensating for the wear. In the rings with radial cuts the segments also move uniformly toward the center in case of wear. For both types of rings the driving force for wear compensation is provided by a garter spring.

Two rings usually form a set that is held together by an interlocking arrangement with overlapping joints to minimize leakage.

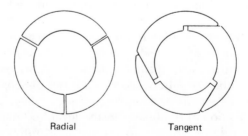

Radial   Tangent

Fig. 5.37 Typical radial cut and tangent cut packing ring set (15).

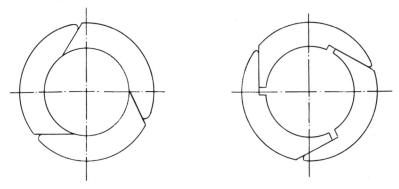

Fig. 5.38   Two types of tangent cut rings [From ASME Design Manual (15), courtesy of ASME, New York, N.Y.].

For satisfactory service the best surface finish possible should be provided for both rod and packing segments with the guarantee that the ID of the packing conforms with the rod surface.

Figure 5.38 illustrates two types of basic metallic segmental rings with two different kinds of tangential cuts. In Fig. 5.38*b* the joint is stepped and tangential to a circle larger than the ring ID. This joint allows a leakage path and must, therefore, be covered by a second ring. In Fig. 5.38*a* the joint is also tangential, but to a circle smaller than the bore of the ring. In this design a leakage path does not exist. This ring is applied in cases where it is not possible to close a leakage path with another ring.

Packing sets are assembled in so-called packing cases. In conventional compressor design a set of packing rings consists of pairs, using a radial and a tangential cut ring. Most major compressor manufacturers consider such a set as a "standard" arrangement for positive sealing in one direction. Such a set is illustrated in Fig. 5.37. The radial ring serves the purpose of sealing the gaps of the tangential ring. One set of rings, radial and tangential, is doweled together to prevent alignment of gaps. The pair of this design is referred to as a single sealing element. For installation the radial ring should always face the pressure. In sets where the tangential ring faces the pressure, the joints of the radial ring provide a passage allowing the gas to bypass the rings.

When the packing has to seal in both directions it is customary to use a pair of tangential rings, as shown in Fig. 5.39.

When rings made of TFE resins are used, a backup ring is required, as shown in Fig. 5.40. They are usually applied for pressures in excess of 750 psi. The backup ring should have zero clearance at the joint but

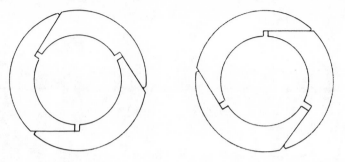

Fig. 5.39 Special double-tangent cut packing ring set [From ASME Design Manual (15), courtesy of ASME, New York, N.Y.].

some clearance with the rod, which may amount to a few thousandths of an inch.

A packing case is intended for housing one set of seal elements for any particular set composition. Several cases then compose a packing case set, as is illustrated in Fig. 5.41, which shows an arrangement of three cases. This design would satisfy a system with moderate pressure for a nonlubricated compressor. Case cavities are not standardized because of the lack of standards for the seal rings. The clearance of the bore against the shaft surface is usually such that the bore is $\frac{1}{8}$ in. or more larger than the shaft diameter.

Systems that require lubrication for the packings use passages with inlets and an outlet for the lubricant. One or more lube points are provided. Fig. 5.42 illustrates how the lubricant supply is introduced to the system and how it passes through the various cases. The design also illustrates how vents are added to the packing. Vents are used to collect hazardous, noxious, or precious gases that otherwise would leak to the atmosphere.

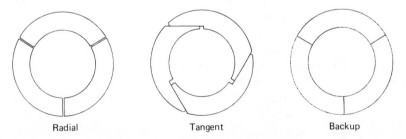

    Radial   Tangent   Backup

Fig. 5.40 Typical radial cut and tangent cut packing ring. Set with backup ring [From ASME Design Manual (15), courtesy of ASME, New York, N.Y.].

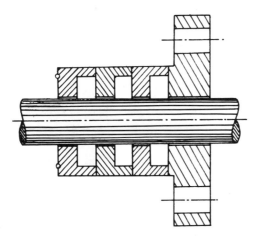

Fig. 5.41 Basic packing case, three-groove design [From ASME Design Manual (15), courtesy of ASME, New York, N.Y.].

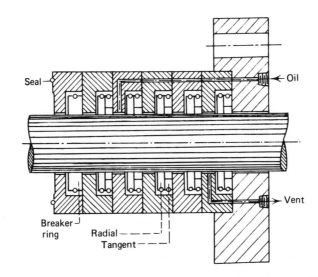

Fig. 5.42 Packing arrangement with casings for lubricated packing sets [From ASME Design Manual (15), courtesy of ASME, New York, N.Y.].

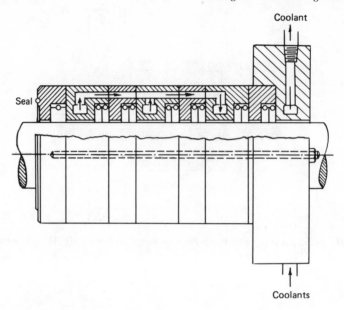

Fig. 5.43 Cooled packing cases [From ASME Design Manual (15), courtesy of ASME, New York, N.Y.].

Packing cases often require cooling to remove heat buildup. Since generation of heat within a packing varies from system to system the design configurations for heat removal often vary considerably. Figure 5.43 shows a packing with cups made as castings, containing cavities for cooling water that direct the coolant to locations where heat is generated.

In Fig. 5.44 the coolant cavities are machined into the cases, a device that is generally applied to compressor designs handling higher pressures.

C.   Special High-Pressure Packing Designs

Sealing of intensifiers and hydraulic presses for superpressures is an extremely intricate problem and requires an absolute understanding of all major sealing principles. Plungers of intensifiers and hydraulic presses usually move with relatively low velocities, reducing the rate of wear. The requirements for seal effectiveness, however, are quite high.

Figure 5.45 shows a plunger of a hydraulic press that was operated at a pressure of 6000 atm (90,000 psi). The plunger is inside packed using leather and brass rings as packing materials. Leather alternates with brass and the rings are beveled toward the cylinder wall.

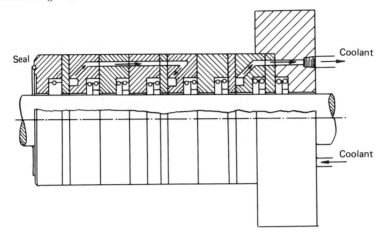

Fig. 5.44 Cooled packing cases [From ASME Design Manual (15), courtesy of ASME, New York, N.Y.].

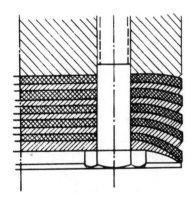

Fig. 5.45 Packed plunger with leather rings for 6000 atm.

A Bridgman plunger seal shown in Fig. 5.46 has been used in an intensifier for extreme pressures. The plunger is inside packed. On the top face of the plunger a homogeneous rubber ring is attached, held in place by a cap with a guide pin. The seal is self-energizing and very effective.

H. Tongue (19) reports also a Bridgman-type plunger seal used in a compressor developed for ICI-Ltd., United Kingdom. The cylinder assembly, shown in Fig. 5.47, was designed for an operating pressure on the order of 2000 atm.

The plunger is provided with four backing rings to support three seal rings, which in turn are attached to three spacer rings. On top of the

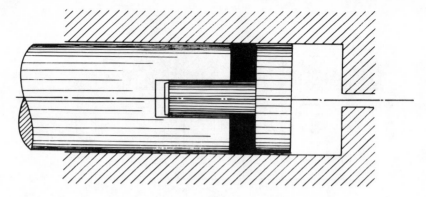

Fig. 5.46   Plunger with Bridgman packing.

packing set is a bearing pad held in position by a hex nut. To prevent leakage along the ID of the packing rings a copper joint ring is used to seal between the top backing ring and the pad.

The plunger moves in a tungsten-carbide sleeve attached to the cylinder by shrink-fit to supply compressive stresses in the sleeve, thus avoiding excessive deformations. The backing rings underneath the seal rings are made of tungsten-carbide for reasons of stability. Tungsten-carbide should also be used for the spacer rings that support the packing seal rings.

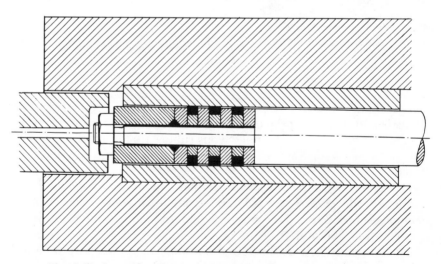

Fig. 5.47   Intensifier plunger in tungsten-carbide cylinder sleeve (19).

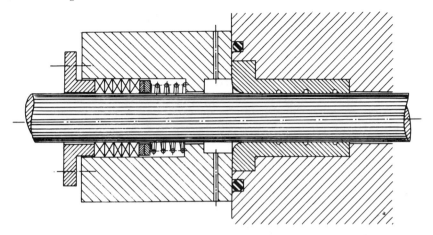

Fig. 5.48 Plunger with hydrodynamic thin-film lubrication as sealing device [From (5), excerpted by special permission from *Chemical Engineering*, November 1951, copyright © 1951 by McGraw-Hill, Inc., New York, N.Y. 10020].

With this kind of seal pressures above the 2000 atm limit can be achieved. The author has used this type of plunger-seal combination successfully for pressures on the order of 4000 atm (60,000 psi).

W. Coopey (5) reports a plunger seal design in which the lubricant utilizes the hydrodynamic sealing principle resulting from an extremely close clearance gap. In the design of Fig. 5.48 an example of a hydrodynamic thin-film lubrication is shown, utilizing the viscosity of the lubricant.

The clearance gap between the oil film packing element and the plunger is reported to be 0.0005 in. The wedge effect formed in the chamber of lubricant supply facilitates the creation of the film along the surface of the plunger. The thin film develops high molecular adhesive forces that suffice to seal the internal gas pressure of the intensifier, which may far exceed 2000 atm.

D. Summary Remarks

In high-pressure service pistons and plungers are used that must be sealed in reciprocating motion. Pistons are characterized as compression devices that carry the seals along during their operation. They are sealed either by piston rings or by inside-packed seals, using U-cups or V-rings. Piston rings are used for pressure levels up to 400 atm. U-cup and V-rings may be used for pressure well in excess of 400 atm.

Plungers are generally sealed by outside-packed devices, using a wide range of packings in the form of plastics, TFE resins and their combinations, soft metals, such as Pb–Sn alloys, and solid metal packings, usually made of hardened steel.

In intensifiers where plungers are used exclusively, the line of packing arrangements is not clear-cut. Commercial machines with large plunger sizes, high pressure levels, and large capacities preferably use metal packings in multiple sets of two with the number of sets increasing as the pressure rises.

Intensifiers of smaller sizes for intermittent or pilot plant service use leather, rubber, and plastics, and their combinations for pressures up to 6000 atm. As a final general conclusion, in modern seal technology there are few problems that cannot be solved satisfactorily with the means available today. It is important that the problem be recognized. By applying suitable sealing principles while meeting the mechanical requirements, most difficulties can be minimized. Most seal problems are created by improper handling of the mechanical prerequisites before the seal begins to function. If these problems are eliminated the seal leaves few factors unsolved.

References

1. Bartheld, J.
   "Spring-Loaded Teflon Ring Seals"
   BAL-Seal Company, Tustin, CA., 1965.

2. Bauer, P., Glickman, M., and Iwatsuki, F.
   "Dynamic Seals"
   IIT Research Institute, Technical Report AFRPL-TR-65-61, Air Force Rocket Propulsion Laboratory, Edwards, CA., May 1965.

3. Boyce, H. L.
   "Selection Factors for Seals and Packings"
   Plant Eng. (February 1973).

4. Buchter, H. H.
   Apparate und Armaturen der Chemischen Hochdrucktechnik
   Springer-Verlag, New York, 1967

5. Coopey, W.
   "How to Pack Reciprocating Rods Against High Pressure"
   Chem. Eng. (November 1951).

6. Coopey, W.
   "How to Solve Soft Packing Problems"
   Chem. Eng. (January 1958).

7. Coopey, W.
   "New Seal for Superpressure Shaft"
   Chem. Eng. (September 1965).

8. Coopey, W.
   "Spring-Loaded Packing Ensures Reciprocating Rod Seal"
   Plant Eng. (May 1966).

9. Coopey, W.
   "A Fresh Look at Spring-Loaded Packing"
   Chem. Eng. (November 1967).

10. Fisher, E. W.
    "Installing and Maintaining Sealing Devices"
    Hydraul. Pneum. (October 1972).

11. Gillespie, L. H.
    "Designing with Seals Made of Teflon Fluorocarbon Resins"
    Teflon by DuPont, J. Mach. Des. (1972).

12. *Handbook of Mechanical Packings and Gasket Materials,*
    Mechanical Packing Association, Inc., Washington, D.C. 1960.

13. Korndorf, B. A.
    Hochdrucktechnik in der Chemie
    VEB-Verlag Technik, Berlin, 1956.

14. McKillop, G. R.
    "Sealing Materials in Compression Packings and O-Rings"
    Mach. Des. (January 1966).

15. ASME Design Manual
    "PTFE Seals in Reciprocating Compressors"
    ASME, New York, 1975.

16. *Mach. Des.*
    Reference issue on seals, 1973–1974.

17. *Mach. Des.*
    Reference issue on mechanical drives, 1975.

18. Norton, R. D.
    "Mechanical Seals for Handling Abrasive Liquids"
    Chem. Eng. (September 1956).

19. Tongue, H.
    The Design and Construction of a High Pressure Chemical Plant
    Van Nostrand, New York, 1959.

Numerous catalogs, manufacturers' literature, papers, and unpublished company reports were also consulted.

CHAPTER 6

Other Configurations of Direct Contact Seals

All seals discussed in Chapters 1 to 5 are members of the category defined in Chapter 1 as interfacial seals. They are characterized in their seal function as establishing a direct contact between the components that must be sealed and the shaft. Felt seals, diaphragm seals, and piston rings are other significant sealing devices that complete in essence the large family of interfacial seals. Felt seals are discussed in Part 1; diaphragm seals in Part 2; and piston ring seals in Part 3.

PART 1. FELT SEAL DESIGNS

Felt seals are sealing devices that are applied in large quantities to prevent leakage of oil or grease from relatively simple systems. They provide a double function by preventing dust and dirt from penetrating into closed systems and preventing lubricants from leaking to the atmosphere.

I. General Functional Purpose

Felt seals are basically designed for the purpose of retaining lubricants. They are always, therefore, installed in close proximity to all kinds of bearings for rotating shafts. They function with relatively little contact with the shaft surface without squeeze and requiring no interference action.

Felt seals are cut from felt sheets. This is a built-up fabric manufactured by interlocking fibers through mechanical work, chemical reaction, moisture, and heat in such a manner that spinning, weaving, or knitting are not required. Felt actually represents a conglomerate of various fibers: wool, animal, vegetable, and synthetic.

II. Design Configurations

Felt seals are of great simplicity and come in a variety of design configurations. They are used for a wide range of shaft speeds and are installed to have close contact with the shaft surface without developing excessive contact pressure. Some of the most common felt seal designs are illustrated in Fig. 6.1.

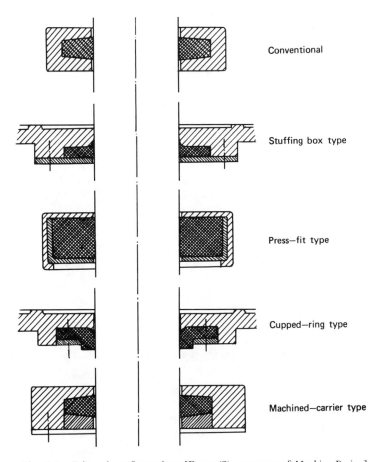

Fig. 6.1   Felt seal configurations [From (2), courtesy of *Machine Design*].

Conventional Design.   In the conventional design the felt ring is inserted in a groove machined in a special casing that is inserted into a cavity of the housing with the shaft. Once the felt ring is inserted the position remains fixed and further adjustment is not possible. Practical running clearance with the shaft and the casing is 0.010 to 0.015 in. The seal can be replaced by the removal from the shaft only.

Stuffing Box Type.   In the stuffing-box-type design the felt ring is inserted in a specially machined groove and then held in place by a thin cover plate. The felt can move slightly toward the oil side, thus enhancing seal effectiveness. The felt can be replaced easily.

Press-Fit Design.   The felt ring is placed in a casing that then is inserted into a cartridge with a press-fit operation. This design is extensively used in low to medium shaft speed applications and is preferably applied where space is a problem.

Cupped Ring Type.   Cupped ring design is similar to the stuffing box-type seal configuration. The ring is installed and operates with interference between shaft surface and felt ring ID. The seal also acts as a very effective exclusion device.

Machined-Carrier Design.   This design is frequently applied to seal ball and roller bearings. For replacement the felt ring is preferably split. This design is very similar to the conventional device.

III. Installation Rules

When radial felt seals are assembled the following basic rules should be considered:

The felt ring should be designed so that its height is larger than its width on the shaft surface. The width on the ring OD is always smaller than the width on the ID.

Great caution must be used when the ring is inserted in the groove. The structure of felt is easily damaged and is susceptible to compression. If handled improperly it disintegrates and when stretched excessively it loses its properties completely.

Best operation can be expected when the ring is made of one single piece without being split.

Whenever maintenance work is done, the seal should be replaced, regardless of condition. However, if the maintenance intervals are too long, the seal should be replaced more often.

IV. Felt Ring Applications

Felt ring seals cover a considerable range of useful applications. They may function as lubricant retainers, filters, exclusion devices, oil wicks, and in many other ways.

A. Oil Retainers

Most ball and roller bearing applications use a felt seal in close proximity as a lubricant retainer.

B. Oil Wicks

Many kinds of machines and equipment are used for intermittent service and develop problems of improper lubrication when they are restarted. Felt seals overcome this difficulty. The felt fibers functioning as capillary systems fill themselves with lubricant after shutdown and hold their contents of lubricity at all times, thus eliminating any period of dry-run.

C. Filtration

Long periods of machine idling are critical for collecting solid particles that may contaminate the lubricant, damage shaft and bearings, and jeopardize operation. Felt ring seals then act as filter elements for eliminating particle sizes down to 0.7 microns.

D. Lubricant Absorption

Felt has a high degree of absorption capacity for oil from the sump. Depending on density, some rings absorb up to 80% of their volume, which then acts as an oil reservoir.

V. Limitations for Applications

Felt seals are available in two basic types, plain and laminated. By impregnating felt with one or more of a variety of suitable materials the range of applications can be broadened significantly. When impregnated with paraffin, petroleum, or colloidal graphite the resistance of felt to water and mud is enhanced. This impregnation further improves resistance to lubricant oils under pressure and decreases friction. Manufacturers recommend that the felt be saturated with an oil or grease of higher viscosity than that used as lubricant.

Dry felt develops a coefficient of friction in the order of 0.22 as an average value. A substantial improvement can be achieved after the felt has been saturated with oil.

Although felt acts as a filter, absorbing abrasive particles from the environment, it has been observed that felt polishes rather than scores a shaft. It seems that the absorbed particles penetrate the felt instead of remaining located at the place of direct absorption.

Felt ring seals can be used through a temperature range of −60 to 250°F. When the felt is replaced by synthetic fiber the resistance to temperature may rise to 400°F. Depending on the composition of the fiber, resistance to corrosion can also be markedly improved.

For chemical compatibility consultation with the manufacturer is recommended.

Felt ring seals are suited for shaft rotational speeds up to 2000 fpm. Synthetic fibers used with adequate lubrication permit shaft speeds up to 4000 fpm and over.

PART 2.   DIAPHRAGM SEALS

I.   Introduction

Dynamic diaphragm seals are used exclusively for reciprocating shaft motions. The elastic diaphragm is an integral part of the reciprocating piston, spanning the gap between the stationary housing and the reciprocating piston, thus separating the operating chamber into two different areas.

Diaphragms may be classified in a number of ways. The most common classification distinguishes between flat and convoluted diaphragms and rolling diaphragms.

II.   Flat and Convoluted Diaphragms

A flat diaphragm is characterized by having either no convolution at all or convolutions amounting to less than 180 degrees. Diaphragms used for dynamic sealing purposes are membranes that operate in the same way as sliding contact packings with zero leakage.

III.   Design Principles

A diaphragm may initially be flat, but convolutions may be formed by fluid pressure during operation. When the diaphragm moves with the piston it will be distorted by the relative motion. During the distortion, elongation of the diaphragm material takes place. This elongation should be avoided, since it will cause future fatigue failure of the diaphragm material. This can be counteracted by providing convolutions that allow the diaphragm to operate under elastic rather than plastic conditions under load.

A. Flat Diaphragms

The diaphragm should be a thin, flexible, and highly elastic membrane. Specific details of a diaphragm design are illustrated in Fig. 6.2. Figure 6.2a represents suitable mechanical stroke limitations to prevent the diaphragm from moving beyond the 90% value of its maximum possible stroke. The sharp corners in Fig. 6.2b will eventually produce surface damage of the diaphragm with subsequent failure. Excessive thickness of membranes will result in wrinkles that will also result in premature failure, as shown in Fig. 6.2c. Flexing requires free space. Excessive space, however, produces fatigue stress in membrane material, as may be observed in Fig. 6.2c. The best diaphragm for long service life has molded-in convolutions that can absorb relatively high extension stresses.

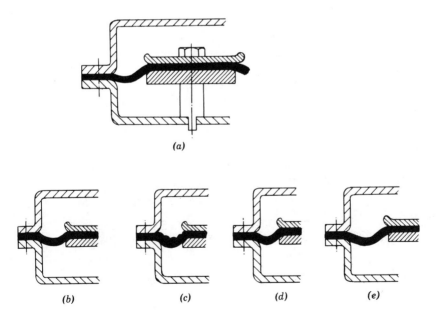

(a)

(b)   (c)   (d)   (e)

Fig. 6.2 Details of diaphragm designs [From (2), courtesy of *Machine Design*].

B. Convoluted Design

The convolution diaphragm contains convolutions that facilitate reciprocating piston motions without producing fatigue failure in the diaphragm. Increased strength is imparted to the membrane by incorporating a fabric structure into the elastomer. Temperature stability is
obtained by the characteristics of the elastomer. Convoluted membranes
work best with short strokes.

IV. Rolling Diaphragms

Rolling diaphragm design is needed when longer operating strokes must
be applied. A longer stroke is defined as a stroke longer than any stroke
that can be accommodated by a flat or convoluted membrane.

A. Mechanism of Operation

The operating mechanism of a rolling diaphragm is schematically illustrated in Fig. 6.3. When the piston moves upward the diaphragm rolls in
the same direction as the piston until the piston side wall is completely
covered by the diaphragm material. On reversal of the motion, the
diaphragm disengages from the piston into the vacant space provided
between the piston and the cylinder wall. The rolling process stresses the
membrane severely in a circumferential direction, resulting in corresponding elongation. In an axial direction no elongation takes place.

   For the rolling motion the diaphragm must be thin, of a thickness
ranging from 0.010 to 0.035 in. The circumferential elongation is made
possible by the special overlay of the fiber, which is oriented so that free
circumferential elongation becomes possible. The rolling diaphragm
design operates with a full 180-degree convolution. During operation
the diaphragm constantly changes in circumference. The number of
possible strokes before failure depends on the extent of these changes,
operating pressure, and temperature.

B. Materials of Construction

Materials for diaphragms must have a high resistance to bursting, a high
tensile strength, and a high modulus of elasticity. A decision must be
made for a suitable fabric and an appropriate elastomer. This latter has
to supply elasticity, flexibility, impermeability, and chemical compatibility. Fabrics have to provide mechanical strength and resistance to abrasion and fatigue.

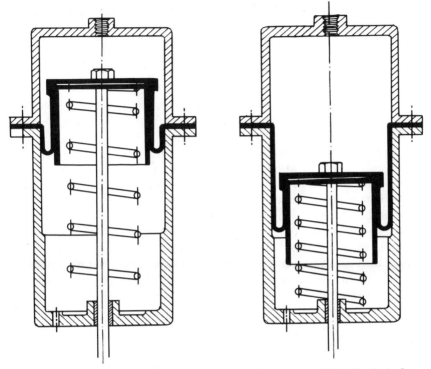

Fig. 6.3 Design of rolling diaphragm [From (2), courtesy of *Machine Design*].

Most commonly used fabrics for diaphragms are polyesters, nylon, and Nomex. Most frequently used elastomers are nitrile, neoprene, epichlorohydrin, silicones, fluorosilicones, polypropylene, and Viton. For detailed information contact with the' manufacturer is recommended.

PART 3. PISTON RINGS AS SEALING DEVICES

I. Introduction

Piston rings are used for a wide range of industrial sealing purposes in systems handling gases, liquids, and, predominantly, hydraulic fluids. They represent very specific sealing devices that provide satisfactory long-term services. This discussion covers the fundamental concepts only. Piston rings have been widely used in internal combustion engines of all types and in aerospace applications to control leakage of hot exhaust gases, to control lubricants in reciprocating engines, and to control liquids in hydraulic supply systems of all kinds.

II. The Piston Ring as a Sealing Component

The piston ring is a sealing component that was first developed in connection with the invention of the steam engine. Its application was continually extended and it was finally applied to Otto and diesel engines on a large scale. Finally pumps, compressors and intensifiers were added to the wide range of applications. The ring was designed, developed, and improved to satisfy many specific requirements and is still undergoing further improvements.

Figure 6.4 illustrates the simplest form of a piston ring seal (a), and a cross-sectional view of the ring, installed in the groove of an engine piston, located in sealing position (b), and in a neutral position (c), but in contact with the cylinder wall. When the ring is in the operating position the external face of the ring contacts the wall of the cylinder because of its built-in tension, which forces the open ends of the ring to stretch

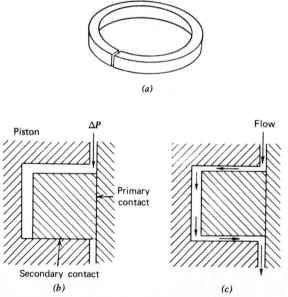

Fig. 6.4 Piston ring. Modes of flow and pressure. (a) Piston square ring. (b) Ring in sealing position. (c) Ring in neutral position [From (3), courtesy of Koppers Company, Inc., Baltimore, Maryland].

outward, thus establishing the so-called primary seal contact. Once pressure ΔP is acting the ring is pushed against the side wall of the opposite side of the piston groove where it reaches its secondary contact position on the downstream side.

III. Basic Sealing Function

In general, the piston ring can be used to satisfy a variety of sealing functions. To simplify the discussion the sealing mechanism of a simple square ring may be described when the ring is inserted in a conventional groove of a piston to seal along the ID of a cylinder wall, as indicated in Fig. 6.5.

During installation into the groove of the piston the ring tends to expand outward because of its built-in spring tension. As a result the external face of the ring establishes contact with the wall of the cylinder; this is referred to as establishing its primary contact. Once system pressure ΔP is applied the ring is pushed downstream to establish its secondary contact with the face of the groove in the piston. As a result of this performance the ring is also referred to as a compression ring. When the pressure is acting the primary contact is increased by the action of the pressure in the back of the ring. The distribution of the forces developed by this pressure is shown in Fig. 6.6. Figure 6.6a shows the ring in sealing position in the groove, and Fig. 6.6b gives the force diagram with respect to the balance conditions.

If the magnitude of pressure ΔP is moderate the primary and secondary contact forces will not be excessive. As the total pressure differential across the piston increases the contact forces will also increase and more piston rings must be added. As the pressure drop across an individual ring becomes excessive, pressure balancing must be applied, as is described in Section IV. Adequate balance design provides a more uniform force distribution for each ring, increasing seal life.

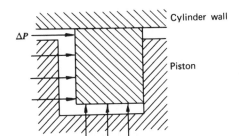

Cylinder wall

ΔP

Piston

Fig. 6.5 Pressure-induced forces acting on compressor-type piston ring. (a) Ring in seal groove. (b) Balance conditions [From (3), courtesy of Koppers Company, Inc., Baltimore, Maryland].

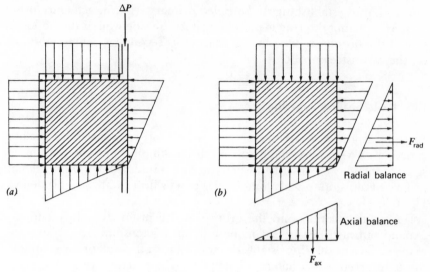

Fig. 6.6 Force distribution and balance conditions on piston ring [From (3), courtesy of Koppers Company, Inc., Baltimore, Maryland].

The total contact seal forces are composed of two components, including the ring tension and the force component, produced by the fluid pressure. In general, the component of the fluid force is the larger of the two. Both components are responsible for the formation of the axial and radial drag forces.

When the fluid pressure is high the influence of the ring tension can be neglected. At very low fluid forces the ring tension may be dominant.

The drag forces are schematically illustrated in Fig. 6.7. Their magnitude can be computed. A computation method is shown in the *Engineer's Handbook*, (3).

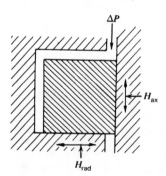

Fig. 6.7 Drag forces on piston ring [From (3), courtesy of Koppers Company, Inc., Baltimore, Maryland].

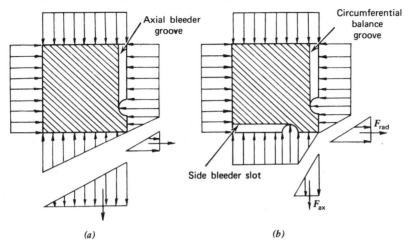

Fig. 6.8 Principles of piston ring balancing [From (3), courtesy of Koppers Company, Inc., Baltimore, Maryland].

IV. Pressure Balancing

In the presence of excessive hydraulic fluid pressure a balancing of the forces becomes necessary. Two typical balance methods are shown in Fig. 6.8, where the axial and the radial forces are balanced to a certain degree. The balance principle is very much the same as shown in Chapter 2 for mechanical end face seals. The balance mechanism in Fig. 6.8 is self-explanatory. The ring of (a) is balanced in the circumferential direction, whereas the ring of (b) is balanced in both the circumferential and axial directions. Balance can be established by providing so-called bleeder slots or grooves.

Pressure balance on piston seal rings is made possible by machining small channels in one or more surfaces of the outer ring. Fluid of the system can then flow into the channels or grooves, thus partially balancing the action of the fluid pressure on a surface opposite the surface along which the channels are located. A typical balanced piston ring is illustrated in Fig. 6.9.

V. Piston Rings and Seal Rings for Lubricated Compressors

Piston rings serve quite a variety of functions. In internal combustion engines they seal gas pressures from the crankcase and simultaneously control the flow of lubricant from the crankcase to the compression

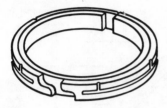

Fig. 6.9   Balanced piston ring [From (3), courtesy of Koppers Company, Inc., Baltimore, Maryland].

chamber. In combustion engines a distinction is made between the compression rings and oil control rings.

In gas compressors the rings serve a similar function when they are installed in the piston. When sealing a double-action motion the rod of the compressor piston must be sealed as well. For rod sealing the rings function in stationary position and seal on the ID of the rings, where they function as contracting rings.

Piston rings used for sealing rotating shafts are installed stationary in a similar fashion in the housing as rings used for sealing reciprocating rods.

A.   Compression Rings

In diesel, aircraft, automotive, and industrial engines the compression rings are exposed to high temperatures, extreme stresses, impact, corrosion, and even abrasion. They must operate with a minimum of lubrication and still provide service at low wear conditions.

Gray iron is the major material for piston rings. In recent years chromium plating, molybdenum coating, and carbide coating have made considerable progress in reducing wear, thereby providing longer cylinder life.

B.   Oil Control Rings

Oil rings are better lubricated than compression rings and they are subject to lower stresses and temperatures. Because of good lubrication and reduced load conditions they are less stressed, and operate under lower friction and temperatures, thus providing a longer service life. The design of the oil control ring is basically governed by the gas pressure in the compression chamber, the shape of the oil-scraping ring edge, and the provision for oil drainage. The oil rings must supply a satisfactory lubricant film along the cylinder wall to lubricate the compression rings continuously. These functions, however, vary from engine to engine; consequently, the oil rings are also subject to variations.

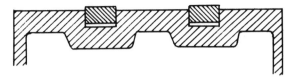

Fig. 6.10 Rings for double-acting compressor piston [From (3), courtesy of Koppers Company, Inc., Baltimore, Maryland].

In compressor design the same basic piston ring practice is used as is customary for combustion engines. Single-acting compressors use oil rings. In double-acting compressors oil rings are not required, since a connection with the oil sump of the crankcase does not exist, because both cylinders are closed, through a crosshead. A piston ring arrangement for a piston of a double-action cylinder of a compressor is illustrated in Fig. 6.10.

Where higher pressures must be handled, multistage design is employed. The final pressure stages use multiring arrangement with built-up ring configurations, as shown in Fig. 6.11. The illustration indicates that spacers are used which provide the actual width of the grooves for the piston rings. The separators are the parts placed as lands between the successive grooves. This design permits the use of rings with heavier walls, usually of the one-piece type. They are also available in the two- or three-piece configuration. The heavier wall is possible because the rings are not installed by the conventional method, being stretched and then slipped over the piston OD.

VI. Rings for Nonlubricated Compressors

Nonlubricated gas compressors must be used where lubricant oil in the gas cannot be tolerated. A typical example is a compressor handling

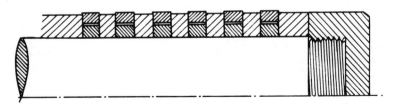

Fig. 6.11 Built-up piston for high-pressure compressor [From (3), courtesy of Koppers Company, Inc., Baltimore, Maryland].

oxygen, where the presence of oil would result in fire or explosion hazards. Other examples are food and drink processing equipment, where oil from lubrication contaminates the end products.

Nonlubricated compressor design varies greatly. The basic feature is the use of a crosshead to separate the crankcase from the gas compression chamber. The oil of the crankcase housing has no chance to be ejected against the wall of the gas cylinder.

Contrary to the conventional compressor design, nonlubricated compressors use nonmetallic piston rings for both piston and rod sealing. The rings usually consist of antifrictional materials, such as PTFE compounds, impregnated with carbon-graphite or molybdenum disulfide.

Fig. 6.12 reflects a typical inboard arrangement whereas Fig. 6.13 illustrates a piston with two outboard rider rings. In the arrangement with two outboard rider rings the number of sealing piston rings between the rider rings depends on the operating conditions of the compressor. In Fig. 6.13 the design illustrates three sealing piston rings. Fig. 6.14 shows an outboard configuration with two sealing piston rings between the two rider rings.

A rider ring is generally fabricated as a one-piece ring, which may consist of a wide range of PTFE compounds depending on service conditions and environment. The material offers a very low coefficient of friction with an excellent resistance to wear without stick-slip and very little drag action (3). The same material is also successfully used for the piston seal rings.

The rider rings are provided with leakage grooves along the surface to prevent the riders from taking the action as seal rings. They function as movable bearings, carrying the weight of the piston along the surface of the cylinder in the form of sliding bearings. To accomplish this function larger clearances must be provided between the cylinder and the piston, compared to oil-lubricated compressors. Taking the action as movable bearings the width of the rider rings must be larger than that of the conventional seal rings.

The one-piece ring with the radial cut joint represents a most fre-

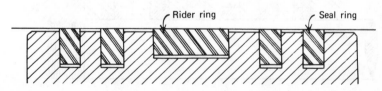

Fig. 6.12 Inboard rider arrangement [From (3), courtesy of Koppers Company, Inc., Baltimore, Maryland].

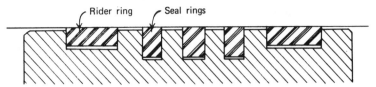

Fig. 6.13   Outboard rider arrangement [From (3), courtesy of Koppers Company, Inc., Baltimore, Maryland].

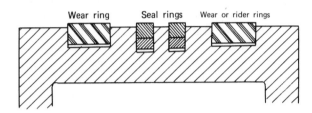

Fig. 6.14   Modified outboard rider arrangement [From (3), courtesy of Koppers Company, Inc., Baltimore, Maryland].

quently used design. It can easily be installed and removed. The solid configuration is difficult to install and cannot be removed without destroying the ring; because it is difficult for the fluid pressure to reach the back side of the ring, back pressure will not develop, which could increase the contact force of the ring and accelerate wear.

A rider ring design configuration of the solid ring type and a ring with a radial cut joint are presented in Fig. 6.15.

For further details on the large variety of piston rings and the many possibilities of application, see reference (3).

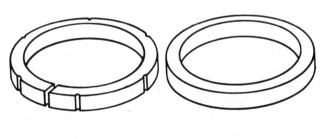

(a) Radial cut   *(b)* Solid ring

Fig. 6.15   Rider rings [From (3), courtesy of Koppers Company, Inc., Baltimore, Maryland].

VII. Final Remarks on the Application of Piston Rings

Piston rings are sealing devices used for sealing pistons, reciprocating rods, plungers, and rotating shafts. In lubricated compressors, combustion engines, pumps, and so on, the rings are generally fabricated from metals. In nonlubricated compressors the rings are of PTFE compounds, impregnated with carbon-graphite or molybdenum disulfide. For rotating shaft sealing the rings are made of similar combinations as those used for nonlubricated compressors. As mentioned earlier, rotating shafts are more difficult to seal than reciprocating shafts. In rotating equipment speeds are generally essentially higher than for reciprocating shafts. Further, the seal rings always contact the same area of the shaft, which makes scorching of the shaft highly probable.

Piston rings are preferably used for sealing gases of all kinds. When they are used for liquid sealing, a certain degree of leakage must be tolerated.

In sealing pistons of reciprocating compressors or diesel engines the maximum pressure that can be achieved with piston rings is on the order of 400 atm (up to 5000 psi). Each ring is capable of sealing a certain pressure differential. By increasing the pressure that exceeds the limit of the differential, the number of piston rings must be increased.

The basic rules are as follows, assuming normal piston speeds and operating temperatures:

p_i up to 300 psi		2 rings
300 to 900 psi		3 rings
900 to 1500 psi	plain face	4 rings
	balanced	5 rings
1500 to 3000 psi	plain face	5 rings
	balanced	6 rings
3000 and up psi	minimum of	6 balanced rings

For conventional piston rings the permissible area contact pressure ranges from 0.3 to 0.7 kg/cm^2. In multistage design the high-pressure stages are designed for 0.7 to 1.50 kg/cm^2.

In multiring applications the inner ring supports the flexibility of the outer ring. For two-piece rings the area contact pressure may reach a value of 5 to 6 kg/cm^2 without exceeding the flexibility of the piston rings. For optimal design the axial (face) stress should not be larger than the radial contact stress with the cylinder surface.

Favorable piston velocities for long-life service of piston rings depend on the pressure range applied. The base rule is high pressure versus low velocities and vice versa. Piston velocities of 500 ft/min in the first stages are considered permissible, using metal piston rings of the lubricated type. For nonlubricated rings of Teflon or antifriction materials velocity should be kept below 400 ft/min. These figures, however, greatly depend on specific design and diameter considerations in conjunction with the influence of the construction material.

References

1. *Mach. Des.*
   Reference issue on seals, 1973–1974.
2. *Mach. Des.*
   Reference issue on mechanical drives and seals, 1975.
3. *Engineer's Handbook of Piston Rings—Seal Rings—Mechanical Shaft Seals*
   Koppers Company, Inc., Metal Products Division Baltimore, 1967.
4. Buchter, H. H.
   Apparate und Armaturen der Chemischen Hochdrucktechnik
   Springer-Verlag, New York, 1967.
5. *Mach. Des.*
   Reference issues on seals, 1973–1974.
6. *Mach. Des.*
   Reference issue on machine drives, 1975.

Numerous catalogs, brochures, and manufacturers' publications were also evaluated and used.

CHAPTER 7

Interstitial Seals

Interstitial seals function by producing a pressure drop in the fluid to be sealed, allowing a minimum amount of tolerable leakage flow in a built-in clearance gap without affecting the rotating motion of the movable parts in the system. Interstitial seals effectively maintain a pressure differential between the interior of a system and the external atmosphere. The pressure differential is achieved by the throttling function of the seal against the leaking fluid.

In interstitial seals rubbing between the seal and the moving machine part does not take place. Consequently, frictional wear is not a problem. These seals, therefore, are distinguished by simplicity in design configuration, durability, reliability in service, and practically no maintenance.

For accurate throttling of fluid a certain minimum flow is required. It is, therefore, impossible to provide a completely leakage-free system by using interstitial seals (with one exception, ferrofluidic seals). Some leakage must be tolerated, and the design is directed toward minimizing this flow.

When the fluid throttling is achieved by steps, using a series of small chambers where a sudden irreversible acceleration with subsequent deceleration of the leaking fluid takes place, the seal device is defined as a *labyrinth seal*. This configuration is preferably used for sealing steam and gas turbines, centrifugal compressors, and other high-speed centrifugal machinery.

When the system fluid is throttled by utilizing viscous friction losses along a small annular clearance gap with constant cross section, the device is referred to as a *bushing seal*. The bushing can be arranged in a fixed position or it can be installed so as freely to follow the shaft motion in a radial direction. Such a seal is called a *floating bushing seal*. When the bushing consists of segments held together by a garter spring with the ring riding on the shaft surface without rotating, the seal is called a *floating bushing seal with segmental rings*.

382

When a separate viscous fluid is pumped into the clearance gap to counteract the leaking system fluid and prevent it from reaching the atmosphere, the device is referred to as a *viscous seal* or *visco seal*. The fluid pumped into the gap to prevent leakage is generally a fluid of a given viscosity. The fluid to be sealed can be either a gas or a liquid.

If the fluid in the gap is magnetic and if a magnet system is arranged to hold the fluid in the gap, the device is designated a *ferrofluidic seal*.

I. Classification of Interstitial Seals

Interstitial seals are available in a variety of design configurations. For discussion in this chapter only those categories are covered that are shown in the classification chart of Fig. 7.1, in which four major groups are considered representative of interstitial seals.

Visco seals may be subdivided into four types of seals: with the screw profile in the shaft, with it in a separate sleeve, with opposing screws, and with a combination of screw-type and other mechanical auxiliary devices.

Labyrinth seals may be divided into straight-through, staggered, stepped, interference, and combined labyrinth seals.

Bushing seals may be classified into fixed, floating, or balanced bushing and floating ring seals.

Ferrofluidic seals may be grouped into devices with single- and multiple-seal arrangements.

When properly analyzed, evaluated, and designed, each of these seal categories functions with a high degree of durability and reliability for extended periods of operating time. Pressure considerations provide the prime parameter for selecting the appropriate seal configuration, resulting in low to mean pressure ranges for interstitial seals as compared to high pressures for the liquid film mechanical end face seals or packings.

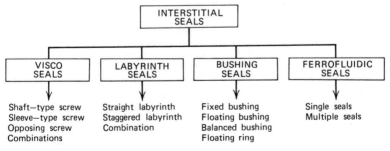

Fig. 7.1 Classification of interstitial seals.

PART 1. VISCO SEALS

I. Introduction

The visco seal is a sealing device in which a pumping action supplies a viscous fluid against a system fluid to be sealed. The seal effect is produced by the viscosity of the pumped fluid moving in the clearance gap between the housing and the rotating shaft surface. The fluid pumped with the desired viscosity is preferably always a liquid, whereas the fluid to be sealed may be either a liquid or a gas.

The seal essentially consists of a threaded portion either on the surface of a shaft rotating in a sleeve or in the sleeve itself. The thread functions like a screw pump, transferring fluid against the system fluid to be sealed, thus establishing the sealing effect. The pumping action is the reason why this design is often designated as a visco pump seal.

II. Operating Mechanism

The visco seal is a hydrodynamic shaft seal that will develop pressure gradients in an axial direction in the clearance gap around the shaft. The helical grooves are either cut in the shaft or machined in a sleeve arranged in a housing through which the shaft must pass, acting as a pump. Figure 7.2a shows a design in which the grooves are machined into the shaft surface. The viscous fluid entry may be considered the neutral line. On the left side of the neutral line the shaft has a right-hand thread, whereas the right side of the shaft has a left-hand thread starting in the system fluid section. When the shaft rotates, the section with the right-hand thread forces the viscous fluid against the system fluid. Each thread section acts as a screw pump. Consequently, the two pumps act against each other and produce a pressure gradient resulting in practically zero flow. This phenomenon is utilized as a sealing device that prevents the system fluid from leaking to the atmosphere through the clearance gap. This configuration is particularly useful for handling certain gases in which the viscous fluid is used to generate the pressure.

Figure 7.2b shows the distribution of the pressures in the clearance gap while the shaft is rotating. The highest pressure is found in the area of the neutral zone where the two fluid streams encounter each other. The thread grooves can either be cut in the shaft or in the sleeve without influencing the pumping effect by reversing the pumping component. Figure 7.2c illustrates a sleeve within which the screwed shaft rotates. Figure 7.2d shows a smooth shaft operating in a threaded sleeve.

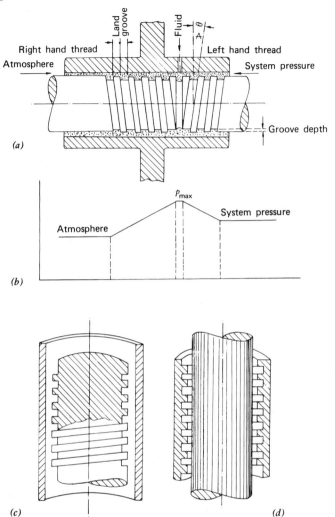

Fig. 7.2 (*a*) Cross section of visco seal. (*b*) Diagram of pressure distribution. (*c*) Screw machined in shaft. (*d*) Screw machined in sleeve.

During rotation of the shaft the sealing fluid fills the annular clearance gap, including the grooves of the threads over the full length. If the system operates under atmospheric conditions, a pressure distribution is observed, as indicated by curve 1, Fig. 7.3*b*, shown as a solid line, reaching a maximum in the neutral region where the sealing fluid meets the system fluid.

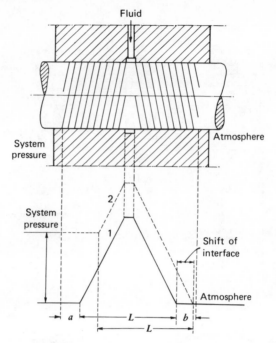

Fig. 7.3   Pressure profiles in dual visco pump [From (3, 4), courtesy of *Chemie Ingenieur-Technik*].

From the atmospheric side of the system (see Fig. 7.3*b*), the ambient air tends to penetrate. It will succeed in doing so over the length designated as (*b*). On the other hand, the system fluid will tend to leak out over the length of the clearance gap, designated as (*a*) in Fig. 7.3*b*.

For a pressure in the system fluid greater than atmospheric the pressure distribution will be as shown as a dotted line in the diagram as curve 2. It will be noticed that the interfaces between system fluid and the atmosphere are shifting in an axial direction.

III.   Static Sealing of Visco Seals

Visco seals are designed to function properly only when the shaft operates at a certain minimum rotational speed. Thus, for very low or zero shaft rotation a secondary sealing device must be provided. A variety of proposed designs for industrial visco seals utilize the centrifugal force during rotation to keep built-in lips from establishing contact with the

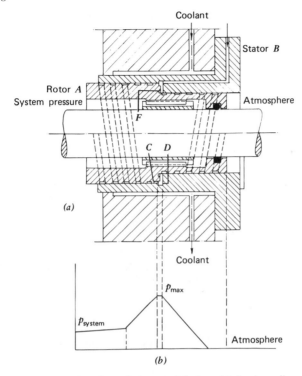

Fig. 7.4 Provision of static sealing of visco seal design. (*a*) Static sealing provision. (*b*) Pressure diagram [From (3,4), courtesy of *Chemie Ingenieur-Technik*].

shaft, but once the shaft stops rotating the lips contact the shaft and establish the necessary seal. This provides a satisfactory seal as long as temperature is not a problem for elastomeric lip materials.

Another design, proposed by Boon (3, 4), suggests a pressure-actuated seal, shown in Fig. 7.4. Rotor *A* is driven by the shaft sealed against the shaft by an elastomeric O-ring (or fluoroelastomer or Kalrez). The external diameter of rotor *A* is larger on the system fluid side than on the face in contact with the external atmosphere, thus providing a recessed shoulder between the two faces that can be used as a sealing shoulder, designated as *F* in Fig. 7.4. The pressure in chamber *C* is a function of the rotating speed. By reducing the rotation the pressure level in chamber *C* decreases and may finally drop below a certain desirable minimum limit value. If the system pressure remains greater than the fluid pressure in chamber *C*, rotor *A* will move toward stator *B*, establish contact with face *D* of stator *B*, and establish a seal with face *F*. The radial clearance between rotor surfaces with the stator are within 15 to 20 microns.

Reports indicate that this device is functioning satisfactorily. One report states that at a gas pressure of 20 atm (~300 psi) at room temperature the seal established a complete shutoff without gas leakage. When the shaft rotation at 1500 RPM came to a stop, no leakage was observed. The sealing fluid used had a viscosity of 70 centistokes.

IV. Analytical Design Considerations

A thorough study of the existing analytical design methods for interstitial seals reveals that for all procedures known so far a considerable range of results can be observed. This is further confirmed by the large number of publications on the subject. The author has developed computer programs for all major publication proposals and has come up with a wide range of deviations of the significant design factors for each particular design configuration. The reason is the large number of assumptions that must be made by the individual designer. Therefore, literature citations for publications processed through the computer are given here. The authors are listed in alphabetical order, showing those papers that provided the closest, respectively narrowest deviation of the particular design results.

The same method of selection of design procedures was used for visco seals, labyrinth seals, and bushing seals. It is left to the reader to make his own approach and then make his own decision. Investigation of these studies will familiarize the reader with the particular processes and provide an understanding of the basic concepts of the problems involved.

For visco seal design the following publications are suggested: Asanuma (1), Bauer et al. (2), Boon (3, 4), Rowell and Finlayson (19, 20), Whipple (28, 29), Froessel (8), and Klüpfel (12). The works of Whipple (28, 29), Woodrow (31), Snell (21), Woodsworth (32), and Whitley (30) concern the development of a spiral groove thrust bearing in the UK and are applicable to the visco seal design.

The intricacy of the analytical design is probably the reason why the visco seal configuration has not gained greater popularity although it has been known now for more than 40 years.

V. Conclusions

In-depth literature studies reveal that the visco seal is sufficiently developed and empirically verified to justify confidence in its performance and efficiency. In many cases it is possible to apply the seal without adding complex systems by utilizing shaft and housing length and existing cooling facilities.

One literature source reports an example in which a gas pressure of 20 atm was sealed through a visco seal design. The shaft rotated at a speed of 1500 RPM. Leakage was practically not observed. The oil used for sealing the system had a viscosity of 70 centistokes. For maximal shaft surface velocity a speed of 23 fps was employed for a shaft with 45 mm OD, rotating at 2950 RPM.

The results of the literature studies permit the following conclusions:

1. The visco pump seal has a wide range of potential applications for effective use in numerous rotating shaft equipment, such as centrifugal compressors, turbines, finishers, and the like.
2. The running clearances are to be designed either equal to or closer than those customarily used for similar types of interstitial shaft seals.
3. The visco seal design can be provided with an auxiliary seal system which is at least not more intricate than those customarily used in buffered-type labyrinth or bushing seals.
4. The visco seal is not restricted to dynamic use. Static sealing for zero shaft motion can be provided by relatively simple means.
5. The visco seal can be equally applied to both laminar and turbulent flow in the clearance gap without jeopardizing sealing efficiency. Optimum seal effectiveness for laminar flow is not the same as for turbulent flow regime. Seals that have to operate in both flow conditions offer greater flexibility when designed with small helix angles.
6. Design analyses for visco seals exist and lead to acceptable results. They must be experimentally verified and the seal behavior can be predicted with reasonable accuracy. Because of the necessity for individual assumptions, each designer will obtain a somewhat different result. This is the primary reason why a specific design theory for general application has not been presented here.
7. The pressure developed in a visco seal can be increased by: increasing the length of the seal, decreasing the clearance in the leakage gap, decreasing the depth of the helical grooves, increasing the viscosity of the sealing fluid.
8. The investment in piping, storage tanks, heat exchangers, pumps, controls, and instrumentation is less than is required for labyrinth or for bushing seals.
9. In case of a power failure secondary sealing fluid can be provided by a small, pressurized accumulator.
10. A major advantage of the visco seal is that the system pressure can vary from atmospheric to design levels while the seal pressure is kept

constant. The movement of the system gas-sealant interface takes place automatically as the value of the system pressure varies.

11. Heat dissipation for the visco seal must be greater than is required for bushing seals.

12. Visco seals usually do not employ heat exchangers for continuous supply of sealing fluid.

PART 2. LABYRINTH SEALS

I. Introduction

Labyrinth seals are characterized as controlled-clearance seals without rubbing contact with the moving parts and with some tolerable leakage. The system fluid to be sealed is a compressible fluid and has to pass through peculiar geometric shapes, representing chambers that force the leaking fluid to accelerate and decelerate through narrow and wide gaps to dissipate pressure energy and thus reduce the flow of leakage. Every step down in this process of pressure dissipation is accompanied by a loss in pressure.

Because of the lack of direct rubbing contact with the moving shaft the labyrinth seal is well suited for sealing shafts operating at high rotational speeds, as in centrifugal compressors and steam turbines. A properly designed labyrinth seal requires neither lubrication nor maintenance. If eventual wear is observed, the only damage is an increase in leak rate.

II. Description of the Labyrinth Design

There are countless design versions of the labyrinth seals. The simplest is the so-called straight-through design with a straight shaft and a straight housing. The bore for the shaft is provided with a sleeve, formed as a rigidly attached ring, containing strips in parallel arrangement, perpendicular to the shaft. The tips of these strips maintain a close clearance with the shaft surface, on the order of about 2 mil/in. of shaft diameter. Figure 7.5 illustrates schematically the design principle of a straight-through labyrinth seal with thin parallel strips attached to the housing ring to form the so-called chambers.

Besides the mechanical factors of clearance gap straightness, alignment, and so on, there are other factors that markedly influence the

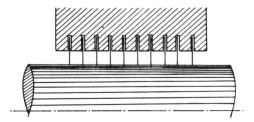

Fig. 7.5 Scheme of simple straight-through labyrinth seal.

effectiveness of the labyrinth design, such as strip spacing, thermo-dynamics of the system fluid, and pressure drop.

III. Sealing Mechanism

The sealing mechanism of a labyrinth seal design may be explained by using Fig. 7.6a, which shows a straight-through configuration with appropriately spaced strips, providing a clearance gap against the shaft, designated as (c). The system pressure has the value of p_1, which must be dissipated through the chamber arrangement of the labyrinth to finally reach level p_2, and eventually atmospheric pressure. The full dissipation of the pressure energy is finally achieved in the form of a complete pressure loss.

When the system gas at pressure p_1 reaches the first flow-restricting strip, facing clearance gap (c), three zones of flow phases can be observed through which the gas has to pass. In zone 1 the fluid accelerates and reaches its maximum possible velocity. In zone 2 the gas jet expands, decelerates, and then dissipates into turbulence, producing noticeable friction. The gas finally fills up zone 3 with relatively low velocity at decreased pressure level. The first labyrinth chamber must be wide enough to allow pressure dissipation in zone 3. If the subsequent strip is too close, zone 3 will not exist as described and the gas will pass directly through as it did in the preceding zone 1. This in turn means that the pressure energy will be carried over without energy loss and fluid leakage will not be controlled. Appropriate labyrinth design must prevent carryover from one chamber to another. A portion of the kinetic pressure energy will always be carried over in straight-through labyrinth design. This is one of the major shortcomings of this simple configuration. The reason for using it is simplicity of design, economy in construction, and low maintenance cost.

The cycle of chamber one continues in every subsequent chamber until the pressure level reaches atmospheric conditions.

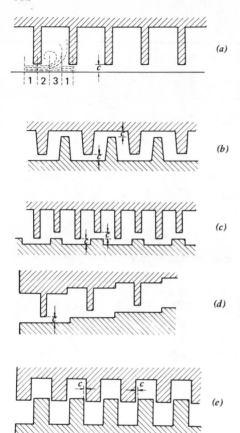

Fig. 7.6 Basic modifications of labyrinth designs.

IV. Design Characteristics

Any improvement of the labyrinth design must aim toward elimination of the carryover flow without pressure dissipation. By modifying the shaft geometry, the leakage path can be changed considerably. The major design objective is to diminish carryover by placing an effective obstacle into the path of the jet stream of the accelerated gas underneath the clearance gap and by modifying the chamber geometry from stage to stage. This may lead to staggered and stepped labyrinth configurations, as shown in detail in Fig. 7.6b. As will be noticed, the straight-through flow line is effectively interrupted and an increase in flow resistance will take place. These configurations are more expensive and more difficult to install. By splitting the housing an entirely new range of design

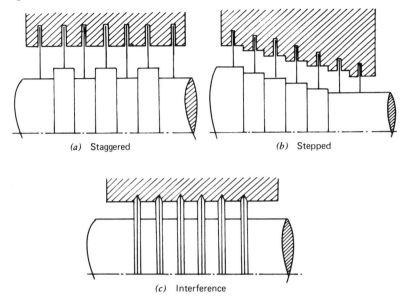

(a) Staggered   *(b)* Stepped

(c) Interference

Fig. 7.7 Various types of labyrinth designs. (*a*) Staggered. (*b*) Stepped. (*c*) Interference. [From J. M. McGrew and J. D. McHugh, *J. Basic Eng. Trans. ASME* (16), courtesy of ASME].

modifications becomes available. This also involves modifications of the shaft and other basic design concepts.

Figure 7.7 shows a variety of gap and housing modifications in connection with variations of the shaft geometry. The housing in Fig. 7.7*a* must be split to assemble the seal unit. Splitting the seal housing is not required in the design of Fig. 7.7*b*. However, larger overall dimensions are mandatory, accompanied by an increased flow area for the system gas to be sealed. The design of Fig. 7.7*c* represents an interference arrangement in which the flow restrictions create their own running contacts with the seal housing.

On machines with very long shafts and where large temperature differentials are likely to occur, it is inevitable that noticeable dimensional variations must be encountered in an axial direction. Stepped or staggered labyrinths are not possible, since close tolerances will not allow significant thermal expansions.

Another important modification, in addition to the length of the labyrinth seal, is the incorporation of annular ports in selected axial locations within the seal where neutral gases can be injected or withdrawn from the chambers before reaching the external atmosphere. An

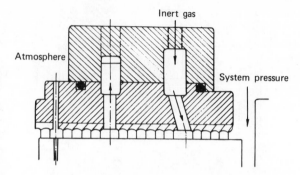

Fig. 7.8 Labyrinth seal with inert gas purge. [From J. M. McGrew and J. D. McHugh, *J. Basic Eng. Trans. ASME* (16), courtesy of ASME].

example of such a design is illustrated in Figs. 7.8 and 7.9, representing a combination of eductor and ejector configurations, as reported by L. B. Sanborn (49). An inert gas is introduced to encounter the system gas in close proximity to where the system pressure enters the seal area. A controlled differential pressure must be maintained, enabling the system to leak the injected inert seal gas to the atmosphere, as indicated in Fig. 7.8.

In the presence of hazardous gases a vent port is preferable to remove the leakage before it reaches the atmosphere. Such a design is illustrated in Fig. 7.9. The pressure differential is kept low to minimize leakage and to establish equilibrium conditions.

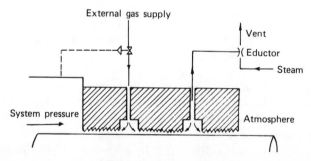

Fig. 7.9 Labyrinth seal with injection-eduction systems [From J. M. McGrew and J. D. McHugh, *J. Basic Eng. Trans. ASME* (16), courtesy of ASME].

V. Design Analysis

Many attempts have been made to develop a reasonable and scientifically justifiable design theory for computing the leakage flow through labyrinth seals. The results have been a wide variety of solutions that deviate widely from each other, as in the situation described for visco seals. Every method proposed requires a series of assumptions that only the designer can make to arrive at an acceptable solution. It was, therefore, decided to recommend those theories that have given reasonable results and let the reader decide which method to prefer.

The approach perhaps most widely used is that of Martin (48). The other methods shown in the literature citations of this section can also be used with confidence, with final selection being a matter of personal preference.

VI. Final Remarks

The design of a labyrinth seal for specific applications is a task best solved on an individual basis. Some methods are available for calculating leakage flow to an accuracy matching the dimensional tolerance requirements of most machines.

Labyrinth seals are generally used for pressures up to several hundred pounds per square inch as long as leakage costs or hazards are not critical. The sensitivity of labyrinth seals to pressure pulsations creates difficulties in compressors controlled by throttling of the suction pressure.

Under certain start-up conditions vacuum may be drawn, leading to air leakage into the system. Such occurrences can be prevented either by using special designs or by injecting a separate buffer gas into the labyrinth.

In conventional labyrinth design leakage is usually limited to an amount less than 0.50% of the weight flow of gas through the centrifugal compressor.

For the calculation of leakage rates a series of methods may be recommended. The most commonly used are Martin (48), Egli (41), and Gercke (42).

The labyrinth design requires computation of the following fundamental characteristics: pressure ratio; p/T conditions of pressurized fluid; physical properties of gas to be sealed; minimal possible clearance, governed by shaft gyration, bearing clearance, alignment, and so on;

thermodynamic concepts; assumption of tolerable leakage; assembly, space, machining cost limitations.

Computations require a variety of assumptions that can be made only on an individual basis.

PART 3.   BUSHING SEALS

I.   Introduction

Bushing seals are interstitial seals that have much in common with the labyrinth seal. In a simplified version the labyrinth represents an orifice, whereas the bushing is a pipe, matching the rotating shaft with a close clearance gap. Considering the mode of laminar flow, the labyrinth follows the orifice flow law and the bushing follows the resistance flow law. The bushing has a distinct advantage over the labyrinth seal, namely, a lower leakage rate per unit seal length. This is particularly true of the floating bushing seal, which permits smaller radial clearances. Although higher in cost, the bushing seal offers better economy because of higher efficiency.

The function of the bushing seal was discussed in Chapter 2 in connection with mechanical end face seals, where bushings may be used to throttle the flow from the pump housing at the stuffing box throat, before the flow reaches the mechanical seal.

II.   Design Considerations

According to the classification chart given earlier, Fig. 7.1, bushing seals may be categorized as fixed, floating, balanced bushing, and floating ring seals. As the name implies, the bushing seal is essentially a close clearance seal or fitting arranged either stationary to the housing or floating with the shaft.

A.   Fixed Bushing Seal

The fixed bushing is a sleeve attached stationary to the housing, surrounding a rotating shaft with an annular gap with relatively close clearance. The system fluid seeps through the narrow gap. Leakage is minimized as a result of the throttling effect when the fluid makes its way through the small annular ring space. The narrow passage imposes an effective restriction on the fluid, producing a high velocity at a noticeable

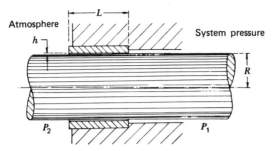

Fig. 7.10 Bushing seal with fixed bushing arrangement (2).

pressure drop as a result of fluid viscous friction. The sealing mechanism is created by the dissipation of dynamic energy through viscous fluid friction. This seal is suited only for sealing liquids. For gases the energy dissipation would not be sufficient and the leakage rate would reach intolerable amounts. A typical bushing seal using a fixed bushing is presented in Fig. 7.10. Fixed bushings are larger and longer than floating bushings. They are simple in design, easy to install, low in cost, and require no maintenance. Because of their length fixed bushings must have larger clearances to compensate for shaft gyrations.

B. Floating Bushing Seal

The floating bushing for seal purposes is designed to follow freely the shaft gyrations; consequently, it can be provided with closer clearances. As an example, the clearance can be smaller by approximately one order of magnitude than is necessary for equivalent fixed bushings.

An example of a floating bushing seal configuration is illustrated in Fig. 7.11. The bushing is free to move radially but is restricted from

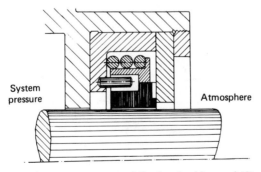

Fig. 7.11 Arrangement of floating bushing seal (2).

rotation by a dowel pin. One face of the bushing is lapped and maintains contact with a lapped corresponding face of the housing, arranged perpendicular to the shaft surface, similar to the arrangement of inter-face mechanical seals. The bushing moves along the contact face of the housing. The two mating faces are well machined so as to be flat and parallel. It is important that they be installed to move in such a way that during radial movements the faces always maintain their parallel posi-tion. For initial stationary sealing a helical spring is employed, establish-ing steady contact of the interfaces. In operation it is the fluid pressure that supplies the contact force. The secondary semistatic seal face of the bushing with the housing requires flatness tolerances of no larger than two to three light bands.

C. Floating Ring Seal

Floating bushing seals are characterized by having the freedom to move in a radial direction relative to the shaft. The motion permitted takes place in a plane perpendicular to the shaft surface. These seals can follow the shaft gyrations as required without generating excessive fric-tion. A disadvantage of this design is the length of the bushing sleeve, because the clearance must be increased proportionally with increasing length of the bushing to minimize friction.

By reducing the length of the bushing the tolerance requirements can be improved and the seal effectiveness increased. A bushing seal with reduced length is referred to as a floating ring seal, characterized by a defined length (L) to diameter (D) ratio. Such a reduction permits closer tolerances and makes the seal more independent. The reduced length of the ring can be compensated, by using two or more rings in any partic-ular seal arrangement. Such a design allows larger misalignments with-out affecting seal effectiveness.

Figure 7.12 compares a conventional floating bushing seal (a) with a design with floating rings (b), using three rings in this set.

D. Segmented Floating Ring Seal

Instead of using solid rings it is also customary to design a floating ring seal as a segmented archbound ring, where the segments are held to-gether by a garter spring, as illustrated in Fig. 7.13. The garter spring provides the required contact force to establish a seal with the shaft surface. The metal retainer is not always necessary. The configuration shown in Fig. 7.13 is primarily designed for sealing gases under moder-ate pressure.

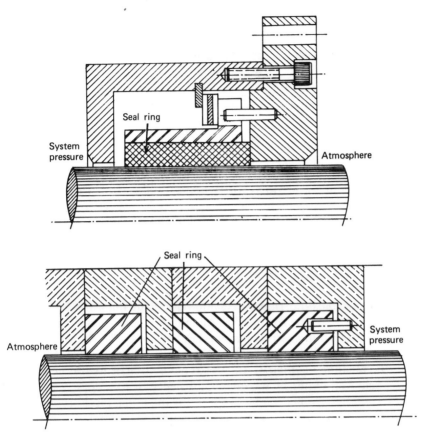

Fig. 7.12 (*a*) Floating bushing seal with one-ring set. (*b*) Floating ring seal with three-ring sets [From (8), reprinted by permission of the American Society of Lubrication Engineers].

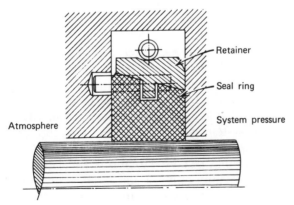

Fig. 7.13 Floating ring seal with segmented ring [From (8), reprinted by permission of the American Society of Lubrication Engineers].

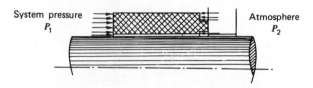

Fig. 7.14 Balance diagram of floating bushing seal.

The ring segments are made of carbon-graphite when used for sealing gases. For oil environments metals like Babbit and bronzes are customary. For sealing water bronzes, stellite, ceramics, and carbon-graphite are the preferred materials.

E. Balanced Floating Bushing Seal

For balancing bushing rings the same basic principles are used as are described in Chapter 2 for mechanical end face seals. A typical example of a balanced bushing seal is illustrated in Fig. 7.14. The diagram for the pressure distribution indicates the degree of balance built in by modifying the bushing geometry. Any desirable degree of hydraulic balance can be achieved either by introducing a sleeve or by stepping the shaft correspondingly. In Fig. 7.14 P_1 represents the initial fluid pressure and P_2 indicates the level after the pressure dissipation through the throttling of the bushing.

F. Multiple Floating Ring Arrangements

Great flexibility and independence from shaft gyration can be accomplished by an arrangement with multiple seal rings of the floating ring design. A series of rings ride separately on the shaft surface while the shaft is rotating. The rings are of the segmental design configuration and garter springs establish the contact force as indicated in Fig. 7.15. All rings run practically dry and wear becomes a factor. The garter springs compensate for wear. The wear rate is primarily dependent on the purity of the system gas to be sealed, surface conditions of the interface contact areas, and the materials involved.

Port A in Fig. 7.15 may be added to improve seal effectiveness. Port B may be added for vacuum applications. The arrangement of multiple rings is not restricted to any given number. The pressure differential decides how many rings per seal unit are required. Manufacturers will readily provide the necessary information. Two examples of designs for floating buffered bushing seals are shown in Fig. 7.16, representing a

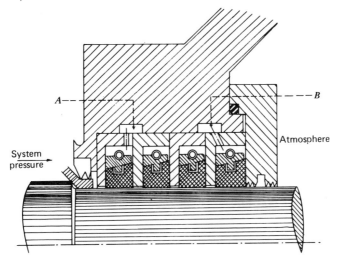

Fig. 7.15 Floating bushing seal with multiple sets of segmented rings [(9), courtesy of ASME].

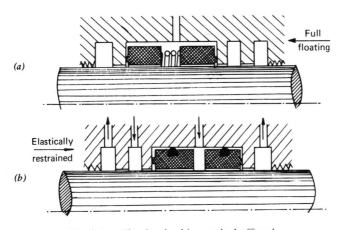

Fig. 7.16 Floating bushing seals, buffered.

full floating design in Fig. 7.16*a* and an elastically restrained design in Fig. 7.16*b*.

III. Design Analysis

Similar to visco and labyrinth seal designs there are also several significant theories for computing the leakage rates through bushing seals. Since bushing and ring seals are used for both liquids and gases, compressible and incompressible flow must be taken into consideration, and each of these may be either laminar or turbulent. In addition, eccentricity and flashing of the liquids play a significant role.

Each of the design theories in use provides different results. It is, therefore, suggested that the reader refer to the references at the end of this chapter. Of a variety of reasonably acceptable methods the paper of Tao and Donovan (69) is perhaps the most useful. They have projected special charts to facilitate computation. Two of their charts are presented as Figs. 7.17 and 7.18.

IV. Materials of Construction

For fixed bushings the best material is an antifriction type, such as carbon-graphite, molybdenum disulfide, or a combination of both with PTFE resins or soft metals used as liners. Where temperatures permit Babbit is a good candidate. For higher temperatures bronzes and aluminum alloys are well suited. Any material chosen must be compatible with the system fluid.

Floating bushing and ring seals are subject to the same basic requirements. Differences in thermal expansion are essential and should be carefully evaluated. With carbon-graphite as a major candidate it is important to consider the low coefficient of expansion compared to steel of the shaft. By shrinking it into a metallic retainer the problem can be minimized. Plasma coatings have proven favorable for high-temperature applications.

The system fluid is an important factor and must be carefully evaluated. It must be clean and should not tend to polymerize or crystallize. When water is used, it should be "soft." High surface velocities may produce erosion effects along the bushing surface.

PART 4.   FERROFLUIDIC SEALS

I. Introduction

The search for effective sealing devices has been a never-ending task and is now becoming a major challenge to the equipment designer, because of the constantly increasing sophistication of both equipment and process operations. A typical example is the extremely complex operation for the start of a space missile. Chemical, petroleum, and related process industries are facing similar problems in sealing equip-

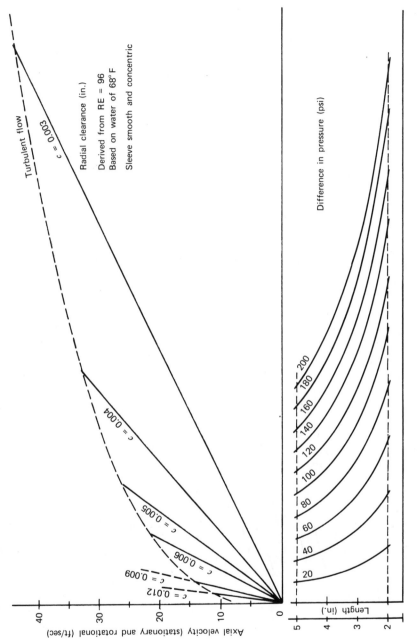

Fig. 7.17 Diagram for pressure breakdown in bushing gap [(12), courtesy of ASLE].

403

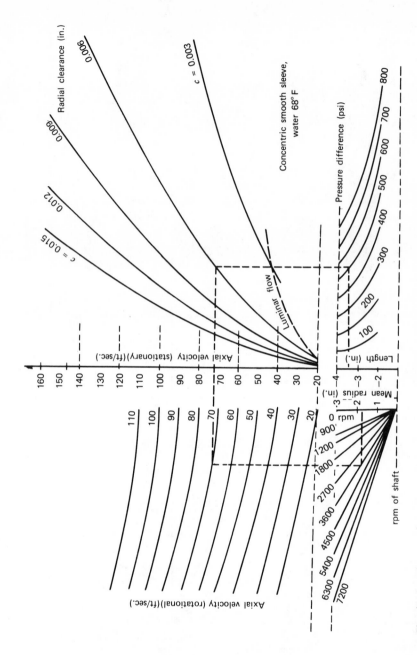

Fig. 7.18 Diagram for pressure breakdown in annular bushing gap [(12)], courtesy of ASLE].

ment that operates under an enormous variety of diversified process conditions, using an extensive spectrum of intricate sealing principles. Billions of dollars are lost and wasted in industry every year as the result of the ignorance of responsible personnel in industrial operations who do not recognize that seals are often the weakest link in the diversified chains of sophisticated industrial equipment.

Fortunately, a new category of seals has entered the industrial scene in recent years that is already competing in many ways with well-established categories. This new member is the ferrofluidic seal. This seal functions as an interstitial seal without rubbing contact with the shaft, using a clearance gap between the seal and the rotating shaft surface. However, it is distinctly different in one way from other interstitial seals, namely, there is no flow of leakage in the clearance gap whatsoever. The gap is filled with a magnetic fluid film held in position by a magnetic field that forces the fluid to remain stationary, thus closing the gap hermetically and providing the ideal seal along the rotating shaft surface.

Ferromagnetic fluids were initially developed by the National Aeronautics and Space Administration (NASA) in the early 1960s for controlling rocket fuels in zero-gravity environments. They were intended to be used for thermomagnetic control systems in satellites. Ferrofluidic components permit applications in seal design for high-speed seals where other types of seals cannot be used.

II. Properties of a Magnetic Fluid

A ferrofluid consists of a carrier liquid that contains ultramicroscopic particles of a magnetic solid, such as magnetite. They are colloidally suspended and then stabilized by physicochemical means. To prevent flocculation even under the influence of a magnet or any kind of a magnetic field the particles are coated, and random collisions (Brownian motion) with the molecules of the carrier liquid keep the particles in colloidal suspension for an indefinite period of time.

The composition of a ferrofluid can best be studied by using an electron micrograph. A ferrofluid is soft, magnetically speaking. When the magnetic field is removed there is hardly any hysteresis effect, as Fig. 7.19 proves. As a magnetic field is applied the magnetic particles of the fluid become oriented. With the removal of the magnet the randomly shaped particles of an average size of 100 angstroms become demagnetized.

At the present time all commercially available ferrofluids are electrically nonconductive in carrier liquids, such as diesters, fluorocarbons,

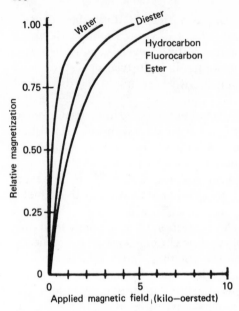

Fig. 7.19 Magnetization behavior of fluids.

hydrocarbons, and aqueous solutions. Chemically and mechanically the magnetic fluid offers the same characteristics as provided by the carrier liquid in which the magnetic particles are colloidally suspended.

It is assumed that the particles are uniformly magnetized, single-domain spheres. For tangent spheres the magnetic energy is proportional to the square of the magnetization and to the cube of the particle radius. When the particles are smaller than 150 angstroms in diameter, thermal agitation alone will prevent magnetic flocculation. Further stabilization can be achieved by coating each particle with a mono-molecular layer, which acts as an elastic cushion. As two particles approach each other compression of the coating provides the elastic repulsion to oppose the attractive magnetic and van der Waals forces. Ferrofluids behave like true homogeneous fluids, but they can be influenced by a magnetic field.

III. Design Concepts for Ferromagnetic Shaft Seals

Fluids that can be magnetized and arbitrarily influenced by a magnetic field are well suited for use in ferrofluidic seal applications.

A. Sealing Mechanism

The design of magnetic seals for rotating shafts perhaps represents the most advanced technological application of magnetic fluids. In addition to providing a virtually perfect seal with negligible friction loss, magnetic fluids can be used to form a seal of practically any geometric size that is absolutely independent of shaft rotational speed or direction of rotation.

In Fig. 7.20 the unique operating principle of a magnetic seal is shown through the use of a ferrofluid in the annular clearance gap between seal and shaft. A ring magnet is placed over the shaft contacting a ring pole on either side. With this arrangement a magnetic field is produced that is enforced in its effect by placing serrations either on the shaft surface or on the ID of the ring poles. The magnetic flux area is complete when the clearance gap between the ring poles and the shaft is filled with a film of magnetic fluid. The fluid bridges the passageway, thus blocking any trace of leakage flow by the system fluid. The shaft is free to rotate without frictional disturbances by the ferrofluid. Without solid mechanical contact no wear is possible and the friction generated in the fluid film is negligible. Even at extreme rotational shaft speeds maintenance is not required.

It is preferable to use a shaft sleeve with thread-type serrations. The grooves simply are filled with the ferrofluid and the tips of the threads easily bridge the gap with the least magnetic resistance. The fluid then forms a solid and retained film, where the serrations act as a barrier against gases that might tend to leak through the clearance gap. The small viscous friction of the fluid with the shaft is negligible.

Manufacturers claim successful operations at speeds in the order of 10,000 RPM in helium atmosphere under pressures and vacuum with-

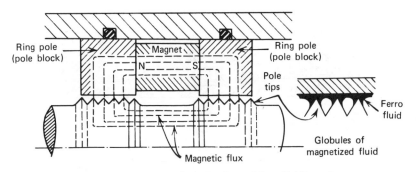

Fig. 7.20 Functional mechanism of ferrofluidic seal.

out a trace of leakage. At the present time the magnetic seal is applied to equipment sealing gases only.

The ferrofluidic seal is birotational and the direction of the pressure is unimportant.

B. Design Configurations

A rotary shaft seal applying ferrofluidic sealing principles is best designed as a cartridge arrangement with single- or double-seal combinations. A seal configuration with one magnet and two ring poles as a single-seal device is schematically illustrated in Fig. 7.21a. The ring poles are arranged concentrically with the peaks of the serrations in the sleeve, covering their full width. Adjacent to the ring pole are two ball bearings, providing a solid basis for true alignment at minimal shaft gyration. Both pole rings are sealed stationary against the housing by O-rings.

A cartridge containing a double-seal arrangement is schematically shown in Fig. 7.21b. The seal between the two ball bearings uses wider pole rings to provide a larger number of serrations. Each serration

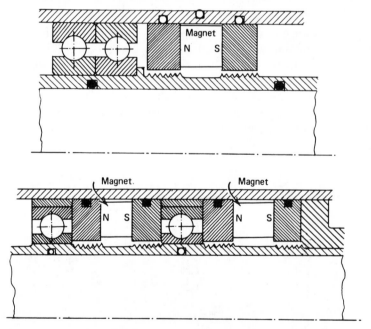

Fig. 7.21 (a) Cartridge with single-seal arrangement. (b) Cartridge with double-seal arrangement.

(valley between two peaks) is referred to as a *stage*, which is responsible for a defined amount of pressure drop to be produced. The number of stages for the total ring pole widths is a measure of the total capacity for pressure dissipation that a single seal with two pole rings is capable of absorbing.

Both units of Fig. 7.21 have the advantage that the cartridges can be completely assembled and adjusted before they are slipped over the shaft for final adjustment. Perfect accessibility is guaranteed.

For small seals the ferrofluid is initially inserted by brushing the surfaces of the ring poles. After an extended period of operation it may become necessary to replace the fluid or to inject more. It is best to inject replacement fluid to avoid disassembly while the seal remains in qpera-tion. The provision of pressure monitoring ports simplifies maintenance and extends seal life indefinitely.

C. Seal Pressure Capability

Commercially available ferrofluids show that at low magnetic solids loading the colloidal particles do not have a large effect on the properties of the carrier liquids, such as density, viscosity, vapor pressure, or pour point. As the fluid reaches magnetic saturation the viscosity rises abruptly, as shown in Fig. 7.22. Most commercial ferrofluids reach their condition of magnetic saturation at 600 gauss.

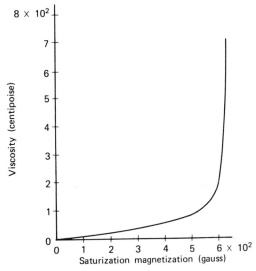

Fig. 7.22 Viscosity as function of magnetic saturation of ferrofluids.

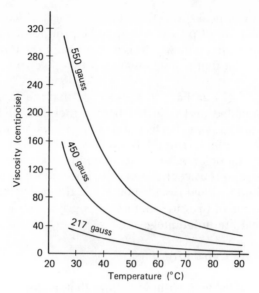

Fig. 7.23 Viscosity of ferrofluids as function of temperature.

For a diester ferrofluid the temperature dependence of viscosity is shown in Fig. 7.23 for various degrees of saturation magnetization (solids loading). The viscosity rapidly drops with rising temperature.

In all magnetic liquid seals the steel focusing structure is driven to magnetic saturation, providing a magnetic field in the film of the gap of 15,000 to 20,000 Oerstedt. Now the modified Bernoulli equation permits prediction of the pressure a single stage (serration) is capable of withstanding without being blown off. At the high field levels applied to hold the fluid the relationship between the differential pressure and the magnetic field intensity is approximated by the relation

$$\Delta p = \frac{M_s \times H}{4\pi}$$

where M_s is expressed in Gauss units, H in Oerstedt, and pressure differential Δp in dynes per square centimeter. This relationship predicts that a ferrofluid that is magnetically saturated at 500 gauss will withstand a gas pressure differential of 7 psi when in a gap of 15,000 Oerstedt of field intensity.

Every single seal is theoretically also a multistage seal. Each stage represents a practical barrier against the passage of leakage flow, reflecting an obstacle of at least 5 psi, considering a certain safety margin against the theoretical figure of 7 psi. Current design calls for 25 stages per inch of serration width, giving a guideline of up to 100 psi pressure differential per inch of serration length as pressure capacity of conventional seal units. Thus a multistage seal utilizes as many stages as are required to maintain the total pressure drop of a given system. Failure of a single stage is not critical or catastrophic, and the ferrofluid is not likely to be blown out of the entire seal. The stages recover quickly as the uniformly strong magnetic field forces the ferrofluid to fill the voids immediately.

D. The Effect of the Radial Gap on the Pressure Capacity

Since seal pressure capability is a function of the magnetic field intensity, the pressure capacity increases with the size of the magnet. Inversely the pressure capacity decreases with the widening of the annular clearance gap.

Radial clearance cannot be increased at random because fringing field effects limit the maximum field intensity attainable from a magnet of a given size. On the other hand, too small a clearance gap creates a mechanical problem that increases the chance of galling when direct contact with the shaft takes place in case of undesirable shaft gyration.

The typical ferrofluidic seal, therefore, has a shaft-to-stationary-element radial clearance of 0.002 to 0.005 in. For rotating structures with large gyration effects the gap may be widened. Some seals have been operated with a radial gap of up to 0.010 in.; however, the pressure capacity decreases correspondingly. There is no precisely defined theoretical limit to the size of the radial gap, since it is determined by the level of the system pressure in combination with the size of the magnet.

IV. Influence of the Operating Conditions

Operating conditions have a distinct influence on the effectiveness of ferrofluidic seals. Drag, speed of the shaft, temperature, and pressure level of the system fluid are some of the factors to be discussed.

A. Drag Torque

Magnetization does not affect the properties of a ferrofluid. Consequently a shaft will rotate without starting friction. The only negative factor is the drag introduced by the seal because of viscous shear. The shaft rotates in a liquid film and contribution to drag can be controlled by using a liquid of the desired viscosity.

With increasing rotational speed the viscosity decreases as a result of the temperature rise. At higher speeds this reflects itself as a reduction of drag torque below the linear relationship otherwise expected.

B.   Speed and Seal Life

Ferrofluids have, like any other liquid, the tendency to evaporate. As the fluid temperature rises, the evaporation tendency increases and this temperature rise is the sole limitation of the shaft speed. Manufacturers claim that seals have been successfully operated at speeds of 120,000 RPM. Temperature rise must, therefore, be controlled by corresponding cooling methods. Such compulsory cooling is mandatory when the fluid temperature reaches the 225°F limit, because the fluid tends to decompose from evaporation.

C.   Optimal Operating Conditions

To achieve optimal operating conditions the following factors must be considered: strong magnetic field, high magnetic saturation, low volatility of ferrofluid, and cooling of seal at high rotational shaft speed. A strong magnetic field is the prerequisite for the achievement of a continuous fluid film in the clearance gap. At present magnets are chosen from a list of materials presented in Table 7.1. The material Rayco XV is still in the developmental phase, although it may soon reach production.

The second factor in achieving optimal conditions is a high degree of magnetic saturation. The length of the seal is also a function of the magnetic saturation and subsequently a function of the system pressure

Table 7.1   Suitable Materials for Magnetic Seals

Materials	Flux Density (gauss)	Coercivity	% Used
Ceramics	3,000	2000	5
Alnico VIII	6,000	1000	30
Alnico V	11,000	500	60
Sumarium (Cobalt)	15,000	2000	5
Rayco XV	21,000	3100	—

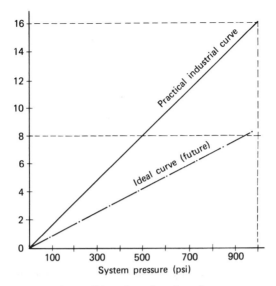

Fig. 7.24 Seal overall length as function of system pressure.

to be sealed. The manufacturer uses a curve that shows the seal length as a function of the internal system pressure, as shown in the graph of Fig. 7.24. The lower curve reflects the ideal case; the upper curve illustrates the actual field situation the designer uses to obtain reliable seal results with an adequate margin of safety. Once a higher degree of magnetic saturation can be achieved, the upper curve will move closer to the ideal curve, which in turn means that the seal can be built shorter. With the present state of the art a seal 4 in. in length should suffice to seal against a pressure of 500 psi. In reality, a length of 8 in. is used to provide a safety margin.

V. Ferrofluids

A ferrofluid is a colloidal suspension of submicron-sized ferrite particles in a carrier fluid with a dispersing agent added to prevent flocculation. When a magnetic field is applied to such a fluid, a body force is developed within it that is sufficient to change radically its gross behavior without altering its fluid characteristics. It is thus utterly unlike a magnetic clutch fluid, whose particles chain together and solidify under the influence of the magnetic field. By judiciously arranging conditions under which it is exposed to a magnetic field, a ferrofluidic liquid can have its internal pressure increased, its velocity increased, or its free surface raised. These phenomena, and many others, have led to a variety of novel applications from accelerometers and attitude control devices for satellites, to stable magnetic inks as well as a method for converting thermal energy directly to mechanical energy.

Ferromagnetic particles measure 100 angstroms and so quadrillions of them can be crowded into the volume of one cubic centimeter. They are kept from clustering together by a monomolecular surface coating. Ferrofluid is an ultrastable suspension of single-domain magnetic particles within a carrier liquid, behaving as a homogeneous Newtonian liquid without losing its initial liquid characteristics. Table 7.2 gives physical properties, composition, and specific details for fluids available at the present time. However, new fluids can be expected to be added to this list.

Any ferrofluid can be established with any number of bases in a series of magnetic intensities. In addition, the manufacturer has the capability in glycerine, silicones, organometallics, polyphenylethers, diesters with low vapor pressure, and many other liquids that allow operation in a diverse number of environments.

Considering the properties listed in Table 7.3 the magnetic saturation ranges from 100 for a fluorocarbon base to 600 gauss for water base fluids. Water base fluids generally contain colloidal particles with average dimensions about 15% larger than in other ferrofluids and consequently permit a higher degree of magnetic saturation. The average saturation capacity amounts to 200 gauss. One member of the hydrocarbon base family provides 400 gauss at saturation.

Hydrocarbon base fluids offer excellent stability with low viscosity and may be blended with numerous other hydrocarbon liquids to achieve suitable properties. Although volatile, they may be used in closed systems at temperatures exceeding the 200°F limit for extended periods of time. Their nominal electrical resistivity is 10^8 ohm-cm at 60 kHz. The dielectric constant is 20 at 1 kHz.

The larger-sized particles in water base fluids lead to a higher initial permeability together with a greater tendency to segregate in a gradient force field. The pH value of these fluids may be adjusted over a wide range without affecting the colloidal stability.

The fluorocarbon base carriers are fluoroalkyl-polyethers, characterized by a wide range in temperature and nonflammability. These carriers are the best candidates when absolute insolubility in other media is a must. They have a high density, high viscosity, and low volatility.

When ester base fluids are formulated with silicate ester carrier liquids they provide serviceability at very low temperatures. They further combine reasonably low volatility with low viscosity, although they offer only a fair hydrolytic stability.

Any of the ferrofluids listed in Table 7.3 may be freeze thawed without subsequent harmful effects.

VI. Applications

Ferrofluidic seals are the most advanced application of ferromagnetic fluids. They provide a perfect and frictionless seal that can be applied to any size shaft. They are ideally suited for reactors with top-entry agitators and will replace a vast number of other seal configurations.

The manufacturer claims that this type of seal is for vacuum conditions at 10^{-11} standard cc/sec of helium atmosphere as required for space flight applications.

A. Ferrofluidic Seal as Exclusion Device for Abrasives

Newly developed ferrofluidic seals have been designed to operate in highly abrasive environments such as grinder, machine tools, mining equipment, construction machinery, and textile equipment. They are capable of excluding abrasives by floating them to the surface of magnetic fluid sealants. The liquid seals have no wear and provide operation at zero leakage. The seals also exclude moisture, mist, and other environments. With abrasive environments containing magnetic constituents, the seals also incorporate any optional magnetic trap.

B. Substitution for Lip Seals

With ferrofluidic seals a new family of exclusion devices has become available, ranging in size from $\frac{3}{16}$ to 12 in. in shaft diameter. They may replace lip seals, scrapers, labyrinths, and other seal configurations. Their totally zero leakage performance makes them eligible for the exclusion of water, vapors, dust, solid particles, oil, and contaminants of all kinds. The magnetic fluid sealant is resistant to—and actually repels—abrasives, such as taconite, pulp fibers, Fiberglas, grinding dust, and other hard materials.

C. High-Speed Seals for Textile Industry

In the textile industry where fibers must be wound in large quantities the high-speed yarn take-up machines are in great need of seals that prevent

Table 7.2 Physical Properties of Ferrofluids

Material Identification	Magnetic Saturation (gauss)*	Density (g/ml)	Viscosity† (centipoise)	Pour Point (100,000 cp)	Vapor Pressure (1 torr)	Initial Susceptibility (M/H)	Surface Tension (dynes/cm)	Thermal Conductivity × 10⁵ (cal/°C-cm-5)	Specific Heat (cal/°C-cc)	Thermal Expansion (5) Coefficient (cc/cc°F)
				Nominal Properties (77°F unless noted)						
DOI	200	1.185	75	−35°F	300°F	0.5				
HOI	200	1.05	3	40°F	170°F	0.4	28	35	0.41	5.0×10^{-4}
HOI	400	1.25	6	45°F	170°F	0.8	28	35	0.44	4.8×10^{-4}
FOI	100	2.05	2500	−30°F	360°F	0.2	18	20	0.47	5.9×10^{-4}

Carrier fluids
Dilution of a ferrofluid with its carrier liquid varies the concentration—doubling the volume reduces the magnetic saturation by half.

Diester base
Low vapor pressure for vacuum and long-life applications, low viscosity for high speed seal applications, excellent lubricating properties for stictionless devices. Viscosity can be specified over a wide range for damping applications.

Hydrocarbon base
Low viscosity, may be blended with other hydrocarbon liquids. Electrical resistivity is 10^8 ohm-cm at 60 Hz and the dielectric constant 20 at 1 KHz.

Fluorocarbon base
A fluoroalkylpolyether, wide temperature range, nonflammability, insolubility in other liquids. An inert

material suitable for reactive environments such as ozone and chlorine.

Ester base

Silicate ester magnetic fluid provides serviceability at very low temperatures. Available at high gauss magnetic saturation (EO1).

Water base

Water base fluid particles about 15% larger than in other ferrofluids, higher initial permeability, same segregation in gradient field. The pH of this fluid may be adjusted over a wide range.

Polyphenyl ether base

Low vapor pressure for hard vacuum, radiation resistance in excess of 10^8 rad.

Any of the ferrofluids described above may be freeze-thawed.

Ester base										
EO3	200	1.15	14	−70°F	300°F	0.4	26	31	0.89	4.5×10^{-4}
EO3	400	1.30	30	−70°F	300°F	0.8	26	31	0.89	4.5×10^{-4}
EO1	600	1.40	35	−80°F	104°F	1.0	21	31	0.89	4.5×10^{-4}
Water base										
AO1	200	1.18	7	32°F‡	78°F§	0.6	26	140	1.00	2.9×10^{-4}
AO1	400	1.38	100	32°F	78°F§	1.2	26	140	1.00	2.8×10^{-4}
Polyphenyl ether base										
VO1	100	2.05	7500	50°F	500°F	0.2				

Source: Ferrofluidics Corporation, Burlington, Massachusetts.

* Field averaged value of $B - H = M$, i.e., $\frac{1}{H} \int_0^H M \, dH$ at saturation.

† Brookfield, shear rate $> 10 \text{ sec}^{-1}$.

‡ Freezing point.

§ 24 torr.

‖ Average over range 77 to 200°F.

Table 7.3 Nominal Properties of Ferrofluids

	Catalog Number	Magnetic Saturation* (gauss)	Quantity	Density (g/ml)	Viscosity† (centipoise)(100,000 cp)	Pour Point	Vapor pressure (1 torr)	Initial Susceptibility (M/H)	Surface tension (dynes/cm)	Thermal Conductivity ×10 (cal/°C-cm⁻⁵)	Specific Heat (cal/°C-cc)	Thermal Expansion‖ Coefficient (cc/°cc°F)
								Nominal Properties (77°F unless noted)				
Hydro-carbon Base	2H03	200	3 oz.	1.05	3	40°F	170°F	0.4	28	35	0.41	5.0×10^{-4}
	2H64	200	Half-gallon	1.05	3	40°F	170°F	0.4	28	35	0.41	5.0×10^{-4}
	4H03	400	3 oz.	1.25	6	45°F	170F	0.8	28	35	0.44	4.8×10^{-4}
Water Base	2W03	200	3 oz.	1.18	7	32°F‡	78°F§	0.6	26	140	1.00	2.9×10^{-4}
	2W64	200	Half-gallon	1.18	7	32°F	78°F§	0.6	26	140	1.00	2.9×1^{-4}
	6W03	600	3 oz.	1.58	225	32°F	78°F§	1.3	26	140	0.99	2.6×10^{-4}
Fluoro-carbon Base	1F10	100	10 cc (approx. 0.34 fl oz)	2.05	2500	−30°F	360°F	0.2	18	20	0.47	5.9×10^{-4}
Ester Base	2E03	200	3 oz.	1.15	14	−60°F	300°F	0.4	26	31	0.89	4.5×10^{-4}

Source: Ferrofluidics Corporation, Burlington, Massachusetts.

* Field averaged value of B − H = M, i.e., $\dfrac{1}{H}\Big|_0^H M dH$ at saturation.

† Brookfield, shear rate $> 10 \ \text{sec}^{-1}$

‡ Freezing point.

§ 24 torr.

‖ Average over range 77 to 200°F.

contamination of the yarn by oil. These machines operate at rotational speeds of 12 to 15,000 RPM and a serious problem develops from the splashing of lubricant oil in the bearings onto the yarn. Ferrofluidic seals on these machines provide the ideal solution.

D. Seals for Large-Scale Reactors with Agitation

Perhaps the most important area for application is the sealing of large-scale reactor shafts with top-entry agitation. Chemical, petrochemical, and related process industries will specifically benefit from this application. The arrangement of ball bearings between the seals is ideal and essential for keeping shaft gyration at a minimum despite a great shaft length. No other seal configuration can be applied in such close proximity to ball bearings.

VII. Final Conclusions

The ferrofluidic rotary shaft seal opens the door for an entirely new concept in sealing technology that is particularly suitable for rotating agitator shaft equipment. Pressures in the systems of up to 500 psi can be safely controlled with no leakage. It may be only a matter of months before higher pressures become possible.

A. Advantages

The major advantages of the ferrofluidic seal may be summarized as follows:

1. No specific surface requirements for the shaft
2. No limitation on shaft rotational speeds
3. Size of clearance gap need not be closer than for average machine shop conditions
4. Certain moderate amounts of eccentricities and shaft runout can be tolerated
5. No measurable friction losses or wear
6. Indefinite life expectancy
7. Only occasional fluid replacement

B. Disadvantages

There are still a series of shortcomings:

1. Not enough variety of system fluids available to cover the desired spectrum in chemical and related industries
2. Process fluid should not be miscible with the ferrofluid
3. Temperature limitations because of evaporation caused by low vapor pressure
4. Ferrofluids not yet available in tank car quantities
5. No safe application for sealing liquids of any kind, vapors only
6. Limitation of system pressure to below 1000 psi
7. No information available concerning behavior of seal in reciprocating shaft motion

It is desirable that use of the ferrofluidic seal be extended to a larger circle of customers and manufacturers to widen competition and to develop incentives for creativity in developing this interesting technological device, which will substantially improve modern seal capabilities.

References

Visco Seals

1. Asanuma, T.
   "Study on the Sealing Action by Viscous Fluid"
   Trans. Jap. Soc. Mech. Eng., No. 60 (1951).
2. Bauer, P., Glickman, M. and Iwatsiki, F.
   "Dynamic Seals"
   IIT Research Institute, Technical Report AFRPL-TR-65-61.
   Air Force Rocket Compulsion Laboratory, Edwards, CA, May, 1965.
3. Boon, E. F., Honing, S. and Van Ryssen, D. C.
   "Some Notes on Seals for Rotating Shafts"
   Proceedings of the Fourth World Petroleum Conference, Paper 1, June 1955.
4. Boon, E. F. and Tal, S. E.
   "Hydrodynamische Dichtung für Rotierende Wellen"
   Chem. Ing. Tech. No. 31 (January 1959).
5. Boon, E. F.
   Waarnemingen aan een Visco Afdichtung in een Transparent Huis
   University of Delft, The Netherlands, June 1968.
6. Fisher, C. F., Jr., and Stair, W. K.
   "On Gas Ingestion and Fluid Inertia Effects in Visco Seals"
   Paper presented at the Fourth International Conference on Fluid Sealing,
   ASLE Meeting, Philadelphia, May 1969.

7. Frössel, W.
   "Hydrodynamisch Wirkende Wellendichtung"
   J. Konstr., **8**, Nr. 11 (1956).

8. Frössel, W.
   "Hochtourige Schmierölpumpe"
   J. Konstr., **12**, Nr. 5 (1960).

9. Gumbel, L. and Everling, E.
   Reibung und Schmierung im Maschinenbau
   M. Krayn-Verlag, Berlin, 1925.

10. Hoek, Van, P. H. F.
    Overzichts Rapport over Reeds Verrichte Onderzoekingen aan Visco Afdichtungen
    University of Delft, The Netherlands, April 1955.

11. Ketola, H. N. and McGrew, J. M.
    "Turbulent Operation of Visco Seal"
    ASLE Trans., **10** (1967).

12. Klüpfel, A.
    "Berührungslose Oelfilm—Wellendichtung für Wasserstoffgekühlte Stromerzeuger"
    D.P.P. (Patent) No. 1,012,999, Germany, August 1, 1957.

13. Lawaczek, F.
    Pumpen und Turbinen
    Springer-Verlag, Berlin, 1932.

14. Leyer, A.
    "Keilschmierung oder Druckoelschmierung"
    J. Schweiz. Bauz. (January 1952).

15. Ludwig, L. P., Strom, T. N. and Allen, G. P.
    "Gas Ingestion and Sealing Capacity of Helical Groove Fluid Film Seal (Visco Seal),
    Using Sodium and Water as Sealed Fluids"
    NASA, TN D-3348, Washington, DC, March 1966.

16. McGrew, J. M. and McHugh, J. D.
    "Analysis and Test of the Screw Seal in Laminar and Turbulent Operation"
    J. Basic Eng. Trans. ASME, Series D, **87** (March 1965).

17. Pai, S. I.
    "Viscous Flow Theory II. Turbulent Flow"
    Van Nostrand, New York, 1957.

18. Pape, J. G. and Vrakking, W. J.
    "Visco Seal-Pressure Generation and Friction Loss under Turbulent Conditions"
    ASLE Trans., **11**, No. 4 (1968).

19. Rowell, H. S. and Finlayson, D.
    "Screw Viscosity Pumps"
    J. Eng., **114** (November 1922).

20. Rowell, H. S. and Finlayson, D.
    "Screw Viscosity Pumps"
    J. Eng., **126** (September 1928).

21. Snell, L. N.
    "Theory of Viscosity Plate Thrust Bearing, Based on Circular Geometry"
    Reprint Atomic Energy (October 1952).

22. Stair, W. K.
"Analysis of Viscoseal"
University of Tennessee.

23. Stair, W. K.
"Theoretical and Experimental Studies of Visco-Type Shaft Seals"
Mech. Aerospace Eng. Rep.
University of Tennessee.

24. Stair, W. K. and Hale, R. H.
"The Turbulent Visco Seal"
Third Int. Conf. Fluid Sealing, Cambridge, England, April 1967.

25. Stair, W. K.
"Effect of Groove Geometry on Visco Seal Performance"
J. Eng. Power, Trans. ASME, Series A, **89** (October 1967).

26. Taylor, E. R.
"Preliminary Experiments with the Visco Seal Concept in Turbulence"
Fourth Int. Conf. Fluid Sealing, ASLE, Philadelphia, 1969.

27. Trutnowsky, K.
Berührungsfreie Dichtungen
VDI-Verlag, Berlin, 1943.

28. Whipple, R. T. P.
"Hearing-Bone Pattern Thrust Bearing"
Atomic Energy Research Establishment, UKAEA-T/M, August 1949.

29. Whipple, R. T. P.
"Theory of the Spiral-Grooved Thrust Bearing with Liquid or Gas Lubricant"
Atomic Energy Research Establishment, UKAEA-T/R, March 1951.

30. Whitley, S. and William, L. G.
"The Gas-Lubricated Spiral-Grooved Thrust Bearing"
Atomic Energy Research Establishment, UKAEA-Report 28 (RD/CA), 1959.

31. Woodrow, J.
"Viscosity Plates Flow and Loading"
Atomic Energy Research Establishment, E/M-31, December 1949.

32. Woodsworth, D. V.
"The Viscosity-Plates Thrust Bearing"
Atomic Energy Research Establishment, E/R-2217, 1952.

33. *Mach. Des.*
Reference issues on seals, 1973–1974.

34. "Mechanical Drives"
Mach. Des., reference issue, 1975.

Labyrinth Seal Design

35. Arnold, F. and Stair, W. K.
The Labyrinth Seal—Theory and Design
University of Tennessee, 1962.

36. Babelay, E. F.
"Proposed Back-Diffusion Test Program"

Proc. URNI. Conference on Shaft Seals for Compressors and Turbines for Gas-Cooled Reactor Application
USAEC Report No. TID-7604, December 1959.

37. Bauer, P., Glickman, M. and Iwatsuki, F.
"Dynamic Seals"
Technical Report AFRPL-TR-65-61, Air Force Rocket Propulsion Laboratory, Edwards, CA, May 1965.

38. Becker, E.
"Strömungsvorgänge in Ringförmigen Spalten (Labyrinth)"
J. VDI, **51** (1907).

39. Bell, K. J. and Bergelin, O. P.
"Flow Through Annular Orifices"
Trans. ASME, **79** (1959).

40. Dollin, F. and Brown, W. S.
"Flow of Fluids Through Openings in Series"
J. Engl., **64**, No. 4259 (August 1937).

41. Egli, A.
"The Leakage of Steam Through Labyrinth Seals"
Trans. ASME, **57** (1935).

42. Gercke, M. J.
"Berechnung der Ausflussmengen von Labyrinth-Dichtungen"
J. Die Wärme, **57** (1934).

43. Heffner, F. E.
"A General Method for Correlating Labyrinth Seal Leak Rate Data"
ASME Paper No. 59-Lub 7, October 1959.

44. Hodkinson, B.
"Estimation of the Leakage Through a Labyrinth Gland"
Proc. Inst. Mech. Eng., **141**

45. Jeri, J.
"Flow Through Straight Through Labyrinth Seals"
Proc. Seventh Int. Conf. Appl. Mech., Vol. 2, 1948.

46. Kearton, W. J. and Keh, T. H.
"Leakage of Air Through Labyrinth Glands of the Staggered Type"
Proc. Inst. Mech. Eng., **166** (1952).

47. Kearton, W. J.
"Flow of Air Through Radial Labyrinth Glands"
Proc. Inst. Mech. Eng., **169**, No. 30 (1955).

48. Martin, H. M.
"Labyrinth Packings"
J. Eng. (January 1908).

49. Sanborn, L. B.
"Centrifugal Compressor Shaft Seals"
Mech. Eng. (January 1967).

50. Stodola, A.
"Steam and Gas Turbines"
McGraw-Hill, New York, 1927.

51. Tao, L. N. and Donovan, W. F.
    "Through Flow in Concentric and Eccentric Annuli of Fine Clearance with and without Relative Motion of Boundaries"
    Trans. ASME (1955).

52. Vermes, G.
    "A Fluid Mechanics Approach to the Labyrinth Leakage Problem"
    J. Eng. Power (April 1959).

53. Zabriskie, W. and Sternlicht, B.
    "Labyrinth Seal Leakage Analysis"
    Trans. ASME, **81** (Setptember 1959).

54. "Evaluation of Helium Seal Selected for Nuclear Reactor Coolant Compressors in Gas-Cooled Power Reactor and Development Program"
    Allis Chalmer Company, Milwaukee, WI, January 1938.

55. "The Labyrinth Packing"
    J. Eng., **165** (January 1938).

56. *Mach. Des.*
    Reference issue on seals, 1973–1974.

57. "Mechanical Drives"
    Mach. Des., reference issue, 1975.

Bushing Seals

58. Agostinelli, A. and Salemann, V.
    "Prediction of Flashing Water Flow Through Fine Annular Clearances"
    ASME Report, December 1957.

59. Bauer, P., Glickman, M. and Iwatsuki, F.
    "Dynamic Seals"
    Tech. Report AFRPL-TR-65-61, Air Force Rocket Propulsion Laboratory, Edwards, CA, May 1965.

60. Bondreau, W. F. and Taylor, E. F.
    "Liquid Buffered Floating Bushing Shaft Seals for Gas-Cooled Reactor Applications"
    Paper C6, First Int. Conf. Fluid Sealing, British Hydromechanic Research Association, April 1961.

61. Brkich, A. and Allen, R. E.
    "Development of Floating Ring-Type Stuffing Boxes for Eddystone Boiler Feed Pumps"
    ASME Report, November 1959.

62. Egli, A.
    "The Leakage of Gases Through Narrow Channels"
    Trans. ASME (1937).

63. Grinnell, S. K.
    "Flow of Compressible Fluid in a Thin Passage"
    ASME Report, 1955.

64. Lenkei, A.
    "Close Clearance Orifices"
    Prod. Eng. (April 1965).

65. Ruthenberg, M. L.
"Mating Materials and Environment Combinations for Specific Contact and Clearance-Type Seals"
Paper 72AM 23, presented at the 27th Meeting of the ASLE, Houston, TX, May 1972.

66. Sanborn, L. B.
"Centrifugal Compressor Shaft Seals"
Mech. Eng. (January 1967).

67. Sence, L. H.
"Floating Ring Seals, Design and Application" (reference to high pressure)
ASME Report, March 1963.

68. Stair, W. K.
"Liquid Buffered Bushing Seals for Large Gas Circulators"
Paper C5, First Int. Conf. Fluid Sealing, British Hydromechanic Research Association,
April 1961.

69. Tao, L. N. and Donovan, W. F.
"Through Flow in Concentric and Eccentric Annuli of Fine Clearance with and
without Relative Motion of the Boundaries"
Trans. ASME (November 1955).

Ferrofluidic Seals

The illustrations and the information in Part 4 are from the Ferrofluidics Corporation in
Burlington, Massachusetts. The author would like to express his sincere gratitude.

Index